THE TECHNOLOGICAL ILLUSTRATIONS OF THE SO-CALLED »ANONYMOUS OF THE HUSSITE WARS«

CODEX LATINUS MONACENSIS 197, I

THE TECHNOLOGICAL ILLUSTRATIONS OF THE SO-CALLED

»ANONYMOUS OF THE HUSSITE WARS«

CODEX LATINUS MONACENSIS 197, PART 1

BY

BERT S. HALL

DR. LUDWIG REICHERT VERLAG WIESBADEN 1979

CIP-Kurztitelaufnahme der Deutschen Bibliothek

Hall, Bert S.:
The technological illustrations of the so-called
„Anonymous of the Hussite wars" :
Codex Latinus Monacensis 197, Pt. 1 / by Bert S. Hall. –
Wiesbaden : Reichert, 1979.
ISBN 3-920153-93-6

NE: Anonymous of the Hussite wars

GESAMTHERSTELLUNG: HUBERT & CO. GÖTTINGEN
PRINTED IN GERMANY

PREFACE AND ACKNOWLEDGMENT

This edition has a dual purpose. On one hand, it is a detailed study and presentation of *Codex latinus monacensis 197, part I,* the so-called »Manuscript of the Hussite Wars' Engineer.« On the other, it is an introduction of the broader study of the history of mechanical technology in the fifteenth century, particularly in Germany and German-speaking parts of Europe. The latter task has presented the most difficulties, since it required the study of many unpublished materials that have been only slightly considered in the existing scholarly literature. I have tried to accomplish the dual purpose to the best of my ability by relating the contents of the individual codex under consideration to the wider context of technical methods and devices that are represented in the numerous additional manuscripts from the fifteenth century. In one sense, any study of any specific manuscript from this group is premature, for the researcher has too little information about all the others to place the text he has selected against a suitable background for study. Conversely, the texts cannot be adequately studied as a group, because no individual manuscript has received the proper amount of detailed attention. The solution to such a dilemma is obvious: one simply begins somewhere and does the best he can.

»Somewhere« in this instance is a manuscript which has been fairly frequently cited in the scholarly literature, but which has never received adequate or full attention. It is of moderate length and difficulty, but it lacks the tortuous problems of language and meaning presented by Conrad Kyeser's more famous *Bellifortis.* Since that work had already received the best efforts of the late Götz Quarg in 1967, Clm 197,I seemed the next most important text on which to work when I began in 1969. It was my original belief that the *Bellifortis* edition, together with this work, plus perhaps studies of one or two minor manuscripts, would exhaust the Germanic manuscripts of the fifteenth century in matters technological. I soon learned that I had been seriously mistaken. As work progressed on this book (originally submitted as a dissertation at the University of California, Los Angeles in 1971), the manuscript bibliography began to mount up at an appalling rate. By the time I was in Europe studying Clm 197,I at first hand (1970), the pages of Jähns and Gille and others had yielded far too many references to unpublished manuscripts to permit them all to be investigated by an impecunious graduate student. Accordingly, I completed my study in Germany, returned to Los Angeles and submitted the dissertation in the best form possible under the circumstances.

Over the next three years I continued to collect the various manuscripts on microfilm and to study them further on two separate trips to Europe. As my understanding of both the German and the Italian developments in technology increased, so did the number of errors I could perceive in my original work. Lines of development that seemed perfectly clear, interpretations of texts that seemed utterly obvious, comparisons of drawings that seemed securely unchallengable, all grew more vexed and confused as the months passed. The time spent in revising the work in accordance with these shifts in perspective, together with the inescapable demands of an assistant professorship, account for the delay in bringing this work to publishable form.

It is my hope that this edition will provide further material for interested scholars to use in broadening our common understanding of technological developments in the pre-modern world. I am keenly aware that one important factor inhibiting research in this field is a lack of suitable editions of major source documents. This seems to me especially true for the later middle ages, a period rich in such documents, but lacking in general interpretive studies or text editions. This lacuna is beginning to be filled. Within the last ten years Kyeser's *Bellifortis,* Francesco di Giorgio Martini's *Trattati,* and Mariano di Jacopo Taccola's works have all reached print, to say nothing of the late Ladislao Reti's monumental edition of Leonardo da Vinci's *Madrid Codices* (which appeared too late for serious consideration in preparing this study). Several other important works have recently appeared or are under way: C.S. Smith and J.G.Hawthorne's *Mappae Clavicula,* A.R.Hall's edition of Guido da Vigevano's *Texaurus,* P.L.Rose's work on Giovanni da Fontana's *Bellicorum instrumentorum* and the edition of Ramelli's *Diverse et artificiose machine* by M.T.Gnudi and E.Ferguson. It is my hope that Clm 197,I can take its place among these works in serving the needs of the historical community. Even more specifically, I frankly wish that this edition might engender more interest than is presently visible among German historians and philologists in the many textual, linguistic and historical problems represented by the northern technological literature of the fifteenth and sixteenth centuries.

The edition consists of three related parts. The introduction is devoted to the usual questions that surround any manuscript, such as date, provenance, authorship, but it also seeks to survey the contents of Clm 197,I in relation to other contemporary treatises and notebooks. It includes a brief history of each general class of devices dealt with in the manuscript, concentrating on fifteenth century developments. Inevitably this requires frequent citation of texts with which the reader cannot be familiar, and I apologize in advance for whatever inconvenience this may cause. Not every machine depicted in the manuscript is discussed in such detail in the introduction, since some of them do not require extensive commentary; in these cases, comparative examples are cited in the apparatus itself.

The apparatus consists of transcriptions and translations into English of all descriptive passages contained in the original together with appropriate philological footnotes. Because more information is conveyed by the drawings than by means of words, there is a detailed commentary on each folio's pictorial contents. These are intended to explain the way each device that is depicted was intended to function, the materials from which it was to be constructed, and, where necessary, discussion of the relations between the drawing and descriptive text. As a rule I have tried to describe each device in depth, but without excessive reliance on a modern technological vocabulary, in the hope that this might assist even readers who are not mechanically inclined to grasp the way each machine was meant to operate.

It would be impossible for me to acknowledge all those who have made important contributions to this work. I would like to thank several agencies under whose financial auspices various parts of the research were conducted: The Woodrow Wilson National Fellowship Foundation, the National Science Foundation, The National Endowment for the Humanities and the Joint Awards Council, State University of New York Research Foundation. Many libraries and librarians have provided invaluable assistance, particularly the Bayerische Staatsbibliothek's manuscript division and above all Dr. Hermann Hauke, the Österreichische Nationalbibliothek, and Dr. Bruno Thomas of the Waffensammlung des Kunsthistorischen Museums in Vienna. Professor Friedrich Klemm of the Forschungsinstitut des Deutschen Museums in Munich deserves mention for the many courtesies and friendly advice that he offered me while I was working there in 1970. My several visits to Munich have been made more pleasant through the kindness of Herr Alexander Pachmann of Lohhof, in whose home my wife and I have frequently been guests.

I would also like to thank several faculty members at the University of California, Los Angeles, in particular Professor Gerhart Ladner, who has taught me more than I have ever learned, and Mr. Albert Hoxie, who assisted me in many ways throughout my undergraduate and graduate careers. I owe much to those who have read this study in various stages of completion and offered their comments: Professors John G. Burke, Arthur B. Groos, Jr., Stephen P. Schwartz and Wilma Iggers. I have a special debt of gratitude to Professor Lynn White, Jr., who first introduced me to the subject of medieval technology and who guided me through the dissertation from which this work emerged. Finally, my deepest thanks go to my wife, Darlene, whose patient support and loving encouragement made it all possible.

B.S.H.
State University of New York, Buffalo

TABLE OF CONTENTS

PART I: THE MILIEU

CHAPTER I

INTRODUCTION

Codex latinus monacensis 197, part I is a technical manuscript from the fifteenth century dealing mainly with machinery. It is frequently cited in general histories of technology[1] and has been the subject of several articles. In spite of this, it is worthy of further study for a variety of reasons. As we shall see, it has been misunderstood by scholars in the past, and current opinion about it needs to be revised. More important, a study of this work serves to clarify the development of mechanical technology during the fifteenth century, particularly when its contents are compared with those of other contemporary technical treatises. In addition, some of the manuscript's folios depict novel devices or techniques, and by identifying them, we can draw new conclusions about their histories.

Modern scholarly interest in Clm 197,I* began in 1891, when the noted French scientist and statesman, Marcellin Berthelot, published a study of it.[2] His article dealt with the manuscript in conjunction with its mate, Clm 197, part II, one of the works of the Sienese engineer Mariano di Jacopo, surnamed »Il Taccola.« Berthelot's study contained little more than a reproduction of some of the manuscript's folios together with often-defective transcriptions and translations of some of the text. It made no attempt to relate the devices shown to the broad history of technological development nor even to similar devices illustrated in other contemporary texts. In addition, Berthelot gave only cursory attention to such questions as the text's authorship and dates; he accepted the traditional, but mistaken, view that the manuscript was the work of one man writing about the time of the Hussite Wars, c. 1430. His article's main value was that it called attention to the manuscript.

---

* For reasons to be made clear in the next chapter, I shall refer to the manuscript throughout in this way rather than by its informal title, »The Manuscript of the Hussite Wars' Engineer.«

1 The manuscript is mentioned by George Sarton, *Introduction to the History of Science*, III (Baltimore: 1948), p. 1551. It is cited frequently in such works as Franz Maria Feldhaus, *Die Technik der Vorzeit, der geschichtlichen Zeit und der Naturvölker* (1914: reprinted München: 1965) (hereafter Feldhaus, *Technik*); *idem*, *Die Technik der Antike und des Mittelalters* (Wildpark-Potsdam: 1931) (hereafter Feldhaus *TdAdM*); Charles Singer *et al.* (eds.), *A History of Technology* (5 vols.; Oxford: 1954–58) (hereafter Singer I, II, III etc.); and Maurice Daumas (gen. ed.), *Histoire general des techniques*, I (Paris: 1962) (hereafter Daumas I). The specific locations in these and other works where the manuscript is mentioned will be cited *infra*.

2 Marcellin Berthelot, »Pour l'histoire des arts mécaniques et de l'artillerie vers la fin du moyen âge,« *Annales de chemie et de physique*, 6me sér., XXIV (Decembre, 1891), 433–521 (hereafter Berthelot).

Berthelot's study caused the German engineer and historian of technology, Theodor Beck, to take an interest in Clm 197 and to devote a chapter to it in his *Beiträge zur Geschichte des Maschinenbaues* (1899).[3] Beck approached the manuscript in the same way Berthelot had. He considered it together with Clm 197, part II, but reproduced only some of the more interesting folios. His transcriptions and translations were better than Berthelot's, but still defective in many cases. More seriously, he treated the manuscript in isolation from other contemporary technical documents, although he attempted to show that some of the manuscript's illustrations reflected traditional engineering practice while others foreshadowed future developments. Beck perpetuated Berthelot's errors regarding authorship and dates. His greatest contribution to our understanding of the work was his incisive analyses of the mechanical operation of some of the machinery shown in it. It is interesting to note that he correctly analyzed several novel features in the machinery, e.g., brakes (see fol. 3v), but did not realize that they were new inventions in the fifteenth century.

The only other major study of our manuscript is by Bertrand Gille. It appeared originally as an article in 1956 and was reproduced in a slightly altered form in his *Engineers of the Renaissance* (1966).[4] We shall consider these two works together. Gille was the first to ponder the manuscript's historical context and to deal with it as part of a developing process of technical activity in the later middle ages. In particular, he insisted that the drawings represented a practicing engineer's view of the technology of his day. We shall consider the accuracy of this view in subsequent chapters. Gille, unfortunately made no serious attempt to reexamine the authorship and dates of the manuscript, and thus he repeated the errors of Berthelot and Beck. Moreover, like previous investigators, he considered only those folios he found interesting, not the entire manuscript; hence he overlooked much data of significance for the history of technology. Gille's attempt to view the manuscript as part of a process of technical activity is laudable, but he concentrated his efforts, especially in *Engineers*, on the Italian groups of technical texts, not the German tradition. He was handicapped in dealing with the German manuscripts by his desire to write a general study of broad developments in technology and by the paucity of monographic material in individual German technical treatises. As a result of all this, Gille misunderstood the meaning of many aspects of the text and its role in the northern tradition of technical writings. His two studies are

[3] Theodor Beck, »Skizzen aus der Zeit der Hussitenkriege (um 1430),« in his *Beiträge zur Geschichte des Maschinenbaues* (Berlin: 1899), 270–92 (hereafter Beck).

[4] Bertrand Gille, »Études sur les manuscrits d'ingénieurs du XVe siècle: Le manuscrit dit de la guerre hussite,« *Techniques et civilisations*, V (1956), 77–86 (hereafter Gille, »Manuscrit«) and *idem*, *Engineers of the Renaissance*, trans. anon. (Cambridge, Massachusetts: 1966), 68–73 (hereafter Gille, *Engineers*). This work will use the above-cited edition throughout. Gille's work appeared originally as *Les ingenieurs de la renaissance* (Paris: 1964). There is also a British edition, titled *Renaissance Engineers* (London: 1966). One other work dealing with the manuscript appeared about the same time as Gille's original study. Lynn Thorndike, »Marianus Jacobus Taccola,« *Archives internationales d'histoire des sciences*, XXXIV (1955), 7–26, published a brief survey of the contents of Clm 197,I. This work is too slight as it concerns part I to need extended consideration here. Where necessary, Thorndike's comments on individual folios will be noted below.

valuable in that they sought to bring Clm 197,I out of isolation, but they fail to provide either a complete view of the manuscript or a correct assessment of its relation to other contemporary documents.

Our manuscript is one of many fifteenth-century technical treatises dealing with military and mechanical technology. In order to place Clm 197,I in its proper context we must examine this literature in a general way. This is not an easy task. Most of the manuscripts have not been edited or even studied. Their authors are often anonymous or very shadowy figures about whom little is known. In many cases the works are badly cataloged and indexed under false and misleading authors or titles. Many of them are bound together with other, similar works in such a way that it is difficult to distinguish one author from another. Once one finds and isolates a particular manuscript, he encounters a new set of problems, one which must have plagued even contemporary readers of such works. In order to transmit technical information accurately, one must have settled conventions of artistic representation and of written description. The fifteenth-century authors often lacked both. Their drawings are usually sadly deficient in proper–or even comprehensible–perspective, and their language –either Latin or the vernacular tongues–had not yet developed a precise technical vocabulary. In an age when printing was in its infancy and copyright laws virtually non-existent, what we would call plagiarism was rampant. This blurs the line of demarcation between one man's work and another's and also confuses the neat modern distinction between a private notebook and a formal treatise. Technical manuscripts seem to have been copied in formal »editions« only rarely. Instead they appear to have passed from one interested party to another, some of whom made full or partial copies, occasionally altering certain drawings to suit their personal whims.

Matters gradually improved as printing became more firmly established as a means of transmitting technical knowledge. The printer produced a more legible and grammatically correct text, and also a standardized set of drawings that did not vary from one copy to another. Copper-plate engraving in particular made possible a precision unobtainable by previous methods, and many of the sixteenth-century works are copiously illustrated with such plates. In addition, changes in artistic conventions tending toward greater realism and better understanding of the laws of perspective enabled the technician to create a better and more comprehensible drawing of a tool or machine. Technical vocabularies also gradually improved, although even in the sixteenth century they never attained a high level of accuracy or brevity. Finally, the printer, if not always the author, of a technical work began to be given some measure of copyright protection for his investment of time and money. This too had a salutary effect on the transmission of technical information. As a result, the technology of the sixteenth century has an accessibility denied to that of the fifteenth. Many sixteenth-century works are familiar to the student of technical development: Biringuiccio's *Pirotechnia* (1540);[5] Agricola's *De re metallica* (1556);[6] Besson's *Theatrum instrumen-*

[5] Cyril Stanley Smith and Martha Teach Gnudi, *The Pirotechnia of Vannoccio Biringuccio* (2nd ed. rev.; New York: 1959).

[6] Georgius Agricola, *De re metallica*, trans. Herbert Clark Hoover and Lou Henry Hoover (London: 1912).

torum (1569?); and Ramelli's *Le diverse et artificiose machine* (1588).[7] By contrast, the only fifteenth-century technician whose works are known outside a narrow circle of specialists is Leonardo da Vinci.[8]

It is strange that the technical works of the fifteenth century should have been so little studied.[9] There are only a few modern editions of individual works and virtually no discussion of such treatises as a general class. Many important questions remain unanswered. Chief among these is why the flowering of this type of literature began only in the fifteenth century. To be sure, treatises on mechanical technology are not unknown before 1400. One needs only call to mind the works of Vitruvius (fl. 1st century B.C.) and Heron of Alexandria (fl. c. 62 A.D.), to say nothing of the poliorcetic writers of antiquity, to recognize that technology was considered, at least on occasion, as a proper subject for literary discussion in the classical world. Yet, though the ancients continued to be read and copied in Byzantium and Islam, the creative example they set was not followed. There were few who sought to extend and develop the technology presented in the ancient texts. For this reason, the flowering of a technical literature in the fifteenth century cannot, I think, be ascribed solely to the example of classical antiquity; it is not a »Renaissance« phenomenon in the accepted sense of that word. Rather, it would seem to have had its roots in the newer cultural and religious milieu of the later middle ages. One aspect of this was »the emergence of a conscious and

[7] There are no modern editions of either work. Jacques Besson's *Theatrum instrumentorum et machinarum* exists in more than thirteen separate editions or translations dating from 1569(?) to 1626(?). Agostino Ramelli's *Le diverse et artificiose machine* (Paris: 1588) is currently being translated by Eugene S. Ferguson, University of Delaware and Martha Teach Gnudi, University of California, Los Angeles.

[8] Leonardo represents a special case. Because of his reputation as an artist rather than as a technician many of his notebooks were edited in the closing decades of the last century. This has given Leonardo a measure of attention denied to his contemporaries and predecessors, which in turn has distorted our view of late medieval and renaissance technology. Gille's *Engineers* is devoted to demonstrating that Leonardo was heavily dependent on the work of other men and that he was genuinely original in only a few areas. Gille's work is a useful counterweight to the uncritical adulation that has so often been given Leonardo, but it overstates the opposite case quite seriously. The appearance of two new Leonardo manuscripts, the Madrid codices (ed. Ladislao Reti, New York: 1974) has already begun to shift the assessment of Leonardo's role in the development of western technology. See Ladislao Reti, »The Leonardo da Vinci Codices in the Biblioteca Nacional of Madrid,« *Technology and Culture,* VIII (1967), 437–45 and »Leonardo and Ramelli,« *Technology and Culture,* XIII (1972) , 577–605. This book focuses on the broader currents of fifteenth-century technology only as they are relevant to our manuscript. Perforce, it must consider Leonardo's works only where they illustrate some phase in the development of a device illustrated in our text. It should not, therefore, be taken as an attempt to deny the genius or originality of Leonardo. For a list of Leonardo manuscripts, their editors and dates of publication, see Gille, *Engineers,* 252–53.

[9] It is still extremely difficult to find bibliographical information in these works. The best recent source is Gille, *Engineers,* 247–53 but his listing is far from complete and at times misleading in its classification of certain manuscripts. Max Jähns, *Geschichte der Kriegswissenschaften, vornehmlich in Deutschland,* Vol. I *Altertum, Mittelalter, XV. und XVI. Jahrhundert* (München and Leipzig: 1889) contains an almost exhaustive listing of German manuscripts and much information regarding foreign ones as well. However, Jähns is very difficult to use because many of his bibliographic references are now badly out-of-date.

generalized lust for natural energy and its application to human purposes.«[10] This new consciousness can be traced back as far as the twelfth century in the thin stream of novel technical treatises that began to appear then. The new »lust for natural energy« demanded a new type of literature dealing with technological processes, and this demand began to be met by such works as Theophilus Presbyter's *De diversis artibus* (1122–23). This was followed about a century later by the earliest of the technological works to carry illustrations, Villard de Honnecourt's architectural notebook (c. 1235).[11] Like Vitruvius' *De architectura* Villard's notebook includes a section devoted to machinery. In the fourteenth century this trend continued and increased. For example Guido da Vigevano's *Texaurus regis Francie Acquisitionis terre sancte* (1335),[12] written to incite Philip VI to a new crusade, contains drawing of military devices to help subdue the infidel. Giovanni de'Dondi has left us a description and drawings of his complex astronomical clock *(astrarium)* which he finished building c. 1362.[13]

The new consciousness of power technology explains why the later middle ages produced works devoted to technology but it does not tell us why the fifteenth century saw such a large number of works of this type. A glance at the manuscript bibliography in this volume will give the reader some indication of how many manuscripts of this literature survive. To understand why there should have been such an outpouring only in the fifteenth century,

---

[10] Lynn White, jr., *Medieval Technology and Social Change* (London: Oxford University Press, 1962), 129 (hereafter White).

[11] For Theophilus cf. C. R. Dodwell (ed. and trans.), *De diversis artibus* (London: 1961) and J. G. Hawthorn and Cyril S. Smith, *On Divers Arts* (Chicago: 1963). Regarding the date see Lynn White, jr., »Theophilus Redivivus,« *Technology and Culture*, V (1964), 230. For Villard see Alfred Darcel, *Album de Villard de Honnecourt* (Paris: 1968).

[12] Paris, Bibl. Nat., Ms. latin 11015, fols. 32–55. A second copy, ascribed to one Martin of Aachen and written on Cyprus in 1375 is now at Yale University. A. R. Hall's treatment of the text is found in his »Guido's *Texaurus*, 1335,« in Bert S. Hall and Delno C. West, *On Pre-Modern Technology and Science: A Volume of Studies in Honor of Lynn White, jr.* (»Humana Civilitas,« Vol. 1: Center for Medieval and Renaissance Studies, University of California, Los Angeles; Malibu: 1976), 11–52. Hereafter Hall and West. This volume appeared too late for inclusion in all notes, but will be mentioned where appropriate and in the bibliography. Cf. also Carlo Promis, »Gl'i ingegneri militari,« *Miscellanea di storia italiana*, XII (Turin: 1871), 418–19; A. R. Hall, »The Military Inventions of Guido da Vigevano,« *Actes du VIIIe congress international d'histoire des sciences*, Vol. III (1958), 966–69; and »More on Medieval Pivoted Axles,« *Technology and Culture*, II (1961), 17–21; Gille, *Engineers*, 28–30; Sarton, III, 846–7. Guido was not the first to consider technology a partial answer to the problems of defeating the Muslims. Hall's edition cited above also notes Aegidius Columna, *De regimine principium* (c. 1285) and Marino Sanudo (Torsellus the Elder), *Liber secretorum fidelium crucis super Terrae Sanctae recuperatione et conservatione* (c. 1306–13) as works which devote some attention to technological methods to assist crusading armies.

[13] Silvio A. Bedini and Francis R. Maddison, *Mechanical Universe–The Astrarium of Giovanni de'Dondi* (»Transactions of the American Philosophical Society,« N. S. LVI, Part V; Philadelphia: 1966) trace the history of the *astrarium* and de'Dondi's descriptive treatise and show modern reconstructions of the clock.

several general factors must be taken into account. The frequency with which technical manuscripts were copied in the fifteenth century is *prima facie* evidence of a new attitude on the part of technicians toward the methods of transmission of technical knowledge. We are sadly lacking in information about how one generation of technicians instructed another prior to c. 1400, but we may safely assume that they did not exchange much information in written form (or at least that, if they did, their works have not survived). This may mean that technicians as a class were largely illiterate in earlier centuries, or it may mean only that they deliberately chose not to commit the rules of their arts to writing. In either case, the situation was changing by the 1400's. By then, the general level of literacy in Europe was considerably higher than it had been previously. Moreover, and more importantly, the level of complexity of technology was also increasing, especially in the relatively new field of firearms and cannon. These two factors taken together seem to have produced a new awareness of the value of written and pictorial treatises in the work of preserving and transmitting technical information. One of the most widely circulated of the technical manuscripts, the primitive gunner's handbook known as the *Feuerwerkbuch*, confirms this argument indirectly. Many manuscripts of this work include a short hortatory section on the duties and responsibilities of the *Büchsenmeister* (»Master gunner«). In addition to fearing God and honoring his prince, the master gunner should also know the art of reading and writing *(der meister sol auch kennen schreiben und lesen)* since without this skill he could not possibly »keep in mind all the elements which belong to this art,« i.e. the various procedures, formulae and recipies.[14] The genuine novelty of this attitude must be emphasized. Former generations of technicians had, as far as we can tell, kept in mind the rules of their arts without frequent recourse to written texts; the *Büchsenmeister* could not be expected to do this. He had to rely on the written word (or drawings) to record the complexities of his profession, and with this reliance went the habit of producing technical treatises of the sort we are now considering.

The geographic distribution and character of the fifteenth century texts also give us clues as to their genesis. Most of the known manuscripts were produced in northern Italy or southern Germany or Austria. Isolated examples are found as far east as the Siebenbürgen (the western part of modern Rumania) and as far west and north as the Seine valley, but there are no manuscripts of the sort we are studying from England, Scandinavia, the Low Countries or the Slavic area. The earliest treatises in Germany and Italy have one dominant theme, warfare. The first of them, Conrad Kyeser's *Bellifortis* (c. 1405),[15] is devoted mainly, but not exclusively, to apparatus for military use. Giovanni da Fontana's *Bellicorum instru-*

[14] The quotation is taken from ÖNB Cod. 2987, fol. 32r. A similar passage is found in ÖNB, Cod. 2952, fol. 35v. It is uncertain whether the same sentiments are repeated in all copies of this work.

[15] Götz Quarg (ed.), *Conrad Kyeser aus Eichstätt Bellifortis* (2 vols.; Düsseldorf 1967). Vol. I is a facsimile reproduction of the best early manuscript, Göttingen, Universitätsbibliothek, Ms. philos. 63. Vol. II contains Quarg's transcriptions, translations and commentary (hereafter Quarg, *Kyeser*, I and II). Quarg's work has been heavily criticized, primarily on philological grounds, by Hermann Heimpel in *Göttingische gelehrte Anzeigen*, CCXXIII, Heft 1/2 (1971), 115–48. He gives an extensive list of corrections to this edition.

mentorum liber (c. 1420, possibly 1420–49),[16] probably the earliest Italian work of this type, shows a similar preoccupation with military matters. The unillustrated *Feuerwerkbuch* that often circulated with the illustrated treatises is devoted entirely to gunpowder-making and gunnery. The new type of warfare of the late fourteenth and early fifteenth centuries played an important role in producing this literature. Significantly, both Germany and Italy were highly fragmented politically, and wars between cities or principalities were endemic. Moreover, these were fought more and more by mercenary armies. War was becoming a specialized business to be left in the hands of competent professional experts. In this context the military engineer was also becoming a specialist, a man who devoted himself to technical aids in waging war. This did not mean that he had no other technical interests. Indeed, the engineer was expected to earn his living in time of peace by assisting in construction projects, designing water systems, making automata to amuse the noble court, or any number of other occupations.[17] Nevertheless, he was clearly becoming more and more a specialized military technician as the fifteenth century progresses. This change is mirrored in the authorship of the texts. The earliest were written by men who could be considered »amateurs« in technology: Guido da Vigevano (c. 1335), Conrad Kyeser (c. 1405) and Giovanni da Fontana (c. 1420) were all physicians, although Kyeser had seen military service with the armies of Sigismund at Nicopolis in 1396. (Physicians, it should be noted, were the only medieval educated professional group–clergymen and lawyers are the other two–whose work by its nature was intimately concerned with natural phenomena.) The texts of the later fifteenth century, particularly in Germany, do not show this »amateur« authorship. When the authors can be identified at all, they almost invariably turn out to be military specialists, men such as Johann Formschneider, a master gunner from Nürnberg, or Philibs Monch, »der Pfalz Büchsenmeister.« This trend toward military authors reached overwhelming

[16] Cod. icon. mon. 242 (hereafter Giovanni da Fontana). The treatise was not titled by Fontana, but by a later owner oι the work. It reflects the predominant character of the manuscript, which is also written mainly in code. There is no edition, but see Alexander Birkenmajer, »Zur Lebensgeschichte und wissenschaftlichen Tätigkeit von Giovanni Fontana (1395 ?–1455 ?),« *Isis*, XVII (1932), 34–53. Birkenmajer thinks the work dates from the latter half of Fontana's life. Marshall Clagett, »The Life and Works of Giovanni Fontana,« *Annali dell'Istituto e Museo di Storia della Scienza di Firenze*, I (1976), 5–28 indicates 1418–20 as the most likely date of composition. I am grateful to Professor Clagett for his showing me the typescript of this study. Cf. also Frank D. Prager, »Fontana on Fountains: Venetian Hydraulics of 1418,« *Physis*, XIII (1971), 341–60. Fontana's work is referred to frequently in the literature. Paul L. Rose has undertaken an edition of the text.

[17] A letter from Leonardo da Vinci to Ludovico Sforza »Il Moro« written about 1482 and preserved in the *Codex Atlanticus*, fol. 319 r lends support to this assertion. Leonardo, seeking employment at the ducal court, advertises himself as a completely competent engineer, stressing his skills at such military engineering feats as siege warfare, bridge bulding, naval warfare, gunnery and non-gunpowder artillery, and sapper operations. He then adds that in times of peace he is skilled in architecture, painting, hydraulic engineering, sculpture and bronze casting. We know also that in actual service to the Sforzas he designed automata and stage settings. Although Leonardo was a remarkable man, there is nothing to indicate that the range of skills he promised was extraordinarily broad by the standards of his day (albeit that his skill at executing them would have been). For a translation of the letter and mention of his automata, see Ladislao Reti and Bern Dibner, *Leonardo da Vinci, Technologist* (Norwalk, Connecticut: 1969), 45–46 and 49.

proportions in Germany during the 16th century. Thus warfare, together with the emergence of technical experts in waging war, was another important factor in stimulating the appearance of the technical literature of the fifteenth century.

Let us now turn to a more detailed consideration of this literature. Although both the German and Italian works initially shared the same impetus, they quickly diverged in their technical concern. We may speak of two schools or »traditions« of technical writing in the fifteenth century, the northern or German tradition and the southern or Italian tradition. The Italian engineers never lost interest in the art of war, but they combined this with other interests, especially architecture and applied mechanics, resulting in a broader set of concerns than is found among the Germans. Recent studies have revealed that Filippo Brunelleschi's work on the cupola of the Florentine *Duomo* (1420–35), during which he invented a number of new construction machines, was the font of this interest in »civil« machinery.[18] Many of Brunelleschi's designs (now lost in their original forms) appear repeatedly in later engineering notebooks and were improved with each new generation.

The main inheritor of the Brunelleschian mantle was Mariano di Jacopo of Siena, surnamed »Il Taccola« (d. 1453–58). Both his earlier work, *De ingeneis* (done 1427–33 but never revised) and his later and more formal *De machinis* (1449) blend classic traditionalism and Brunelleschain novelty.[19] His treatises are interesting for the ways in which they outline the path the Italian tradition was beginning to take. His fellow Sienese, Francesco di Giorgio Martini (1439?–1502) continued the same tradition after Taccola–some Taccola manuscripts have annotations by Francesco–but added new and original designs in many areas. His main works, collectively given the modern title *Trattati de architettura, ingegneria e arte militare* (1482–1501) are filled with interesting material.[20] Francesco's work was used by Leonardo da Vinci and was repeatedly plagiarized during his lifetime and in the sixteenth century.[21] It is interesting to note that the Italian tradition, in combining military engineer-

[18] Frank D. Prager and Gustina Scaglia, *Brunelleschi: Studies of His Technology and Inventions* (Cambridge Massachusetts: 1970), drawing on other recent studies, demonstrate this clearly.

[19] The manuscripts of Taccola's works are scattered. Books I and II of *De ingeneis* occupy part II of Clm 197, Books III and IV are Florence, Biblioteca Nazionale, Ms. Palatinus 766. The best manuscript of *De machinis* is Clm 28800. Copies include: New York Public Library, Spencer Collection, Ms. 136; Paris, Bibliothèque Nationale, Ms. lat. 7239; and Venice, Biblioteca Marciana, Ms. lat. VIII 40 (2941). Cf. Prager and Scaglia, xi and Paul Lawrence Rose, »The Taccola Manuscripts,« *Physis*, X (1968), 337–46. There are two new editions of Taccola's works: James H. Beck (ed.), *Mariano di Jacopo detto Il Taccola: Liber tertius de ingeneis ac edifitiis non usitatis* (Milano: 1969) and Gustina Scaglia (ed.), *Mariano Taccola: De machinis* (Wiesbaden: 1971). Frank D. Prager and Gustina Scaglia, *Mariano Taccola and His Book* **De ingeneis** (Cambridge, Massachusetts: 1972) falls between a study and an edition, but it reproduces numerous folios from Taccola's earlier work. I have relied on Scaglia, Prager and Scaglia, *Taccola*, and my own study of Clm 197, II in this work.

[20] The *Trattati* exist in two main manuscripts; Florence Biblioteca Laurenziana, Ms. ex-Ashburnham 361 and Turin, Biblioteca Reale, Ms. Saluzziano 148. Both have been combined in Corrado Maltese (ed.), *Francesco di Giorgio Martini: Trattati di archittetura, ingegneria e arte militare* (2 vols; Milano: 1967) (hereafter Maltese).

[21] See Ladislao Reti, »Francesco di Giorgio Martini's Treatise on Engineering and Its Plagiarists,« *Technology and Culture*, IV (1963), 287–98. Leonardo possessed and annotated the Laurenziana 361 manuscript.

ing and architecture, retained more of an amateur flavour for a longer time than did the German. Taccola was a notary and an artist by profession, while Francesco di Giorgio was an architect-artist who also did some sculpture. Leonardo, naturally, was perhaps the most astounding amateur of all time (though he often earned his living as a military engineer). Much of the breadth of the Italian tradition (which has made it so interesting to historians) is a result of this »amateurish« character. The exception to this is Roberto Valturio (1413–84), whose *De re militari* was composed about 1455 and printed at Verona in 1472. Valturio's contribution is more limited, more conservative, and, as its title indicates, more military in its concerns than either Taccola's or Francesco's. Nevertheless, it acquired a contemporary fame far larger than any other. There are over twenty-two manuscript copies extant; Valturio's drawings were plagiarized countless times; and he was the only fifteenth century technical writer to be printed during his lifetime.[22]

The German tradition is less continuous, more variegated than the Italian, but also considerably more restricted with respect to its subject matter. German authors do not seem to have been as closely connected with each other as were the Italians, perhaps because there were more of them spread over a wider area. Yet at the same time, there is a much greater tendency among the Germans to produce direct copies of someone else's original. The Germans seem to have specialized somewhat earlier than the Italians, and with specialization came a heavy (though not exclusive) concentration on military technology. In the entire fifteenth century German tradition, one sees the powerful influence of two works: Kyeser's *Bellifortis* and the anonymous *Feuerwerkbuch*. Kyeser in particular made an enduring impression north of the Alps. He provided the stylistic model for most of the pictorial works produced before c. 1480, and the range of Kyeser's technical concerns seems to have defined the range of later authors to a considerable extent. Fortunately, that range was broad enough in the beginning to allow for considerable variation in later years. The full degree of Kyeser's influence has not yet been adequately assessed. In addition to the numerous whole or partial copies of *Bellifortis* still extant, one finds illustrations from this work, or recognizable variants, adorning treatises of the sixteenth and even the seventeenth centuries. Although *Bellifortis* was not printed in its entirety until 1967, some of its illustrations (together with some of Valturio's) graced and early sixteenth-century printed edition of Vegetius.[23] There is a tendency within Kyeser's work to embellish technological information with astrological or magical symbols, images of past heroes—especially Alexander the Great—and grotesque figures of men or beasts. The ultimate meaning of this linkage between technology and magic will require further study,[24] but its appearance in Kyeser helps to elucidate the same

[22] For a list of manuscripts, see Gille, *Engineers*, 250 and Erla Rodakiewicz, »The *editio princeps* of Robert Valturio's *De re militari* in Relation to the Munich and Dresden Manuscripts,« *Maso Finiguerra*, V (1940), 15 ff.

[23] See the manuscript bibliography in this volume for a listing of *Bellifortis* manuscripts. The lists in Quarg, *Kyeser*, II: xxx ff. and Gille, *Engineers*, 247–48 are limited to the principal copies.

[24] One can speculate that it certainly was not mere whimsey. Kyeser was a physician, which in the fourteenth and fifteenth centuries meant that he was necessarily adept at astrology. Moreover, the distinction between

tendency in later works, as well as the frequent appearance of technological works in conjunction with treatises on such arts as numerology, divination, or alchemy. Kyeser's blend of magic and technology, of fact and fancy, was echoed strongly at first within Germany, but this gradually died out during the course of the century. Perhaps it fell victim to the growing tendency toward professional military authorship of technical works. As we shall see, there are traces of Kyeser's influence on Clm 197, I.

The *Feuerwerkbuch* is the other staple item in the fifteenth-century German tradition. The commonly accepted date of its initial composition is c. 1420, but the earliest dated copy is 1429, and there are numerous undated manuscripts, some of which may prove to have been earlier.[25] *Feuerwerkbuch* is thoroughly practical and didactic in nature; it contains none of the magical or fantasy elements found in Kyeser. Part of the work contains a simple set of theoretical generalizations regarding the art of ballistics, but the bulk is given over to pyrotechnic recipies of considerable variety and complexity. Some of the copies I have examined show signs of spills and burns on a few pages, which indicate that these copies may have been used like a cookbook in the preparation of gunpowder mixtures. The full extent of *Feuerwerkbuch's* influence is still unclear. The work appears in a host of manuscripts–over thirty are included in the bibliography–and it continued to be copied into works in military technology throughout the sixteenth century. It reached print for the first time only in 1529, but it received two editions in that year.[26] Because of the nature of Clm 197, I, *Feuerwerkbuch* was less directly influential on it than *Bellifortis*, but the range of concerns it shows was defined in part by the milieu that produced *Feuerwerkbuch* as well.

The practical, prosaic, unillustrated *Feuerwerkbuch* and the fanciful, wide-ranging, lavishly illustrated *Bellifortis* together constitute the two main poles of the fifteenth-century Germanic tradition. Most other works can be defined in terms of their relation to one or the other. It should be noted here again, however, that there is much room for innovation, variation and development of the basic themes laid down in these works. *Bellifortis* copies, for example, often mix additional, original material with Kyeser's drawings; in some cases

manipulation of the physical world through technology and manipulation of the cosmos through magic was not always clear to Kyeser's contemporaries. Indeed, even J. B. della Porta's late sixteenth-century treatise on chemical and household technology bore the title *Natural Magic*. Finally, it is possible that Kyeser himself had a personal belief in magic that was extraordinary even for his day; he may even have been a sorcerer! See Lynn White, jr., »Kyeser's Bellifortis; The First Technological Treatise of the Fifteenth Century,« *Technology and Culture*, X (1969), 436 ff. See also *idem*, »Medical Astrologers and Late Medieval Technology,« *Viator*, VI (1975), 295–315.

[25] The only modern edition of this work merely reprints and comments on the Augsburg printed edition of 1529. It makes no attempt at a critical edition or collation of the extant manuscripts. Wilhelm Hassenstein, *Das Feuerwerkbuch von 1420* (München: n.d. 1941–43). Hassenstein's work shows strong traces of National Socialist propaganda themes and should be used with extreme caution.

[26] Augsburg: Heinrich Stainer under the title *Hye nachvolget vonn Büchsen geschoss, Pulver, Feuerwerck...*; Strasbourg: Christian Egenolphen under the title *Büchsenmeÿsterei. von Gschoss, Büchsen, Pulver, Salpeter und Feuerwerck* etc. ... In the absence of a fully collated modern edition, the precise relationship between these printed editions, as well as their relationship to the numerous fifteenth-century manuscripts, is uncertain.

later copies modify a Kyeserian original into something entirely novel. In addition, there are certain texts which seem to blend the style of Kyeser with the subject matter of *Feuerwerkbuch*. This group is designated by the term »Primitives.« The term was coined by Gille in the belief that the earliest examples of this type ante-date *Bellifortis*.[27] This can no longer be maintained, but the word is still useful if understood to refer only to an artistic style of drawing which is fairly crude and habitually uses a rubricated border around each illustration. The oldest of this group, Nürnberg, Hs. 25,801, and Cgm 600, are very short and are concerned exclusively with saltpeter and gunpowder production. Later members, ÖNB Cod. 3069, Waffensammlung, Ms. P5014 and Zürich, Zentralbibl., Ms. Rh. hist. 33b retain the earlier drawings (and the crude style) but eliminate more and more of the text while adding other illustrations, some from *Bellifortis*. The original concept of an illustrated treatise is pushed to its logical extreme here; text is abandoned altogether in favor of drawings alone. (Some *Bellifortis* copies show this same tendency; it was not illogical in view of the extreme difficulties involved in expressing technical information in fifteenth-century language.) It is interesting to note that the last of this group, Waffensammlung, P5014 (c. 1435), contains a pictorial dedication to Emperor Sigismund; it is entirely without text, but is done on parchment in what is obviously a deluxe manner for presentation to an important person, probably Sigismund himself. This progression from crude, short original to elaborate, stylized presentation copy within about thirty years is illustrative of a repetitive tendency within the Germanic tradition. One frequently finds a similar group of drawings repeated in increasingly more elegant artistic fashion until they at last appear in compendia of technical drawings that are obviously intended for the libraries of a noble or royal patron. Some of the drawings in Clm 197,I were accorded this same treatment.

Until the latter third of the fifteenth century the majority of German technological works continued to be only copies or revisions of *Bellifortis* and *Feuerwerkbuch*, or, naturally, mixtures of both. Two types stand out as exceptions to this, however. The first is represented by only one manuscript, Waffensammlung, Ms. P5135, which dates from c. 1410–c. 1430. The inevitable sections of *Bellifortis* and *Feuerwerkbuch*, are included, but a major portion of the whole appears to be an original treatment of military technology. Since this work has not been studied in any depth at all, final judgment about its contents must be reserved. The second exceptional type appears in five manuscripts from the fifteenth century:

1. Cgm, 734
2. München, Deutsches Museum, Ms. 1949/258
3. Nürnberg, Hs. 719
4. Sibiu, Mun. Arch., Ms. Varia II 374
5. Weimar, Nat. Forsch. u. Ged., Ms. qu. 342

[27] Gille, *Engineers*, 56–57. Watermark evidence contradicts this, although it is barely possible that Nürnberg, Hs. 25,801 might have been slightly earlier that the original *Bellifortis*. I know of no evidence to support Gille's opinion that these stem from a common *Urtext* of the late fourteenth century.

The dates of all these manuscripts are still very tenuous, but they appear all to stem from the third quarter of the century. The only name that has emerged as a candidate for authorship of this group is that of one Johann Formschneider, a *Büchsenmeister* from Nürnberg (fl. c. 1460–c. 1480). Formschneider may have been either author or a plagiarist of the text and drawings, but for the sake of convenience we shall call these the »Formschneider« types. »Formschneider's« knowledge was apparently largely restricted to gunnery and related matters, and, as one would expect, he seems to have known *Bellifortis* quite well. His drawings and descriptions of cannon and siege engines appear to be an attempt to improve and update Kyeser by modifying the latter's designs in light of advances in cannon mounting and aiming techniques. The two Münchener manuscripts and the Nürnberger are brief copies (fragments?), but the Weimarer is a deluxe parchment edition, containing also the *Feuerwerkbuch*. The Sibiu manuscript is the most interesting of the group. It was written by successive master gunners in the service of that city, (formerly Hermannstadt) and contains *Feuerwerkbuch* plus a sixteenth-century treatise on rocketry by Conrad Haas as well as the »Formschneider« segment, which is included as part of a collection of drawings from German and Italian sources.[28] The author of this portion is identified only as an anonymous Nürnberger. »Formschneider's« work seems to have acquired some fame, since drawings taken from it appear in a French seventeenth-century compilation of older technical authors (Paris, Bibl. Nat., Ms. franc. 20006).

From roughly 1470 onwards the German tradition produced several works containing a greater amount of new, fresh material. Generally, this period also saw a lessening of the direct influence exerted by the earlier major works. In part, the new appearance of technical works was linked with the emergence of a new artistic style that was based on Italian and Flemish models, a style that was most highly developed within Germany in the area of Nürnberg. The best example of this artistic style in a technological manuscript is found in the so-called *Mittelalterliches Hausbuch* (c. 1480) by the anonymous »Master of the Amsterdam Cabinet.« The *Hausbuch* has been fairly well studied by art historians, particularly because its style anticipates Dürer's in many ways.[29] The illustrations in the *Hausbuch* contain many astrological figures (whose relationship to Kyeser's is yet unkown), garden and workshop scenes, and machinery of both the military and domestic sort. The text is largely composed of recipes for pyrotechnics, but its treatment of these themes, technologically as well as artistically, is quite novel. A further example of this same tendency is seen in Cgm 356 (c. 1492–c. 1512). Here the traditional material is again reproduced, *Feuerwerkbuch* and drawings from Kyeser and Valturio, but this is supplemented by a substantial

[28] This manuscript is the subject of a study by Doru Todericu, *Preistoria rachetei moderne: Manuscrisul de la Sibiu* (Bucharest: 1969). A shorter, but more readily available discussion is *idem,* »Raketentechnik in 16. Jahrhundert: Bemerkungen zu einer in Sibiu (Hermannstadt) vorhanden Handschrift des Conrad Haas,« *Technikgeschichte,* XXXIV (1967), 97–114. As the titles indicate, Todericu focuses heavily on the sixteenth-century portion of the text.

[29] Helmut Theodor Bossert and Willy F. Storck (eds.), *Das mittelalterliche Hausbuch* (Leipzig: 1912) (hereafter Bossert and Storck).

amount of new material from as yet unidentified sources. Another way this codex breaks with past models is that it reproduces no known work, but gives instead entirely new drawings and text, including part of a little-studied tractate by one Martin Merz, *Kunst aus Büchsen zu schießen*. Two manuscripts from Heidelberg are exclusively military in their concerns: Cpg 126 and 130. The former, done in 1496 by a master gunner in the Rhineland named Philibs Monch, is a fresh approach to cannon and siege engines (possibly showing some Italian influence). The latter, done by Ulrich Bessnitzer of Landshut c. 1485, is a pictorial inventory of an imaginary(?) armory. The advantages of pictorial inventories for keeping track of ordnance are obvious, and such works at times reached respectably high levels of artistry, e.g. Cod. icon. Mon. 222, a *Zeugbuch* done by Bartholomeus Freysleben for Emperor Maximilian I c. 1495–1505.

The turn of the century saw the beginnings of what would later become an encyclopedic literature on military technology. In its developed form, this literature is characterized by its comprehensive character, the large size of its manuscripts (generally 250 fols. or more per copy), the elegant quality of its volumes, (most of which were obviously produced under aristocratic patronage), and by its indebtedness to fifteenth-century forerunners. The sixteenth-century literature has scarcely been studied at all, and this work will, of necessity, have to ignore most of them.[30] We should note in passing only that two of the most frequently copied reproduce verbatim portions of *Feuerwerkbuch* and contain as well illustrations that are recognizable descendants from Kyeser or Valturio. These are Franz Helm's treatise on military techniques, which originated in the 1530's and an anonymous work beginning »Ein Buch durch einen gelehrten Kriegsverstendigen zusammen gezogen aus vielen probierten Kunsten und Erfahrungen ...« The latter's date of origin is uncertain, but it appears to be older than Helm's work. Texts of this sort were foreshadowed by a different type of encyclopedic compilation which is even more directly related to the fifteenth-century tradition. Two extant manuscripts, Erlangen, Universitätsbibl., Ms. 1390 and Weimar, Nat. Forsch. u. Ged., Ms. fol. 328, are large collections of illustrations taken from fifteenth-century sources. The Erlangen work, which runs to 305 fols., was done c. 1500 by Ludwig von Eybe zum Hartenstein (the Younger). The Weimarer collection is anonymous and undated, although a dubious tradition holds that it once belonged to the Albanian hero Skanderbeg (Georgius Castrioti).[31] It is a parchment codex with very fine pen and ink drawings, some of which may be Italian in style. To judge from the drawing style and the machinery and costumes depicted, the manuscript must date from the early years of the sixteenth-century, or, at the very oldest, the closing decade of the fifteenth. Since the Weimarer text includes extensive borrowings from Clm 197,I, we shall refer to it frequently. Both these

30 The only comprehensive discussion of the sixteenth-century German military texts I have found is still that of Jähns, Vol. I. (see note 9).

31 A discussion of the manuscript's history and the Skanderbeg attribution is found in Vulpius *(sic)*, »Sogenannten Skanderbegisches Ingenieur Kunst- und Wunderbuch,« *Curiositäten der physisch-literarisch-artistisch-historischen Vor- und Mitwelt*, X, Stück iv (1824), 289–308. See E. Marx, »Bericht über ein Dokument mittelalterlicher Technik,« *Beiträge zur Geschichte der Technik und Industrie*, XVI (1926), 317–21.

manuscripts, of course, will prove extremely valuable for future study of fifteenth-century German technology, since they reproduce drawings that have otherwise disappeared from the record.

Finally, two additional manuscripts from the sixteenth century are worth noting. Both depart somewhat from the mainstream's concentration on matters military. Pr. Staatsbibl., Ms. germ. fol. 132 is a collection of drawings and descriptions of tools, weapons, torture devices, animal traps, household apparatus, and several recipes. It was done in 1505, most probably in the vicinity of Nürnberg, and it may be related distantly to the *Hausbuch* whose style it superficially resembled.[32] The extremely sophisticated pictorial technique of the unknown draftsman should allow the historian a very detailed glimpse of the devices it illustrates. Cgm 973 is a collection of texts and drawings produced by a minor Münchener artist, Christof Seselschreiber, in 1524. It includes many items, e.g. bridges and fountains, that Seselschreiber had seen during his travels, but there is also a large number of illustrations taken from older works as well. Included in this latter category are several drawings taken directly from Clm 197,I, which Seselschreiber must have seen at some point.[33] Cgm 973 is of relatively little importance in trying to understand the overall development of late medieval technology, yet its rather unimaginative, derivative character makes it a useful collection through which to see the works of other men.

In closing this brief introduction to the texts on technology from fifteenth-century Germany, a cautionary note should be sounded. Manuscript studies have been hitherto an esoteric and often neglected aspect of the history of early technology. Because of the relative paucity of information available, writers of general works have been driven to use whatever secondary literature was at hand (the alternative being to spend a lifetime studying the sources themselves). All too often this has meant that weak, obsolete or incomplete secondary studies have been quoted from again and again with little or no check on the accuracy of the information they present. For this reason, anyone who presumes to say something about the sources, the manuscripts themselves, bears a heavy responsibility to make as clear as possible the limits of accuracy of what he writes. In the final analysis, the only way that a complete and adequate study of the German manuscripts here discussed can be pursued is through the laborious cross-comparison of each text with all others of its type, with the full application of all the skills and techniques available to the investigator. Nothing of the sort has been done yet for these manuscripts. To do so would be a very long-term project involving extensive use of microfilm collation, sophisticated indexing programs, and perhaps computers as well. Until such a work is undertaken (and I am presently planning a project of this type), all statements about the manuscripts are to be taken as provisional, tentative

[32] There seems to be no mention in the learned literature of this text, which has now perished in its original. The Staatsbibliothek Preussischer Kulturbesitz possesses a photocopy.

[33] A discussion of Seselschreiber's biography and Cgm 973 is found in a rare and inaccessible journal that was published in conjunction with an exhibition of handicrafts in München in 1926–27. See Otto Hartig, »Christof Seselschreiber and sein Buch über Büchsenmeisterei and technische Erfindungen,« *Kultur des Handwerks,* Heft 9 (July, 1927), 278–84.

judgments, subject to revision upon further study. They are as accurate and as complete as the present state of knowledge about the subject permits, but because studies in this field are still in their infancy, the level of both accuracy and completeness is fairly low. The situation is analogous to that which faced the early sixteenth-century cartographers in attempting to construct maps of the Americas from fragmentary and often inaccurate explorers' reports. In seeking to relate each isolated bit of information to a comprehensive picture of the whole, the cartographers often produced ludicrous results. A similar danger awaits us here. Clm 197,I can be understood only against the background of the texts here described, and yet this background is visible to the researcher only in vague and shadowy outlines. On the level of specific details the situation is somewhat more favorable, since a reasonable amount of information can be gained from the individual folios of the texts discussed, but the larger task, that of producing a comprehensive picture of the entire Germanic tradition, will have to await future research efforts.

CHAPTER II

THE MANUSCRIPT

Description

Clm 197, part I is an anonymous, paper, illustrated manuscript that depicts and describes military devices and some civilian machinery as well. It occupies the first forty-eight folios of the codex. Part II consists of Books I and II of *De ingeneis* by Mariano di Jacopo Taccola. The drawings in part I are crude, but usually accurate if interpreted within the artists' conventions of representation. They are colored and executed on both sides of the folio. A few drawings cover two adjacent pages. Nineteen of the ninety-six pages contain some text, ranging in length from one or two lines to a paragraph. All the passages are early New High German prose and all describe devices on the same of a facing page. None is a paraphrase or variant version of passages from any other known work.

The codex is a folio volume in a modern, eighteenth-century(?) binding. Pagination, in a modern hand, is continous, with no lacunae or duplications in part I. The quires are as follows: A^7 (fols. 1–14); B^4 (fols. 15–22); C^3 (fols. 23–28; these may have been part of B at one time); D^3 (fols. 29–34; but 30–31 and 32–33 folded separately); E^1 (fols. 35–36); F^1 (fols. 37–38); G^1 (fols. 39–40); H^4 (fols. 41–48). Folios 1–28 measure 310 × 215 mm., while folios 29–48 measure 325 × 220. The manuscript is in excellent condition, with only normal wear and a few repaired tears.

The manuscript is not a unified set of drawings from one pen. It is composed of two parts, hereafter called section A and section B. Section A comprises folios 1–28 and section B folios 29–48. This follows the quires and paper sizes noted above. Watermarks also confirm such a division. The fact that there are different papers quired separately does not in itself prove that there must have been two separate authors. Many manuscripts are composed in just this manner. In this codex, however, subject matter of the two sections, the styles of drawing and the implements used in drawing all indicate that the manuscript was the work of two different men.

The subject matter of the two sections differs greatly. Some items that appear in both sections are treated in radically different ways. For example, both sections show cannon. Section A's all taper off to a long, tail-like breech and appear to be muzzle-loaders. Section B's are all of a type equipped with a removable powder-chamber that fits into a cut-out segment of the breech, i. e., they are breech-loaders. A pile driver appears once in each section, but with extreme contrast in drawing techniques and design. The example from section A (fol. 8r) is badly drawn in particularly poor perspective and relies on a slip-hook to release the driver-weight to fall on the pile and a windlass to raise the weight. The example from section B (fol. 37v) is much better drawn, correct in its perspective, but fastens its lifting rope directly to the driver weight and requires a gang of men pulling on ropes attached to the lifting rope, rather than a windlass, to raise the weight. Ratchet devices appear in both sections, but they are treated differently, section B's being more carefully designed and drawn than A's (cf. fols. 1v, 32v–33r, and 42v). A screw cannon-lifting device is figured in both sections (fols. 9r and 37r), but the latter drawing is artistically superior to the former. Portable shelters for a besieging army are sketched in both sections, but section B includes more examples and shows them in considerably more elaborate forms than does section A. Finally, X-extention frames or Nuremberg shears shown in both sections differ in design and illustration (fols. 16r, 17r, and 45v).

Other differences abound. Gearing, which appears thirteen times in section A, appears only once in section B. All gearing systems in the former section are shown as part of a complete machine engaged in some work. The lone example from the latter section is shown *in abstracto* as a study in gearing isolated from a specific machine. There are differences in emphasis on particular types of machinery. For example, grain mills operated by human muscle power appear eight times in section A, but only once in section B. The latter section includes many more fixed edifices, either siege structures of wood (see fol. 41r) or permanent stone buildings, than section A. Crossbows appear only in section B, never in section A. Finally, all folios having textual explanations occur in section A; section B is completely textless.

The general drawing styles of the two sections are at considerable variance with one another. The drawings in section A are cruder, duller in color and done with blunter pens than those in section B. Section A's drawings appear to have been made by someone with little training, equipment or talent. Most of the drawings in section B, on the other hand, give the impression of having been done with a straight-edge or by a skilled freehand draftsman. Both sections distort proper artistic perspective, but the distortions of section A are naive and often internally inconsistent, while section B's techniques of representation are usually less severe, more consistent, and often formalistic. In depicting grain mills, for example, A never follows the usual medieval convention of showing the millstone in a horizontal plane and the millwork in a vertical, even in cases where this would make a drawing more comprehensible. B does this, even though it is not really required for clarity.[1]

[1] This explanation focuses on elements of disparity which can be verified from photographic reproductions. The contrast is much more apparent when viewing the colored folios of the manuscript at first hand. One may fairly ask why this division of part I was not noted in previous studies. The modern catalog of the Staatsbibliothek was first printed in 1868 and revised in 1892. See Karl Halm and Georg Laubmann, *Catalogus codicium latinorum bibliothecae regiae monacensis* (München: 1892), p. 41. Both this work and the 1868 catalog note the division of Clm 197 only into part I, anonymous German, and part II, Taccola Latin, and report only one paper size, 320 × 220 mm. for part I. This size is a rough average of the two sizes present.

In general, then, the two sections of part I differ from one another strongly enough in technical concerns and artistic style to be considered the work of two different individuals. It is possible, of course, that we may have here the work of one man done at widely separated points in time, but there is no evidence to support such an assumption. It is more reasonable to assume separate authors, particularly in view of the fact that many manuscripts of this type contain the works of separate individuals bound together. Clm 197 is itself a case in point, containing as it does these German sections bound together with a Taccola manuscript (see appendix for other examples). Since many of the technical treatises are of relatively few folios, it would have been costly to have bound each one separately. As a rule two or more related or even unrelated manuscripts were placed together to make a codex large enough to bind economically. This habit, of course, varied with the interests and financial abilities of the individual collectors and libraries. It is not surprising, therefore, to find bound together in our codex a twenty-eight folio manuscript, then one of twenty folios, both followed by a larger manuscript on a related topic. As we shall see shortly, it is probable that Clm 197 as we know it was assembled for the convenience of a former owner. There is no reason to assume common authorship merely from the juxtaposition of sections A and B.[2]

Dates

The manuscript contains little textual material with which one can date it precisely. The traditional date for both sections A and B of the manuscript (formerly thought to be a unit) is c. 1430.[3] The first reason for this dating is the mention of the siege of Saaz (often Sacz or Satz in other documents; the modern name Žatec will be used hereafter) on fol. 16v. This operation took place in late September and early October of 1421 during the second incursion of the Emperor's armies into Bohemia to combat the Hussite heresy. The second reason is the mention on fol. 23v of a wagon-fortress together with a crude drawing representing such a fortress. Author A says that he has depicted a wagon-fortress »from which the Hussites

Berthelot, Beck and Gille seem to have followed the catalogs in this matter. Also, major shifts of style and the inclusion of extraneous material are not uncommon in manuscripts of this type. By following the catalog, an investigator who is accustomed to shifts in style could easily overlook the divisions of part I, particularly if he failed to inspect the codex carefully for quires and watermarks.

[2] Since, as we shall see shortly, there is no ground on which to connect either section with the Hussite Wars of 1419–34, we shall have to dispense with the designation »Manuscript of the Hussite Wars' Engineer« and Feldhaus' »Anonymous der Hussitenkriege« as a name for its authors. Hereafter, »author A« will signify the author of section A, and »author B« the author of section B. Where either author interchangeably or both collectively are meant, »the authors« will be used.

[3] This is the date assigned to part I in Halm and Laubmann, *Catalogus*. Their reasons for this date are not stated. All subsequent investigators have followed the catalog in this matter; cf. Berthelot, 434; Beck 270 and 283; Gille, »Manuscrit,« 79; Jähns, 259. Beck also notes that the verb *vechten* (»fight«) is in the present tense in referring to the Hussite wagon-fortress on fol. 23v. From this he assumes that the text could not have been written after the conclusion of the Hussite Wars, c. 1434.

fight.« The »Hussite Wars« in the common modern understanding of that term were over by c. 1434; thus the manuscript was reasonably considered to date from the period between 1421 and 1434, or c. 1430. Since there is no other textual or pictorial evidence which *prima facie* contradicts such a conclusion, this date has never been questioned.

Other evidence, however, conflicts with so early a date. There are four watermarks in sections A and B. These form similar pairs, and each pair appears only in one section of the manuscript.[4] The marks from section A (my numbers 1 and 2) show a scale inscribed in a circle,

[4] The following is a table of the distribution of the watermarks throughout the manuscript:

	Fol. number	*Watermark number*	*Remarks*
Section A	1	1	This type somewhat more elongated than Briquet's 2504.
	2	1	
	6	2	Corresponds well with Briquet's 2504.
	8	2	
	10	1 (?)	Mark obscure; identification difficult.
	11	2	
	12	2	
	15	2	
	17	1	
	18	2	
	21	2	
	24	1	
	26	2	
	28	1	
Section B	30	3	Similar in shape to Briquet's 2571, but differs in size by 1–3 mm. in places.
	32	3 (?)	Mark somewhat obscure.
	34	3	
	36	3	
	37	4	Corresponds closely to Briquet's 2571; mark slipped slightly from chain line, rotated 1–3 degrees clockwise.
	40	3	
	41	3	
	42	3	
	44	3	
	46	3	

having round-bottomed pans, a loop or bulge about halfway up the handle, and a loop at the top. Mark 1 measures 76×42 mm. Mark 2 measures 71×42 mm (see figs. 1 and 2). The marks in section B (my numbers 3 and 4) also show a scale inscribed in a circle, but having flat-bottomed pans and lacking a loop or bulge halfway up the scale handle. The scale has a loop at the top of the handle. Mark 3 measures 74×56 mm. Mark 4 measures 76×55 mm. (see figs. 3 and 4). The similarity between marks 1 and 2 is strong enough to permit the con-

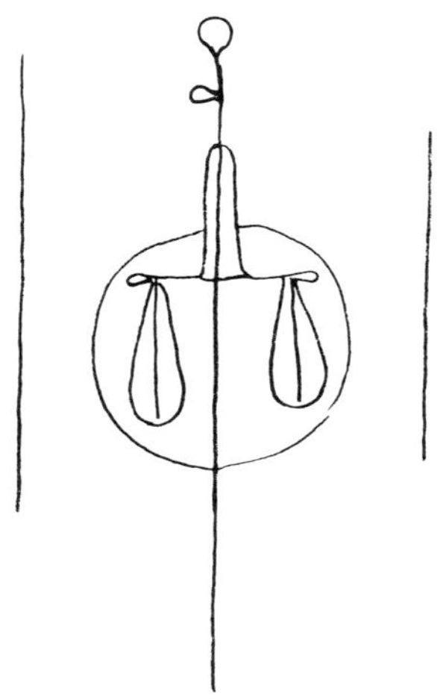

Figure 1.
Watermark 1 from section A of the manuscript.

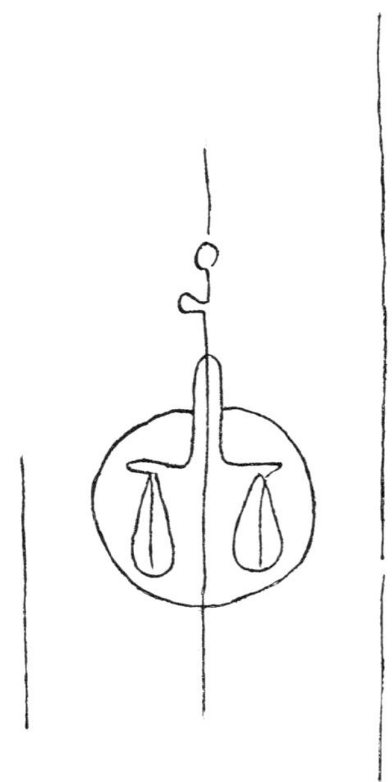

Figure 2.
Watermark 2 from section A of the manuscript.
It agrees well with Briquet's 2504.

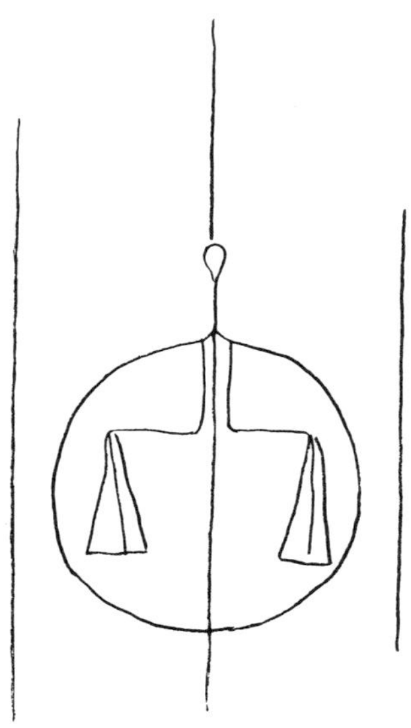

Figure 3.
Watermark 3 from section B of the manuscript.

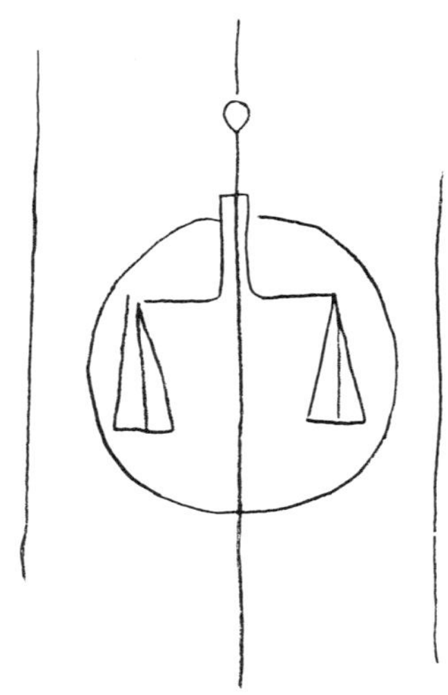

Figure 4.
Watermark 4 from section B of the manuscript.
It agrees well with Briquet's 2571.

clusion that they are pairs or mold-pairs, i.e., that they are attempts to represent the same mark from the same mill on different molds. The same is the case for marks 3 and 4. Therefore mark 1 must be considered contemporary with mark 2, while mark 3 must be contemporary with mark 4.

The first mold-pair (marks 1 and 2) must date from 1472–90, and it has been noted in German and Austrian archives only from the years c. 1472–74. The second mold-pair (marks 3 and 4) is found only in the period 1486–92. These dates are derived not only from the older catalog done by C.M. Briquet, but also from the more extensive, and still partially unpublished, Wasserzeichenkartei Piccard in the Hauptstaatsarchiv Stuttgart.[5] There can be no

[5] For marks 1 and 2, see Charles Moïse Briquet, *Les Filigranes: Dictionaire historique des marques du papier des leur apparition vers 1282 jusqu'en 1600* (2nd ed. 4 vols.; Leipzig: 1923) Vol. I, fig. 2504. Briquet's citation dates this mark's first appearance at an unknown date in a series of documents terminating in 1472. This series is the *Fridericiana* in the Vienna Haus-Hof- und Staatsarchiv; the mark, according to Briquet appears in T. XII of this collection. The *Fridericiana* is a collection of about 1100 acts from the time of the Emperor Friedrich III. (1442–93) and is today organized into ten fascicules. Cf. L. Bittner *Gesamtinventar des Wiener Haus-, Hof- und Staatsarchivs* I (Wien: 1936), 357. At the time of Briquet's research, the *Fridericiana* was organized in cartons and the document in question was probably part of the then undated section of the collection, carton 9. (Dr. Richard Blaas, Direktor des Haus-, Hof- und Staatsarchivs, communication of November 16, 1970, in response to my query). Although Briquet's citation is in error, therefore, regarding the exact location of the document in question, it is unlikely that his dates are in error. The other printed catalogs containing this mark, or reasonably close variants of it, all place the mark later than Briquet's citation, i.e., 1475, 1480–90, and 1482. Cf. Domenico Urbani *Segni di cartiere antiche* (Venice: 1870), plate I, nos. 11 and 19; also Ferdinand Ongania (ed.), *Early Venetian Printing Illustrated* (Venice, London, New York: 1895), plt. 64, no. 5. The latter work is cited by Briquet as *L'Art de l'imprimière pendant la renaissance à Venise.* It is unlikely that the watermark was in use earlier than Briquet's citation would indicate. Any lingering doubts on this point are removed by the evidence from the Wasserzeichenkartei Piccard, in which marks 1 and 2 are represented from the following sources:

a) Staatsarchiv Nürnberg, Rep. 18a, 1807 Ellingen; date: 1472
b) Staatsarchiv Königsberg (Pr.) = Staatl. Archivlager Göttingen, Ordens-Briefarchiv Elbing; date: 1474

Subsequent versions occur in:

c) Archivio di Stato di Mantova, Gonzaga 514, Rothenburg o.d.T.; date: 1474, Feb. 20
d) Archivio di Stato di Brescia, Archivio notarile Brescia 3; date: 1476
e) Archivio di Stato di Ravenna, Archivio notarile 55; date: 1476.

(I am grateful to Dr. Gerhard Piccard of the Hauptstaatsarchiv Stuttgart for his kind assistance in providing me with this information from the unpublished portion of the collection.)

For marks 3 and 4, see Briquet, fig. 2571, which he dates at 1488. The document cited by Briquet also comes from the *Fridericiana,* T. VII. This dating has been confirmed by Blaas in the above-cited correspondence. Briquet also notes a similar mark appearing in Munich in 1491–92. The Wasserzeichenkartei Piccard contains one further example of these marks: Stadtarchiv Nördlingen, *Missive* (Nürnberg, 1486).

The paper used in both section A and section B is of north Italian provenance according to Piccard, as are all papers with encircled scale marks. The use of Italian papers in German documents should not be surprising. Although few studies have been made of the size and extent of trans-Alpine paper trade, one Italian city's production which has been studied, Fabriano's, reveals a healthy trade with the northern countries, including Germany. Cf. A. F. Gasparinetti, »Paper, Papermakers and Paper-Mills of Fabriano,« (Italian essay of 1938,

doubt as to the accuracy of the dates given, and there is no indication whatsoever in any catalog of watermarks that the mold-pairs of Clm 197,I were in use as early as the 1430's. Thus, on the basis of watermark evidence alone, we may conservatively estimate the date of composition of section A as lying between 1472 and 1475. On the very same grounds, section B is roughly eleven to twenty years younger, 1486–92. The degree of confidence we may place in the revised estimate of the manuscript's age depends in part on the general level of credence we may wish to give to watermarks as dating evidence–and historians differ widely on this point. Still, in this case, the watermarks are the best available evidence for dating, since the internal evidence on which the older estimate of the manuscript's date was based is at best equivocal. Indeed, it need not conflict at all with the date suggested by the watermarks.

Examining the sections separately, we find that section B represents a very simple case. There is nothing in it to confirm or to contradict a date *circa* 1486–92. Author B's drawings of weapons, machines, buildings and people show nothing that may be used as evidence for dating. In the absence of any other data, we must conclude that the watermarks are the strongest available evidence and that section B was done between the mid-1480's and early 1490's.

The dating of section A is more complex. All the evidence that was thought to indicate c. 1430 as the manuscript's date appears in this section. The *terminus post quem* of September, 1421, the date of the siege of Žatec, is clear.[6] However, the mere mention of the battle is no reason to assume that A was a participant in the siege or that it occurred within his lifetime. The way in which he alludes to the event indicates that he knew of it only through the accounts of others (see Chapter IV and fol. 16v Commentary). The *terminus ante quem* of c. 1434–36 is open to challenge. It is based solely on the identification of A's wagon-fortress illustration with those of the early Hussites during the period 1419–34.

The wagon-fortress, in fact had been used long before 1419 and it continued in use for many years after 1434. It was employed by a number of central European and other peoples throughout the fifteenth, and in some cases well into the seventeenth century. Pictures of such fortresses and discussions of their tactical use appear in works on military matters long

---

revised 1953) printed in *Zonghi's Watermarks* (»Monumenta chartae papyraceae historiam illustrantia, Vol. III; Hilversum, Holland: 1953), 63–82. Moreover, a brief survey of Briquet's citations for marks produced in Italy, e.g., the encircled scale, certain floral marks, etc., shows that 16–22% of the citations occur in German-Austrian archives. Piccard indicates that Venice-Nuremberg was most probably the trade route for the paper of Clm 197,I. A final note should be added here. While I was engaged in preliminary research on this aspect of the study in 1970, Frank D. Prager and Giustina Scaglia were preparing their *Mariano Taccola and His Book* **De ingeneis** (Cambridge, Mass.: 1972). They were naturally concerned with Clm197, most especially with part II. Quite independently of my work, they noted that the watermarks of part I were similar to Briquet's nos. 2445–2608, and that these suggested part I was »middle or later Quattrocento, more recent than is usually assumed« (*Taccola*, 26, n. 4). I was unaware of their opinion until I had the pleasure of meeting Prager late in 1971.

[6] See Franz Palacky, *Geschichte von Böhmen*, Vol. II. pt. 2 (Prag: 1851), 252–54 for an account of the battle.

after 1434.[7] Moreover, the »Hussite Wars« that ended in 1434–36 were only the first of many »Hussite Wars« in successive decades, when conflicts between Catholic and Hussite Bohemians and between Bohemia and surrounding kingdoms were endemic. Indeed the period from the late 1460's through the mid 1480's was one of constant Hussite struggles

[7] They were used by the Visigoths at the Battle of Adrianople, 378. Cf. White, 7. The *Bellifortis* shows a schematic drawing of wagons in a marching formation. Cf. Quarg, Kyeser, I: 16v–17r; II: 18, where he speculates that this may be a wagon-fortress. The *Kriegsordnung* of Hajek von Hodjetin (1413) gives instructions for wagon-fortresses. See Max von Wulf, *Die hussitische Wagenburg* (Inaugural-Dissertation; Berlin: 1889), 7.
There are many examples of post-1434 texts that deal with wagon-fortresses, e.g., Philip von Seldeneck's *Kriegsbuch* of 1480–93. Regarding this cf. Jähns, 323–28 and Kurt Neubauer, *Das Kriegsbuch des Philip von Seldeneck vom Ausgang des 15. Jahrhunderts* (Inaugural-Dissertation: Heidelberg: 1963). Among later texts, Frans Helm's *Buch der probierten Kunst* (1535?) contains an extensive *Wagenburgordnung*. Cf. Jähns, 611–14. This feature was preserved in copies of Helm's work as late as the 1570's. Cf., e.g., Munich, Universitätsbiblíothek, Fol. cod. ms.602, done in 1573. Cf. also Theodor Hampe, »Ausrüstung einer Wagenburg im 15. Jahrhundert,« *Mitteilungen aus dem Germanischen Nationalmuseum*, (1897), 79–80. Jan Durdík, *Hussitisches Heerwesen* (Berlin: Deutscher Militärverlag, 1961), p.256 (hereafter Durdík, *HH*) cites some other examples as well. The military reformer Lazarus von Schwendi discussed *Wagenburg* tactics favorably in his *Kriegs-Discurs* (1578). Cf. Eugen von Frauenholz, (ed.), *Lazarus von Schwendi: Der erste deutsche Verkünder der allgemeinen Wehrpflicht* (Hamburg: 1939), 217–19.
Several older authorities deny that wagon-fortresses were used after 1436. Charles Oman, *A History of the Art of War in the Middle Ages* (2nd ed. rev.; London: 1924), Vol. II, 370 and Hans Delbrück, *Geschichte der Kriegskunst im Rahmen der politischen Geschichte* (2nd ed., 1923; reprinted Berlin: 1964) Vol. III, 503–25, two of the most widely-read scholars dealing with military history, both deny that the wagon-fortress was used after 1434. They consider the appearance of such fortresses in later texts as mere theoretical speculation. However, von Wulf, 8–9 notes the appearance of wagon-fortresses not only in theoretical treatises but in military ordinances from Frankfurt in 1444, from Nuremberg in 1450 and under Duke Albrecht III, Achilles of Brandenburg in 1462, 1475 and 1477. Durdík, *HH*, 253–57, lists a number of instances in which wagon-fortresses were used after 1434 by peoples as divergent as Spaniards and Ukranians, the latter of whom used them well into the seventeenth century. The Turkish *tabur*, used in the sixteenth century, was a form of wagon-fortress. Cf. von Frauenholz, 234.
Drawings of wagon-fortresses appear in some of the illustrated manuscripts dating from after 1434. One of the best is in ÖNB 3062, fols. 147v–148r. The army in this drawing has the Chalice on its tents and banners and a flag with a goose (Czech – »hus«) on it flies over the wagon-fortress. It is a Hussite army. The date of this manuscript is not precisely fixed. Unterkirchner, *Inventar*, dates it about 1450, while Quarg, *Kyeser* II, xxx, claims that it is probably first half of the fifteenth century. He admits that he has not seen this manuscript personally. The date of 1437 as given by Fairfax Downey, *Cannonades: Great Artillery Actions of History* (Garden City, N.Y.: 1966) caption to ill. facing p.96 is unsupported by any evidence. Josef Macek., *The Hussite Movement in Bohemia* (2nd ed.: Prague: 1958), n. p.[151] and Frederick G. Heymann, *John Žižka and the Hussite Revolution* (Princeton: 1955), ill. facing p.178 also reproduce this drawing. Other illustrations of wagon-fortresses appear in Karlsruhe Durlach 11, fol. 16v; Hans Schermer, *Ueber Basteibau* (Heidelberg, Universitätsbibliothek, Ms. palat. germ. 562), fols. 4r–4v; and Cgm356 (1492–93, p. *[sic]* 59. An exceptionally fine double-page drawing of a wagon-fortress appears in *Das mittelalterliche Hausbuch* (Bossert and Storck, figs. 60–61 and p.21). It is depicted under the banners of the Emperor Friedrich III. Clearly if such fortresses were to be used by the Emperor himself, they were not confined to armies of poor peasants fighting for heretical causes.

against Catholic powers, foreign and domestic. Both sides in these wars used wagon-fortresses with considerable effectiveness.[8]

All this shows the error of assuming that A's drawing of a *Wagenburg* dates section A prior to 1434–36, and it gives us the framework in which we can interpret the drawing. Author A drew a wagon-fortress because such fortresses were in use in his day. He described it as a »Hussite wagon-fortress« because the Hussite Bohemians were using them and because the Hussites had re-introduced them into European warfare with impressive successes a half-century earlier. His use of the present tense verb *vechten* (»fight«) in his comment indentifies his drawing not with the war of 1419–34, but with the Bohemian wars under George of Poděbrady (d. 1471) and Vladislav II (d. 1516), struggles which were present conflicts in his day. We should establish as the *terminus ante quem* the Peace of Olomouc (1478) or the Treaty of Kutná Hora (1485) rather than the Battle of Lipany (1434) or the ratification of the Compacts of Prague (1436).

The interpretation given here is consistent with the available evidence in all ways. There remains, of course the possibility that the manuscript as it has come down to us is a copy, done in the 1470's, of an original dating from the 1430's. Assuming a lost original is a convenient way of accounting for the apparent discrepancy between the references to Žatec and the Hussites and the unequivocal watermark evidence. Such an assumption, however, runs afoul of the principle of logical economy, and would in any case involve a burden of proof greater than that undertaken here. There is simply no reason to postulate a lost original unless we stubbornly insist that the reference to the early phases of the Hussite Wars must indicate that the manuscript was composed during the 1430's or slightly earlier. Until and unless an *Urtext* is discovered, we must accept the dates suggested by the watermarks as compelling.

To sum up, the watermarks in both sections provide the strongest evidence of dates. They give us a range of 1472–90 for section A and of 1486–92 for section B. We may narrow section A's dates on the general grounds that marks 1 and 2 appear in German archives only in the 1470's and also by assuming that author A's present-tense description of a Hussite wagon-fortress was probably done well before c. 1485. The period 1472–1475 is the most reasonable estimate. Section B should be given a slightly wider range of dates, roughly mid-1480's to early 1490's. The manuscript as a whole is about forty to sixty years younger than was formerly thought.

---

[8] One of the most complete accounts of the conflict is still Palacky, Vol. IV, part II (Prag: 1860), cf. esp. p.53 for an account of the Battle of Turnau, June 2–4, 1468, in which both the Hussite and the Catholic League army used wagon-fortresses. Recent works in English on this period include Otakar Odložilík, *The Hussite King: Bohemia in European History* (New Brunswick, New Jersey: 1965), who alludes to such fortresses only once, and Frederick G. Heymann, *George of Bohemia, King of Heretics* (Princeton: 1965), who, pp.457, 492, and 495, makes some reference to wagon-fortresses but does not deal extensively with their role. Durdík, *HH*, 156 asserts that such wagons were used extensively and offensively, and cites the Battle of Mähren in 1468, where George of Poděbrady personally commanded the field army and its fortresses. These wars, which involved Matthias Corvinus, King of Hungary as well as Bohemian Catholics, did not end until George of Podě-

Authorship and Origin

Given the scanty evidence this manuscript offers, questions of authorship and origin cannot be answered conclusively. We can only localize and characterize our authors in a general way. For the always-simple case of section B, there is little to be said. The provenance of the watermarks in this section, as given by Briquet and Piccard, is Austro-Bavarian or Franconian. The general drawing style is similar to other contemporary texts from this area, e.g., the Formschneider texts from which B copied one drawing (fols. 35v–36r). Since there is no evidence contradicting the watermarks, we must conclude that this section was done by an artist or engineer active in south Germany or Austria.

Section A presents not only pictures, but also a set of written texts. Although drawing conclusions from palaeographic evidence is difficult and often inconclusive in dealing with fifteenth-century manuscripts, comparison of the hand with published specimens from the period reveals the strongest similarities with *bastarda* as written in the middle decades of the century in south Germany and in Moravia.[9] More significant than this similarity, however, is the variation between different passages within section A. The hand alters noticably from one folio to another, ranging from the neat precision of folio 1r to the hurried, blurred and brief line of 5v *infra* to the cramped and inverted text of 16r. There is not enough difference in the formation of individual letters to lead to the conclusion that the various passages were written by different persons, nor is there any discernible pattern to the differences in appearance between passages. This suggests that the written comments, and probably the drawings as well, were done not all at once, but over a period of time, perhaps several months or years, under varying conditions and in a number of different settings. This unevenness in drawing and in writing is precisely what one would expect if section A is a notebook based on its author's readings or personal observations.

The dialect used by author A would be an important clue to the area where this section originated. Its examination is rendered difficult by the shortness of the individual passages, their highly specific and technical character and the paucity of place or personal references

---

brady's successor, Vladislav II, signed the Peace of Olomouc in 1478, ending war with the Hungarians, and the Treaty of Kutná Hora in 1485, ending internal Hussite-Catholic conflicts.

[9] In descending order of similarity, the hand resembles the following:

1. Joachim Kirchner, *Scriptura gothica libraria* (München: 1966), fig. 61b. Vienna, Österreichische Nationalbibliothek, Cod. 2828, done in 1464 at the Carthusian foundation in Königsfeld in Moravia. *Bastarda cursiva*.
2. *Ibid.*, fig. 57a. Clm 11444, done in 1446 in Erringen in Bavaria by a native of Kaufbeuren (also in Bavaria). *Cursiva currens*.
3. Ernst Crous and Joachim Kirchner, *Die gotischen Schriftarten* (Leipzig: 1928), Plt. 20, fig. 33. Preussische Staatsbibliothek, Ms. Germ. Fol. 19, done in 1448 by a scribe from Werdenberg, Canton St. Gall, Switzerland. *Oberrheinische Bastarda*.
4. *Ibid.*, Plt. 22, fig. 38. Preussische Staatsbibliothek(?), Ms. germ. fol. 459, done in 1432 possibly in western Upper-Bavaria. Transitional. *Oberrheinische Bastarda*.

to serve as guides. The most revealing data are author A's references to Munich (fol. 1v), Nuremberg (fols. 2v and 15r), Erkinger (von Seinsheim, a fiefdom in the Schwartzenberg area, not far from modern Würzburg) and Žatec (fol. 16v). The area surrounding these four points includes parts of northern Bavaria, East Franconia, Bohemia and eastern Swabia. It is in this general area that we should expect to locate author A's dialect.

Unfortunately, the language used by A is not distinctive enough to bear close comparison to any of the known fifteenth-century dialects from this region. The text contains a mixture of words, spellings, and usages from several dialect areas. Certain words, for example, *sayllen* (»ropes«), *untz daz* (»once« adv.), and *kunst* (in the sense of »device« or »technical artifice«) are exclusively south-eastern, i.e., Bavarian, Bohemian, or East Franconian.[10] Other words, however, such as *korn* (»grain«), *vall prugk* (»siege-ladder«), *leder* (»ladder«) and *auff leyt* (»supports«) are generally associated with Low German.[11] Moreover, the spelling of certain sounds varies greatly. Author A uses Upper German »p« and Low German »b« interchangeably, e.g., *pewart–bewart* (fols. 5v and 21r), *pruck–brugk* (fols. 5v and 26v). At times he uses the Upper German »ch,« e.g., *chumpt* (fols. 21r and 26v), *chomp* (fol. 18v). In other passages he prefers the Low German »k« for the same sound, e.g., *kunst* (fol. 5v), *kumt* (fol. 16v) and *kamp* (fol. 20r). Vowel sound interchanges are frequent throughout the text, for example »u« and »w,« »ei« and »ai,«, »ai« and »ay,« as well as »i« and »y.« Some of these interchanges are doubtless insignificant orthographic variations; the problem is to distinguish these from significant differences, and then to assess the meaning of the real differences in the context of a fifteenth century engineer's training. This is an extremely difficult task.

It is impossible to identify author A's language precisely. It resembles West Middle German, but has occasional Low German traits as well as Upper German characteristics.[12] What sense are we to make of this? The similarity of A's dialect to West Middle German is consistent with the broad geographical region indicated by the place names alluded to in the text. We

---

[10] *Sayllen* (fols. 21v, 26v *supra*). Cf. Gerhard Ising, *Zur Wortgeographie spätmittelalterlicher deutscher Schriftdialekte* (2 vols.; Berlin: 1968) Vol. II, 28–29. *Untz daz* (fol. 21r). *Ibid.*, I, 90; II, 66–67. *Kunst* (fol. 5v). See Johann Andreas Schmeller, *Bayerisches Wörterbuch*, ed. G. Karl Fromann (2nd ed., 2 vols., 1872–77; reprinted Leipzig: 1939), Vol. I, col. 1266.
In addition to Schmeller, the following works were consulted throughout this study: George Friedrich Benecke, Wilhelm Müller, Friedrich Zarncke, *Mittelhochdeutsches Wörterbuch* (4 vols.; Leipzig: 1854–66); Matthias von Lexer, *Mittelhochdeutsches Handwörterbuch* (3 vols.; Leipzig: 1872–78); Jacob Grimm and Wilhelm Grimm (eds.), *Deutsches Wörterbuch* (16 vols.; Leipzig: 1854–1954); Alfred A. G. Götze, *Frühneuhochdeutsches Glossar* (5th ed.: Berlin: 1956). For meanings of technical terms Walter Taenzler, *Der Wortschatz des Maschinenbaus im 16., 17., und 18. Jahrhundert* (Inaugural-Dissertation; Bonn: 1955), proved useful.

[11] *Korn* (fol. 17v). Cf. Ising, II, 44–45. *val prugk* (fol. 5r) *leder* (fol. 26v) and *auff leyt* (fol. 7r) were identified as being of Low German origin by Professor Stephen P. Schwartz, Department of Germanic Languages, University of California, Los Angeles.

[12] I am grateful to Professor Schwartz and to his colleagues, Professors Franz H. Bäuml, Eli Sobel and Arthur B. Groos, Jr. for their assistance in this aspect of the dissertation.

could account for either the Upper or Low German elements in the text by assuming that A was a native speaker from one or the other of those dialect areas and that he was trying consciously to write in a dialect not native to him. Such an assumption would not account for the presence of traits from both dialects, however, and it seems strained as well. There are other assumptions we could make to account for the strange traits, assumptions that run counter to those ordinarily made by philologists, and to explain these requires a brief excursus.

In dealing with technological texts, historians are obliged to apply, consciously or unconsciously, the standards and assumptions of scholars of literature, who are, after all, the acknowledged masters in this general area. In some cases these assumptions do not serve the historian of technology well. For example, in dealing with a literary or philosophical text, one can normally assume that its author possessed at least a modicum of literacy and skill at assembling words. In fact, in most cases such skills were fairly highly developed; we have little experience in dealing with the words of the semiliterate. Yet the writers of technological works were, in many cases, men educated in the workshop, not the grammar school or the university. Their mastery of language was often on a fairly rudimentary level. Even the great Leonardo called himself a »man without letters,«[13] and this should not be taken as a mere literary *topos* indicating humility, similar in spirit, let us say, to Gregory of Tours' famous apology for his bad Latin. Indeed, Leonardo's Italian is rude, often ungrammatical, and even confused at times. Author A's prose is plagued with similar problems; syntactical difficulties abound, forcing the reader–as with Leonardo–to refer to his drawings to follow what is written. Author A seems to have spelled many words as he probably heard them, without regard for the niceties of formal orthography. At times, in fact, his language skills are just inadequate to describe lucidly what his drawings show clearly. To take one example out of several, on fol. 10v, author A depicts a stamp, the three pestles of which are held upright by a board with three holes. The text, however, assures us that the pestles all fit in one hole. If we assume that A was a man of the workshop, then his verbal confusions (including his use of several dialect forms) become easier to comprehend.

A second literary assumption is inappropriate to technological texts: the assumption of stable dialects. If a medieval vernacular work was the product of a member of the lower orders of society, then the scholar who studies it is generally on safe grounds in thinking that its author was the speaker of a single dialect. Medieval peasants simply did not travel widely. But in those instances in which we can trace the career of a late medieval engineer, it inevitably turns out that he worked and traveled far from his place of birth. Conrad Kyeser of Eichstätt (near Nuremberg) died in exile in Bohemia, after having served several major German, Austrian, Bavarian and Bohemian princes, with a brief sojourn in Lombardy as well. Francesco di Giorgio Martini's life centered on Siena, but he spent time at Rome and Turin (1458–64) and also Urbino (1480–86). Leonardo's career was especially migratory, including Rome, Florence, Venice and Milan; he died, of course, in France while in the

[13] Gille, *Engineers*, 122–23.

service of Francis I.[14] While we cannot know how widely author A may have traveled, it is reasonable to assume that his career took him to different parts of Germany. Upon being exposed to different dialects, he may have acquired some traits of language foreign to his natal tongue. If he was none too sure of what constituted »proper« German, then he could easily have reproduced these traits in his writing.

Another literary assumption that serves the historian of technology poorly concerns what one might call the »essence« of any literary text. That is to say, a work of literature is always seen as an entity, unique unto itself, with a definitive form and content. Usually these characteristics are determined by the *Urtext*, against which other copies are judged; copies which stray too far from the original are considered »false,« »variant,« »partial,« etc. Technological literature cannot always be viewed from the same perspective. With some notable exceptions, (e.g. *Bellifortis*), such works are compiled, not composed, and their authors drew on numerous sources. Once a given work had reached a given form at the hands of one author, it became in turn the source for other writers, who freely modified, added to, deleted from and rearranged the »original« to suit their needs and fancies. Naturally, this blurrs beyond recognition the neat modern distinction between »original« and »copy.« In the literal sense of the word, no fifteenth-century work on technology is »original;« there are only different degrees and types of copying. Some works copy from life; others from different written sources; most mix both. When we examine a technological work, we must discard the literary scholar's notion of originality, and ask instead how the author came to know of the devices and process he includes. No doubt, the range of sources for any given author varied widely, from written treatises to oral tales to personal observation. Any technological work is the end result of a chain of information transmission of which we can (in most cases) observe only the end product.[15]

The fact that this information chain is more hidden than usual in the case of Clm 197,I section A (i.e. that we assume it to have been »original«) does not imply that author A »created« his work in the same sense that a work of literature is created. As noted above,

[14] Cf. Gille, *Engineers*, 58 for Kyeser, 103–05 for Francesco di Giorgio, and 121ff. for Leonardo. The lives of the sixteenth-century engineers illustrate the same trends on a wider scale. Jacopo Aconcio, for example, fled from Milan to Switzerland and thence to England in 1559. He died there in 1566–67. See Lynn White, jr., »Jacopo Aconcio as an Engineer,« *American Historical Review*, LXXII (1967), 431–39. The still-mysterious Milanese engineer Olgiati worked in Vienna, the Balearic Islands, Flanders and Hungary as well as his native city. *Ibid.*, 442–43. Similar examples could be multiplied indefinitely.

[15] A somewhat similar note is sounded by C. S. Smith and J. G. Hawthorne in their introduction to the *Mappae Clavicula*, a treatise on various workshop practices compiled from the ninth to the twelfth centuries. They insist that the work is a »compilation of compilations,« part of a complex literary tradition in which the relationship between what is written and what was done can become strained at times. They warn against a naive view of medieval technological works as literal reflections of technical realities. Although the *Mappae* shows these tendencies in a more extreme form than our manuscript, their warning is appropriate. We shall discuss these matters in more depth in Chapter IX below. Cyril Stanley Smith and John G. Hawthorne, *Mappe Clavicula: A Little Key to the World of Medieval Techniques* (»Transactions of the American Philosophical Society,« N. S. Vol. LXIV, Part 4; Philadelphia: 1974), 14–18.

there is evidence that it was compiled over a fairly long period of time. By the same token, however, it lacks the informality that characterizes the purely »private« notebooks of Leonardo da Vinci. Author A never makes personal reference to himself, nor does he write in a fragmentary, broken fashion (only one incomplete sentence occurs in the entire manuscript, fol. 20v). There are no drawings that shows signs of alteration or re-drawing; none are quick sketches, but instead tinted, »formal« drawings. Clearly then, author A's work was in the »public domain« in the double sense that he drew upon extrinsic sources for information and that he sought to pass this information on in an organized fashion to others. Hence, we may reasonably assume that, although A's words appear to us to be his own, some of them may not be entirely so. In the process of absorbing information, A may well have absorbed descriptive words along with the devices he depicts, and in presenting these drawings, he may have reiterated words, spellings and syntactical forms that were not part of his natal dialect.

To sum up, the area in which section A was composed is difficult to determine from the dialect employed. The best indication lies in the mention of or allusion to four place names: Munich, Nuremberg, Seinsheim (near Würzburg) and Źatec (Bohemia). The basic dialect of the text appears to be West Middle German, which is consistent with the general region suggested by the place names. The lack of distinctive characteristics in A's language, as well as his occasional employment of both Upper and Low German forms, can be accounted for if we assume that author A, like most fifteenth-century technicians, lacked much formal education, that he traveled fairly widely, and that he absorbed certain descriptive words and phrases from other regions, which he reiterated in the process of compiling his work.

Copies and Provenance

No complete copy of Clm 197,I was ever made, but portions of it were excerpted and included in two other works, one of which is useful in determining provenance. The library of the »Nationale Forschungs- und Gedenkstätten der klassischen Deutschen Literatur in Weimar« (formerly Thüringische Landesbibliothek) possesses an exquisite parchment manuscript, Hs. fol. 328, that contains deluxe drawings of subjects taken from most of the known illustrated works on technology from the trans-Alpine region. It has been nicknamed the »Ingenieur Kunst- und Wunderbuch,« but in fact almost nothing is known about its dates, provenance or purpose of compilation. To judge from its fine draftsmanship, careful lettering, and subjects depicted, the »Wunderbuch« must have been done between c. 1490 and c. 1525 as an encyclopedic presentation copy for someone of high rank and wealth.[16] The unsolved problems surrounding this manuscript make it difficult to draw too many conclusions from its inclusion of drawings from Clm 197,I. We can assume that, since the

[16] The tradition that it once belonged to the Albanian hero Skanderbeg (Georgius Castroiti) is apparently a pure fable. I am grateful to Dr. Konrad Kratzsch of the Nationale Forschungs- und Gedenkstätten for his assistance in allowing me to examine the manuscript and for providing photocopies of selected folios.

Weimar copyist re-drew twenty-four illustrations from our manuscript, the latter must have gained some fame by the early sixteenth century. Also, since drawings from sections A and B are interspersed, it is likely that both existed as a unit when the copying took place. Unfortunately, the nature of the Weimar codex makes it impossible to determine whether the copyist had before him a more complete version of our manuscript than that which now exists. Some of the other Weimar drawings are similar in style and content to those in Clm 197,I, but there is no other positive evidence to link these drawings to our manuscript.

Drawings from the manuscript appear in one other work. In 1524 a minor Bavarian artist named Christof Seselschreiber copied four drawings from it in his own collection of technical drawings (Cgm 973).[17] We know very little about Christof Seselschreiber, but from scattered references in his work we may gather that he had contact with, or worked for, both the imperial house and Wilhelm IV, Duke of Bavaria (1508–50). Also, a certain »Painter Christof« appears in tax records in Munich in 1525–26, but it is not certain if this is a reference to Seselschreiber.[18] However, since Seselschreiber's work is written in Bavarian dialect, was later in Wilhelm IV's possession[19] and may have been done for the ducal house, we should assume that Seselschreiber wrote and drew his manuscript in Munich or central Bavaria. This would mean that Clm 197,I was in the same area at roughly the same time.

The manuscript next appears as part of the estate of the famous German orientalist, Johann Albrecht von Widmannstetter (1506–57). Von Widmannstetter's career was divided between warfare and scholarship. He fought as a young man with the armies of the Emperor Charles V in Italy (c. 1527) and later, in 1546, led a Spanish regiment in the Schmalkaldic War, for which service he was knighted by Charles on March 2, 1548.[20] He had however, studied philosophy, law, and languages (Hebrew and Greek) at Tübingen and possibly also at Heidelberg and Basel.[21] In 1539 he entered the service of Ludwig X, who shared ducal honors with Wilhelm IV from 1516–45, and in 1542 von Widmannstetter married the Duke's natural daughter. He was one of the most noted scholars and book collectors in

---

[17] Otto Hartig, »Christof Seselschreiber und sein Buch über Büchsenmeisterei und technische Erfindungen,« *Kultur des Handwerks*, Heft 9 (July, 1927), p. 284 notes this copying and also some Valturio drawings appearing in Seselschreiber's work. In a work like Seselschreiber's, obviously copied from a number of sources, it is extremely difficult to identify positively the origins of drawings. Nevertheless, since many of Seselschreiber's sketches are similar, but not identical, to examples found in Clm 197,I, A, it is likely that he had the original text before him as he worked, not merely a copy of the four folios identified above. His comments indicate that he considered it a source for old-fashioned machines and methods.

[18] Hartig, 280; the Seselschreiber family had some fame. Christof's father, Gilg Seselschreiber (fl. c. 1482–c. 1518), was chosen to prepare a monumental sarcophagus for the Emperor Maximilian I, but after a number of delays his commission was revoked in 1518. Christof's manuscript betrays an acquaintance with a number of south German and Austro-Hungarian cities, but centers on the Bavarian-Upper Frankish area.

[19] *Ibid.*, 280–81.

[20] Max Müller, *Johann Albrecht von Widmannstetter, 1506–1557; Sein Leben und Wirken* (Inaugural-Dissertation, München; Bamberg: 1907), 14 and 57–58 respectively.

[21] *Ibid.*, 13.

Table

Comparison of Folios Excerpted from Clm 197,I

Clm 197,I	*Weimar, Hs. fol. 328*	*Cgm 973*
2v	138v	–
5r	273r *(inf. dex.)*	–
5v	136v	–
12v	–	126r
13r	–	139v
13v	134r	140r
14r	–	135v
15r *(supra)*	272r	–
15r *(inf. sin. et dex.)*	273r	–
16r	135v	–
16v	138r (and 271r?)	–
17v *(supra)*	137r	–
17v *(infra)*	271v	–
18r *(inf. sin.)*	270v *(sup. sin.)*	–
18v	272v	–
20r	270r	–
20v	275v	–
21r	274v	–
23v *(infra)*	139v	–
25r	139r	–
27r *(supra)*	270r *(supra)*	–
27r *(infra)*	137v *(infra)*	–
31v	134v	
34r	134r	–
34v	134r	–
39r	134v(?)	–
40v	135r	–

Note: Several drawings from Clm 197,I are grouped together on some folios of Weimar.

Bavaria.[22] A year after his death, 1558, his family was obliged for financial reasons to sell his library to the reigning Duke, Albrecht V (1550–79). Clm 197 came to the ducal Hofbibliothek with this transaction.[23] It is reasonably certain that the manuscript arrived in the ducal library in substantially its present form, although the library's records are not specific

[22] Otto Hartig, *Die Gründung der Münchner Hofbibliothek* (»Abhandlungen der Königlichen bayerischen Akademie der Wissenschaften: philosophische, philologische und historische Klasse,« Vol. XXVIII, Abhandlung 3; München: 1917), 9–10. Von Widmannstetter was an early exponent of the Copernican system. In 1533 he delivered a lecture on the subject, based on Copernicus' *Commentariolus* of 1530, before Pope Clement VII in the Vatican gardens. Leopold Prowe, *Nicolaus Coppernicus* (2 vols.; Berlin: 1883–84), Vol. I, pt. 2, 273–74. Von Widmannstetter was also responsible for a Syriac edition of the New Testament printed at Vienna in 1555.

[23] The signature of Von Widmannstetter's library is found on fol. 74r [77r] of part II of Clm 197.

on this point. The earliest catalog, done in 1582, records what appears to be Clm 197 together with Giovanni da Fontana's *Bellicorum instrumentorum liber* (now Cod. icon 242) under a shelf or »station« number, *Stat.* II. 44.[24] Unfortunately, there is insufficient information in this notation to allow a full description of the manuscript to be reconstructed from it. At some undetermined later date the Fontana codex was separately shelved. Only in 1806 do we find, in a new catalog done by I. Hardt, any indication of the two-fold character of our manuscript. Hardt describes clearly only the Taccola portion (our part II), but he notes that the codex is inscribed with the (still visible) tag *Liber machinarum et mechanica quaedam*. Presumably the latter phrase refers to the 48-folio German segment that we call part I.[25] Once the manuscript had been studied and described by Jähns (1889) and Berthelot (1891), the library's printed catalog began to distinguish clearly between parts I and II.

Despite the silence of the library's catalogs, it is nevertheless probable that the Taccola part and the German part of the manuscript were already joined under von Widmannstetter. He is known to have visited Italy several times in his career, and he was in Siena five times between 1533 and 1542. There he studied law under Marcello Biringuccio (a relative of the famous technician Vanoccio Biringuccio) and under Allesandro Sozzini (a descendent of one of Taccola's admirers). He also was a member of the order of S. Jacopo, to which Taccola had formerly belonged.[26] He probably acquired the Taccola manuscript (an autograph) through these connections. Since part I was already in or around Munich by c. 1524, von Widmannstetter must have acquired it separately, possibly only after he settled there permanently in the 1540's. It would have been reasonable for him to have gathered together relatively brief works on related subjects under one set of covers as a measure to reduce the costs of binding. His financial means were apparently not excessive (his family was in some difficulties after his death) and technology was not in any case a large portion of his total library, the bulk of which was dedicated to his linguistic interests. Consequently, in the absence of any evidence to the contrary, we should assume that it was von Widmannstetter who brought Clm 197 to its present condition, save only for its later rebinding and the rearrangement of some pages in part II.

In conclusion, we can confidently assume that sections A and B of part I were brought together by an unknown person in the period around 1500. In this form the codex circulated in south Germany, probably Munich, and acquired sufficient fame to encourage the copyist of the Weimar codex and Christof Seselschreiber to excerpt certain drawings for their compilations. Then, sometime between 1524 and 1557, the manuscript became the property of J. A. von Widmannstetter, who probably joined to it those portions of Taccola's *De ingeneis* that he had acquired in Italy. In this form Clm 197 passed to the ducal Hofbibliothek, now the Bayerische Staatsbibliothek, where it has remained since 1558.

---

[24] *Cod. bavar. cat.* 61; see Hartig, »Gründung,« 358. The same codex later appears under a »running« *(Laufende)* number, 98, and still later under the designation »Cim 134.« *Ibid.*, 375.

[25] *Cod. bavar. cat.* 65, Vol. I, 513 ff., quoted in English by Prager and Scaglia, *Taccola*, 29.

[26] *Ibid.*, 28, n. 15, following Müller.

CHAPTER III

LIFTING MACHINERY

As Bertrand Gille remarks, »Lifting apparatus takes pride of place in this notebook.«[1] Indeed, there are more examples of lifting machinery than of any other sort, section A being particularly devoted to displaying them. This is hardly surprising in view of the interest shown by other texts in similar machines and their obvious importance. In spite of the large number of lifting machines shown, they can be divided readily into three basic classes, cannon hoists, construction cranes and lifters and bucket hoists for wells.

Cannon Hoists

Lifting heavy objects was a problem dating back to the dawn of engineering. All of the basic mechanical elements necessary to perform this task had been developed by the Graeco-Roman period,[2] but the use of one of them, the jack-screw, does not seem to have become general until the middle ages. The most common machine for raising cannon was a windlass in which, in its simplest form, three or four masts rise from ground level and meet to form a tetragon or a pyramid. They are joined at the top, and a pulley or block is hung from this point. A rope (or ropes) is passed over this pulley and terminates in a winch fixed between two legs of the structure. This type is shown five times in section A, fols. 1v, 2r, 5r, 6r, and 25v, and once in section B, fol. 42v.

There is a strong general similarity between this class of machines and examples used in antiquity. The *Mechanica* of Hero of Alexandria discusses hoists of this type, listing single-, double-, triple- and quadruple-masted variants.[3] Vitruvius' *De architectura* discusses a two-

---

[1] *Engineers*, 69.

[2] The most important of these elements, the sheave or pulley, is first found in an Assyrian relief dating from the reign of Ashur-nasir-pal (883–859 B.C.). Recent research has turned up an Assyrian sheave specimen contemporary with the relief; cf. Joseph W. Shaw, »A Double-Sheaved Pulley Block from Kenchreai,« *Hesperia*, XXXVI (1967), 391 and note 6. Shaw also refers to other (Roman) examples of sheaves and blocks, which, although rare, are extant particularly from later periods. The earliest Chinese sheave dates from 147 A.D., Joseph Needham and Wang Ling, *Science and Civilisation in China*, Vol. IV, pt. 2 (Cambridge: 1965), 95. A.G. Drachmann, »A Note on Ancient Cranes,« Singer II, 658–62 discusses this subject in general, noting that ancient cranes did not use geared wheels. Cf. also Franz Maria Feldhaus, *Die Maschine im Leben der Völker* (Basel & Stuttgart: 1954), 112 and 130 and fig. 69 for other hoisting equipment (hereafter Feldhaus, *DM*).

[3] Beck, 272; Drachmann, Singer II, 660–61; A.G. Drachmann, *The Mechanical Technology of Greek and Roman Antiquity* (Madison: 1963), 97–103 (hereafter Drachmann, *MT*). That these types of hoists were in

masted hoist similar in style to Hero's and stabilized by guy-ropes.[4] Such hoists seem to have served throughout the middle ages as heavy-duty lifters. Particularly in the fifteenth century many manuscripts, both Italian and German, show them.[5] They were engineering commonplaces. Neither of our authors shows cannon hoists that are different from those of his contemporaries, but several features are worthy of comment because they illustrate normal late medieval practice in building and using such hoists.

In contrast to classical models, author A, along with most of his contemporaries, avoids the problems of erecting three or more separate masts by having one element of the hoist prefabricated. One mast is shaped like an inverted Y with a windlass held securely between the two shorter members (see fol. 1v for example). At ground level such a hoist has three or four feet for stability, while at the apex there are only two or three masts to fasten together. Hoists made in this way could be set up or dismantled more easily than one whose elements were all separate. This was an important advantage, since many of these hoists were meant to be portable and to accompany an army on the march.

Another common feature in these hoists is the extensive use of iron as a reinforcement at points where stress would be concentrated. Such iron parts can always be identified by their color.[6] Iron bands are frequently used to reinforce wooden beams, for example at the junction of the three legs in the inverted-Y shaped masts. Iron sheave blocks and pulleys are also frequently used, and one can identify sheave axles, ratchets and lifting and anchor hooks as being of the same material. The extent to which iron is used on a hoist varies directly with the load it is expected to lift; the heavier the intended load, the more iron parts are used.

---

practical use by the Romans is confirmed by several reliefs showing such devices in operation. Cf. Shaw, figs. 78b, c, e.

4 Drachmann, Singer II, 659; Drachmann, *MT*, 142–45. It is difficult to say whether the renewed popularity of Vitruvius in Italy in the fifteenth century might have influenced hoist construction in any way. The list of Vitruvius' manuscripts as given by Carol Herselle Krinsky, »Seventy-Eight Vitruvius Manuscripts,« *The Journal of the Warburg and Courtauld Institutes*, XXX (1967), 36–70 shows that German copies of *De architectura* were much earlier. It is more likely that Vitruvius' crane designs, which reflect Roman practice, remained traditional throughout the middle ages, as did many of his architectural ideas. For an example of medieval use of architectural proportions deriving directly from Vitruvius, cf. Kenneth J. Conant, »The After-Life of Vitruvius in the Middle Ages,« *Journal of the Society of Architectural Historians*, XXVII (1968), 33–38.

5 For example, Cgm734, fol. 61v; Cgm973, fol. 6r; Cpg126, fol. 18v; ÖNB 3069, fol. 93r; ÖBN 10,855, fol. 89v; Vienna, Waffensammlung, Ms. P. 5014, fol. 34r; *Das mittelalterliche Hausbuch* (Bossert and Storck, figs. 69–70). Bayerische Staatsbibliothek, Ms. icon. 222, (c. 1500) still retains this type. Italian examples include Taccola (Clm 197, II), fol. 124v [133v]. For Valturio's drawing of the same, see Giovanni Canestrini, *Arte militare meccanica medievale* (Milano: n.d. [1945]), 256, fig. XCI. For Leonardo's, see Fritz Toussaint, *Lastenförderungen durch fünf Jahrtausende* (Mainz: 1965), 38 and Arturo Uccelli, *Storia della technica dal medio evo ai nostri giorni* (Milano: n.d. [1944]), 113, fig. 62.

6 The coloration is constant throughout section A and different, but also constant, in section B. Iron is always gray, wood yellow-brown, rope yellow-tan, leather red-brown. The standardization of colors is of great assistance in analyzing a given device.

But all the cannon hoists show some parts made of iron. One method of spreading stress over a wider area and economizing on the use of iron elements was to mount the pulleys in tandem in the structural members of the hoist (see fols. 25v and 42v). This gives the hoists an odd appearance by modern standards, for we prefer to group pulleys side-by-side and to mount them in the same block.[7] The tandem mount offers the distinct advantage of permitting the use of a simple iron collar to reinforce the structural member where it is pierced by the sheave axle. A complicated and expensive iron block is not required.

Other features are dictated by the hoist's function. Author A shows all his hoists with X-handles to turn the windlass rather than cranks. Although one can lift a load faster with a crank than an X-handle, cranks are useless where the load is so heavy that speed is out of the question, and X-handles give the workers a better grip. Both authors use block-and-tackle arrangements that are overly-complicated by modern standards. This is because medieval engineers, like their classical predecessors, seldom employed gearing in heavy-duty hoists. Gearing would have given a desirable mechanical advantage, but it was expensive relative to pulleys, and it could be rendered useless if a tooth were to break. Without gearing a heavy duty hoist using only manpower must have a complex tackle or it cannot lift its burden.

Another, more novel type of cannon hoist also appears in our manuscript. This uses a screw around which a nut is turned to raise the screw. Hooks are fitted on the screw's bottom which hold a cannon. There are two examples of this type, one in section A, fol. 9r, and one in section B, fol. 37r. The general use of screws for fastenings and to transmit motive force dates at least from classical antiquity.[8] In particular, screws were used in wine-presses, but they occasionally appear in machines to lift heavy weights.[9] Also, of course, the Alexandrian mechanics used powerful endless screws engaging spur gears (»worm and spur« drive) to achieve large gear reduction ratios.[10] Most of these devices were used in the middle ages.

---

[7] Some ancient texts describe blocks and such blocks have been found by archaeologists *in situ*. However, the diagrams accompanying some texts, e.g., the Arabic ones of Hero of Alexandria's *Mechanica*, show pulleys arranged in tandem. It is difficult to say whether these were drawn in this fashion merely for clarity or whether such an arrangement was actually used. In any case, both blocks and tandem arrangements were used by the authors and their contemporaries, including da Vinci, *Codex Atlanticus*, fol. 104v, b; cfl. Toussaint, 39.

[8] The best general survey is Rudolf Kellermann and Wilhelm Treue, *Die Kulturgeschichte der Schraube* (2nd ed.; München: 1962) (hereafter Kellermann and Treue), which covers the subject from antiquity to the twentieth century. The original edition was by Treue alone and covered only down to the eighteenth century: *Kulturgeschichte der Schraube von der Antike bis zum achtzehnten Jahrhundert* (München: 1955). For classical antiquity, cf. Franz Kiechle, »Zur Verwendung der Schraube in der Antike,« *Technikgeschichte* XXXIV (1967), 14–22 and Alfred Mutz, »Römische Bronzegewinde,« *Technikgeschichte*, XXXVI (1969), 161–67, who puts forth the provocative thesis that the bronze screws found on medical specula at Pompeii were turned on a lathe. Cf. also Feldhaus, *Technik*, col. 981 ff.

[9] For example, Hero of Alexandria's *Mechanica* II: 5 shows a primitive jack which uses a peg fitted into the threads of a vertically mounted screw; as the screw turns the peg is raised or lowered. See Drachmann, *MT*, 57–59, who explains the use of a peg in terms of the difficulty in cutting a female screw or nut.

[10] E.g., see *ibid.*, 60–61.

Screw presses are frequently illustrated from at least the eleventh century onward[11] and lifting screws in the form of jacks appear in the thirteenth; Villard de Honnecourt depicted one in his notebook.[12] Worm and spur drive appears in the early fifteenth century in Brunelleschi's work.[13]

The screw-lifts for cannon shown in Clm 197,I are similar in design and operation to a number of contemporary examples appearing in German and Italian manuscripts.[14] The origins of these hoists are unclear,[15] but since they appear only shortly before Brunelleschi began to design screw-powered hoists to use in building the cupola of the Florentine *Duomo* (c. 1417),[16] these two developments may be related in some way. All of the drawings of screw-lifts for cannon save one show a machine in which a nut is rotated about a screw. Only Taccola shows the converse system in which the screw itself is turned in a fixed nut to lift the cannon.[17] This involves several disadvantages: not only does the capstan turning the

[11] The earliest one I have noticed appears in Utrecht, Archepiscopal Museum, Ms. 3 (*Evangelarium* of St. Bernulf, done at Reichenau c. 1000), fol. 51. Cf. *Bulletin de la société français de reproductions de manuscrits à peintures*, XV (1931), 111–12 and fig. XXXVI. Bertrand Gille, article »Machines,« Singer II, 647 (hereafter Gille, Singer II), notes extant examples of French screw presses from the thirteenth century and claims that the screws of oil-presses in southern France were being produced on a standard pattern at this time.

[12] Darcel (ed.), 145 and fig. XLIII. Gille, Singer II, 646–47 is of the opinion that such a use of screws was a medieval innovation and that Villard expresses »astonishment« at the power of the screw-jack. This is not supported by the notebook's text.

[13] Gustina Scaglia, »Drawings of Machines for Architecture from the Early Quattrocento in Italy,« *Journal of the Society of Architectural Historians*, XXV (1966), 90–114 (hereafter Scaglia, *JSAH*) claims that Brunelleschi »re-invented« worm and spur gearing during the course of his work on the cupola (c. 1417–36). It appears frequently in later fifteenth-century Italian treatises such as Francesco di Giorgio's, but especially in the *Zibaldone* by Bonaccorso Ghiberti (Florence, Biblioteca Nazionale, Ms. BR 228, cf. fols. 122v, 123r, 124r, 125r, 126v and 127v; Scaglia, *JSAH*, figs. 12, 13, 15, 17, 19 and 20 respectively). The *Zibaldone*, Scaglia argues, is based in part on Brunelleschi's now-lost original designs. Cf. Gustina Scaglia, »Drawings of Brunelleschi's Mechanical Inventions for the Construction of the Cupola,« *Marsyas*, X (1960–61), 45–68 (hereafter Scaglia, *Marsyas*). Cf. also, Frank D. Prager and Gustina Scaglia, *Brunelleschi: Studies of His Technology and Inventions* (Cambridge, Mass.: 1970), 65ff. (hereafter Prager & Scaglia, *Brunelleschi*). There is no evidence that worm and spur gearing was used earlier in the middle ages, and I can find no German Text which depicts it before Cpg 126, fol. 4v.

[14] ÖNB, 3069, fol. 92v; Cgm 499, fol. 15r; Cpg 126, fol. 22r; Canestrini illustrates two Italian examples, p. 198, fig. LXIX (unidentified, possibly di Giorgio), p. 264, fig. XCV (Taccola). The enduring popularity of this device is attested to by its appearance in Seselschreiber, (Cgm), fols. 63v–64r, Bonaiuto Lorini's *Le fortificatione* (Venice: 1609), 209–12 and in the work of Diego de Prado y Tovar (1603). Cf. A. Rupert Hall, article »Military Technology,« Singer III, 368, fig. 236.

[15] ÖNB 3069, fol. 92v is the earliest I have seen (c. 1411).

[16] Such cranes also appear in the *Zibaldone*, e. g. fol. 104r (cf. Scaglia, *Marsyas*, fig. 4) and fol. 129r (cf. Scaglia, *JSAH*, fig. 4). Scaglia identifies these cranes with Brunelleschi's designs. Cf. Prager and Scaglia, Brunelleschi, 65ff.

[17] Taccola (Clm 197,II), fol. 22v. Similar machines appear in other Taccola manuscripts; cf. Canestrini, 258, fig. XCIII.

screw rise along with the screw and thus grow ever more difficult to reach, but the cannon being lifted will rotate along with the screw as well. This would pose a serious problem for the operators of such a lift. The rotating-nut system allows workers easier access to the handles by which the nut is turned and also prevents the cannon from rotating. Why Taccola chose to illustrate an obviously inferior system is a mystery, particularly since Brunelleschi's screw-powered cranes use rotating nuts.

Hoists of the screw-lift type were considered alternatives to block-and-tackle hoists for use where especially heavy cannon were involved. Our manuscript shows us nothing novel in respect to such hoists, but gives us representative pictures of how they were constructed. It is interesting to note that the example on fol. 37r is designed to be dismantled for transport, indicating that screw-lift hoists were used in the field as well as in armories.

Construction Machinery

Other machines designed to lift lighter burdens comprise a different class of devices in the manuscript. In number this is the principal class, having a total of fifteen folios devoted to it. Morphologically and functionally these can be divided into three main groups: 1. cranes properly speaking, i.e., devices for general lifting purposes; 2. earth-moving devices; and 3. pile drivers. All the examples in these groups are designed for construction or assembly operations. Taken as a whole, they are illustrative examples of general purpose construction machines from the period.

The general history of cranes is long and complex and much of it is irrelevant for our purposes.[18] The basic form of construction crane used throughout most of the middle ages (which is not shown in our manuscript) evolved by Roman times at the latest. This was a long mast or masts guyed to maintain an angle of 45–60 degrees relative to the horizon, with a more or less sophisticated block-and-tackle for lifting. Power was supplied either through a winch operated by handspokes or, for heavier work, through a tread-drum or »squirrel cage« large enough for several men and mounted directly on the winch-axle. The best known Roman representation of such a crane is found in a relief on the sarcophagus of the Hatarii now in the Lateran Museum.[19] Illustrations of lighter-duty cranes always show

---

[18] The most recent general survey is Toussaint's cited above. A brief survey in English is given by F. R. Forbes Taylor, »The Winch from Well-Head to Goliath Crane,« *Chartered Mechanical Engineer*, IX (1962), 200–09. Paul Brandt, *Schaffende Arbeit und bildende Kunst* (Leipzig: 1927) Vol. I: *Im Altertum und Mittelalter* contains some illustrations and discussion of cranes during the periods.

[19] For illustrations, cf. Taylor, 207; Feldhaus, *DM*, 142, fig. 102; Shaw, pl. 78a; Brandt, fig. 175; Drachmann, Singer II, fig. 603; Drachmann also illustrates another such relief from Capua, fig. 578. Somthing of the continuity of this form may be gathered from a miniature showing one in a fourth- or fifth-century Vergil manuscript (Rome, Bibliotheca Vaticana, Cod. 3225), reproduced by Hugo Th. Horwitz, »Technische Darstellungen aus alten Miniaturwerken,« *Beiträge zur Geschichte der Technik und Industrie* (now *Technikgeschichte*), X (1920), 175–76, and fig. 1.

the bi-, tri-, or quadrupedal types used in the middle ages as cannon hoists, and Vitruvius informs us (10:2:10) that for loading or unloading ships upright and horizontal cranes in pivots were used.[20] Since we lack pictorial evidence to show us precisely what these looked like, we might assume that they resembled the lighter construction cranes shown in the manuscript.

Medieval crane design departed from the classical in many important respects, however. The elements of ancient crane construction were mixed to yield a number of new forms, the best known of which is the loading crane common to medieval towns and docks. These were often tread-drum powered, but housed and pivoted like a windmill.[21] Similar cranes were used to some extent in construction, but the more usual form was a smaller one which could be mounted on the higher parts of the edifice under construction and re-mounted to best advantage as the building rose. These were powered either by windlasses, occasionally by cranks, or by smaller tread-drums. Particularly in the thirteenth century and later, these began to look like some of the cranes found in Clm 197,I.[22] By the fourteenth century and in the Bohemian area, this similarity had become very strong. Also, some of these Bohemian cranes are obviously »slewing« cranes, designed to pivot to assist in the precise placement of their loads.[23] Slewing cranes also appear in Italy in the early fifteenth-century, in a very sophisticated form which could adjust to shift a suspended load laterally.[24] The cranes of the manuscript, while not so highly developed as the Italian examples, are related to the types frequently seen in the Bohemian illustrated texts, and unequivocally do pivot.[25]

All of the examples shown in our manuscript resemble cranes shown in other contemporary treatises.[26] Most of them are alike in general structure, single-masted cranes with a single

---

[20] Drachmann, Singer II, 660; Drachmann, *MT*, 148.

[21] Cf., e.g., Gille, *Engineers*, ill. opp. p. 66 for a fifteenth-century specimen from Lüneburg which survived until modern times. Feldhaus, *DM*, 216–17 dates the original of this at 1337 and notes similar cranes in Strasbourg in 1385 and at the same date in Frankfurt a.M. Trier had one shortly after 1413. By 1405, Feldhaus claims, Frankfurt had a professional »Kranwender« and by 1430 he was entitled »Kranenmeister;« cf. *ibid.*, 207–08.

[22] Cf., e.g., two thirteenth-century examples reproduced by Toussaint, 26–27 and 30.

[23] Uccelli, 24 notes the first illustration of a pivoted crane in the famous Heidelberg »Minnesinger Handschrift« dating from 1310. Also a miniature from the Velislav Bible (c. 1340) shows a tread-drum crane which may pivot. Cf. Brandt, 245, fig. 330. The Wenceslas Bible (c. 1389–1400) contains several such cranes engaged in building the Tower of Babel, some of which must pivot to function. Cf. *ibid.*, 251, fig. 339 and Václav Husa, Joseph Petrán, Alena Šubrtova, *Homo Faber* (Praha: 1967), fig. 90.

[24] They appear in the *Zibaldone* as cited above, fol. 107v and 114v; cf. Scaglia, *Marsyas*, 57–58, who claims these too are Brunelleschi's inventions.

[25] Scaglia, *JSAH*, 98 claims that Clm 197 shows only rigid, non-pivoting cranes; this is not the case for part I.

[26] ÖNB 3062, fols. 200–07, shows several examples similar to Clm 197's. A crude one was illustrated by Francesco di Giorgio. See Maltese, fig. 89. The *Ingenieur Kunst- und Wunderbuch* shows one. See E. Marx, 320, fig. 12. Similar cranes used in iron foundries appear in *Das mittelalterliche Hausbuch*. See Bossert and Storck, fig. 39. Some more complex versions are illustrated by Leonardo. See Bertrand Gille, »Léonard de Vinci et son

windlass and a simple pulley system (see fols. 1r, 4v, 6v, 8v and 38r). Author B's (fol. 38r) is the most elaborately drawn; it is shown mounted in a frame with a vertical winding drum turned by a capstan. Author A's are all much simpler and are shown without stands or frames. The reason for this is that cranes of this type were used in a variety of construction operations and could be mounted in a number of different positions on a building as it grew progressively higher. Author A did not show frames to hold the cranes because such frames would have varied depending on where and how the crane was mounted at various stages of construction operations. He wished to focus on the details of the cranes themselves without dictating to the reader the manner of their emplacement. Since such cranes were common items, A's readers could visualize for themselves the mounting.

By far the most interesting crane is shown on fol. 2v, the »double-acting crane of Nuremberg.« It has two masts and a common winding drum with one rope. As one hook ascends, the other descends and each turn of the winding drum is made to do useful work. Cranes like this first appear in the fourteenth century and are also depicted in some of the other fifteenth-century treatises.[27] Although they appear elementary by modern standards, they were nevertheless the first western machines other than bucket lifts in wells to make use of the »double-acting« principle, which later became important in machine design in quite a different way.[28] It is also interesting to note that A claims this crane is powered by horses, whereas all his other cranes are man-powered. We can doubt the veracity of his assertion for a variety of reasons (see fol. 2v Commentary), but the fact that he made such a claim is indicative of the value his milieu placed on machines that substituted animal for human labor. The fifteenth century saw a rise in the use of animals in machines requiring rotary power, and this trend continued well into the eighteenth century.[29] A's crane is not an efficient machine for exploiting horses to generate rotary motion, but some of his others are (see fols. 21r–21v). His concern with replacing human- with animal-power is praiseworthy even though his drawing shows a crane that could not be powered efficiently by horses.

temps,« *Techniques et civilisations*, II (1952), 79–81 (hereafter Gille, »LdV«). Agricola used them as well. Cf. Hoover (ed.), 508 and 514; and Toussaint, 53.

27 The earliest example I have seen of what appears to be a double-acting crane appears in the *Kasseler Weltchronik* (1385). See Brandt, 274, fig. 333. Taccola illustrates them frequently in Clm 197, II, fols. 22v [32v], 23r [33r], 24r [34r], 65v [68v]. German examples include Cpg 126, fol. 10r and the *Wunderbuch*. Cf. Marx, 318, fig. 3.

28 For a discussion of this principle as it applies to machines for gas and fluid flow in the Italian Tradition and China, see Ladislao Reti, »The Double-Acting Principle in East and West,« *Technology and Culture*, XI (1970), 178–200.

29 Although animal-powered devices, »gins,« are quite ancient, horse-gins are not frequently seen in post-Roman Europe before the fifteenth-century, according to Carl Graf von Klinckowstroem, »Zur Geschichte des Göpels,« *Technikgeschichte*, XXXIV (1967), 30. Frank Atkinson, »The Horse as a Source of Rotary Power,« *Transactions of the Newcomen Society*, XXXIII (1960–61), 32, does not specifically agree with this late a date, but notes that harness problems delayed the wide-spread use of horses in this context, and that one sees a steady increase in the number and variety of horse-powered rotary machines from the sixteenth through the eighteenth century.

In construction it is necessary to move not only discrete blocks of material but also quantities of loose material such as earth, rubble or mortar. In their simplest form earth-lifters are nothing more than cranes with buckets in the place of hooks, but to move larger amounts of material faster or over longer distances, different designs are required. Our manuscript shows four machines that serve as earth-movers, fols. 3r, 3v, 25r and fols. 35v–36r (one drawing). Each of these presents a different sort of solution to the problems of earth-moving, and two of them are especially worthy of discussion.

Fol. 3v shows a large, pivoting, double-acting bucket lift to use in demolition operations. The loaded bucket lifts an empty one as it descends, and the problem author A must face is how to slow the loaded bucket's descent and bring it to a halt. In essence this is the same sort of problem that a weight-driven clock must overcome, and it is not surprising to find author A using a device from clockworks, the fan-escapement, as a braking device.[30] He was not the only engineer to apply clockwork elements to other purposes. Weight-driven grain mills appeared in Milan in 1341,[31] and Kyeser shows a rather fanciful elevator in which one's descent is slowed by a fan escapement.[32] Author A takes an additional and important step, however, by adding a brake to his machine. His brake is crude: it consists of a loop of rope weighted at one end and a lever, both acting on a ribbed pulley. Its origin is unclear. Although claims have been advanced for brakes in ancient cranes, A's is in fact the earliest drawing of a brake to appear in a western manuscript.[33] This is all the more surprising because one would expect windmills to have some sort of brake, yet the earliest depiction of windmill brakes is in da Vinci's work.[34] Windmill brakes may have derived from brakes like the one A shows, for the appearance of brakes in our manuscript marks the emergence of a concern for controlling power as well as generating it.

[30] The fan-escapement appears early in the history of clockwork. White, 121 (and fig. 10) cites one from c. 1250. They were the preferred means of regulating bell-striking trains in medieval clocks, where a moderately high gear velocity was required.

[31] A. G. Keller, *A Theater of Machines* (New York: 1965), 13, mentions these mills, as does White, 124, n. 5.

[32] Quarg, *Kyeser*, I: 134r; II: 100. Kyeser also shows a windmill-powered lift (I: 83r; II: 52–55). Quarg notes in respect to the latter that it may have been derived from the use of fan-escapements in construction machinery, but he cites no examples of this and I have seen none ante-dating Clm 197 or the *Bellifortis*. Kyeser's machines probably derive from windmills and clockwork. I cannot agree with R. J. Forbes, article »Power,« Singer II, 620, who claims that fol. 3v is based on the windmill. Clockwork is a much more likely source.

[33] Taylor, 207, and G. A. Harper, »Controlling Power,« *Chartered Mechanical Engineer*, XIV (1967), 120–21, claim that brakes were used in antiquity in cranes and cite the Hatarii sarcophagus relief as having »external, self-wrapping rope brakes.« This is untrue. The relief shows the normal ancient technique, described by Drachmann, *MT*, 52, of not allowing excessive turns of rope to build up on a winch axle, but instead leaving the rope unanchored and having a workman pull off the excess under tension. This was done while the crane was in operation, and just enough rope was left on the axle to prohibit its slipping. This maneuver, which relies on friction between the rope and the axle, looks like a braking action, but its purpose is, in fact, exactly the opposite.

[34] Our manuscript is the earliest to depict the interior of a windmill (see fols. 19r and 19v) and it shows no brakes of any kind. Windmill brakes must have been developed in the last decades of the fifteenth century.

Another interesting earth-moving machine is shown by author B on fols. 35v–36r. He illustrates a rather awkward railway along which a cart moves. It is used in construction earthworks for fortifications and can be moved as the construction progresses. Author B's drawing is not original, but a copy from the group of manuscripts we have designated as the »Formschneider« type.[35] Neither B's copy nor the original depicts a well-designed machine, but these drawings nevertheless represent early attempts to utilize a cart running on rails to transport material.[36] The use of rails became more widespread in the sixteenth century, particularly for hauling ore from mines, and machines similar to B's are shown at work building fortifications in treatises as late as the early seventeenth century.[37]

The final type of construction machinery to involve lifting is the pile driver. Such machines are presumably quite old, possibly Neolithic, as Feldhaus claims,[38] but the earliest western text to mention them is Vitruvius' *De architectura*, which, unfortunately, gives no details on their form or operation.[39] Roughly contemporary with Vitruvius is a Chinese device very much like a pile driver for punch-drilling deep deposits of water or brine.[40] Pile drivers were also a normal part of medieval construction equipment, and, by the fifteenth century at least, in Nuremberg they were owned and operated by the city itself for municipal purposes.[41]

---

[35] The same illustration appears in Cgm 734, fols. 76v–77r and Munich, Deutsches Museum, Hs. 1949/258, fol.s 76v–77r.

[36] The origin of this device seems to have been the mines of central Europe, where wooden tracklines were used in transporting ore from an early date, possibly the twelfth century. Unfortunately, we know very little about the details of such installations. Illustrations of mining activity from the late fifteenth century show wooden carts and tracklines. See Charles E. Lee, »Some Railway Facts and Fallacies,« *Transactions of the Newcomen Society*, XXXIII (1960–61), 3 and Gille, »LdV,« 72. For more detailed discussion of mining railways, see Michael J. T. Lewis, *Early Wooden Railways* (London: 1970).

[37] Lee, 4ff. discusses these and reproduces several. The best-known example is Agricola's; Hoover (ed.), 156. A. Wolf, *A History of Science, Technology and Philosophy in the Sixteenth and Seventeenth Centuries* (2nd ed.; London: 1950), 512 erroneously claims that Sebastian Munster's *Cosmographia Universalis* (1550) depicts the earliest railway. Lorini, 220 shows a fortification construction machine that uses rails.

[38] Feldhaus, *DM*, 32, but he cites no evidence for this view.

[39] *Ibid.*, 105.

[40] Needham, 52, fig. 360 shows one such pile driver in use in China in 1943. It is remarkably similar in design to the example on fol. 37v.

[41] Feldhaus, *DM*, 197–98 cites the instance of Eberhardt of Cologne, who in 1238 began a building project for the cloister of Bloomkamp. For a pile support he employed ten pile drivers, each worked by a crew of four men, with another four-man crew in reserve. Endres Tucher, *Baumeister* of Nuremberg, 1461–76, inventories pile drivers among municipal property and describes the duties of the city's *Hoyermeister* or »pile-driver-master.« See Matthias von Lexer (ed.), *Endres Tuchers Baumeisterbuch der Stadt Nürnberg*, introduction and apparatus by Friedrich von Weeck (Stuttgart: Bibliothek des literarischen Vereins in Stuttgart [= LXIV of this series], 1862), pp. 47 [13a] and 45 [12a] (bracketed numbers refer to folios of the original manuscript as given in this edition; hereafter Tucher).

Given the well-established position of the pile driver in medieval engineering, it is not surprising to find many examples of such devices in fifteenth-century technological treatises.[42] As a rule these treatises show pile drivers that resemble the one on fol. 8r, i.e., machines operated by a windlass with a slip hook to release the driver head and guide rails to direct its fall. Some of them show automatic triggers which release when the driver-head is lifted to a certain height. The only other example in the manuscript, fol. 37v, is of a much simpler type, possibly much older than the former example. Since it requires a four-man crew, I suspect that it is identical to the thirteenth-century one mentioned in note 41, and it is strikingly similar to the Chinese type noted above. Author B's drawing is the earliest western illustration of such a pile driver. The only other drawing I have seen of this type is a Dutch engraving of c. 1700, which shows it mounted on a barge and used in maritime construction.[43]

Bucket Lifts for Water

The final type of lifting apparatus encountered in the manuscript is the bucket lift for raising water from a domestic well. Such an elementary device may seem out of place in an engineer's notebook, but we should note that, although devices to draw water from wells are quite old, the application of cranks and gearing to domestic, as opposed to irrigation wells, was delayed until late in the middle ages. The crank, for example, does not seem to have been applied to the task of drawing water from a well until c. 1425.[44] Perhaps this lag was caused by the very humbleness of the problem; drawing water, after all, was woman's work. In any case the fifteenth century saw the proliferation of new techniques to ease this arduous task. Taccola presents a number of ingenious solutions to the problem, but one suspects from their very elegance that he is attempting to display his virtuosity rather than to provide a practical device for everyday use.[45] Ramelli's treatise of 1588 has this same tendency with respect to well pumps.[46]

---

[42] The earliest fifteenth-century sketch of a pile driver I have seen is Taccola's (Clm 197, II), fol. 51v [54v], which shows a large *recumbent* hammer. Although the object of the hammer's blows is not shown an obvious copy of the drawing in the »Bonino Manuscript« (U.C.L.A. Ms. 170/358, fol. 7r) shows it sinking a pile. This unusual style did not find much use, and even Taccola shows a version like section A's, but better developed, later in the Munich manuscript (fol. 119r [128r]; cf. Beck, 290–91). Francesco di Giorgio Martini included a similar design. See Maltese, fig. 44; cf. *Das mittelalterliche Hausbuch* (Bossert and Storck, fig. 44). The persistence of this type is evident from its appearance in Lorini, 234.

[43] Adrien Pouderoyen, »Les polders aux Pays-Bas,« *Techniques et civilisations,* III (1954), 24, fig. 8 and Johann van Veen, *Dredge, Drain, Reclaim: The Art of a Nation* (5th ed.; The Hague: 1962), 81 reproduce this engraving, but neither gives its source.

[44] White, 167, notes that all the well hoists in the Princeton Index of Christian Art (to 1400) are equipped with handspokes, and that the earliest illustration of a cranked well hoist he has found dates from 1425.

[45] One solution (Clm 197, II, fol. 16r [15r]; cf. Berthelot, 475, fig. 28a) involves an axle with a wheel mounted on it. The bucket's rope winds onto the wheel, which is turned by a falling weight whose rope winds about the

The examples shown in our manuscript, on the other hand, are the essence of practicality. Both of them (fols. 9v and 10r) are constructed along the same general lines: a spoked wheel turns an axle about which a rope is wound. Both ends of the rope are attached to buckets, and lifting the filled one lowers the empty one. This is merely an elaboration of the normal medieval method of drawing water from a well with two buckets, in which a single rope is passed over a pulley suspended over the well. This type appears in one of the tales of Reynard the Fox and can be traced back to the late eleventh century.[47] The earliest illustration I have seen of it dates from the early fourteenth century.[48] It is interesting to note that double-bucket wells thus provide the first application of the double-acting principle which A and his contemporaries utilized in a number of different ways.

---

axle itself. Taccola has given the bucket, not the weight, the mechanical advantage here, in order to reduce the length the driving weight must descend–thus that weight must be considerably heavier than the bucket it is to lift. Another solution (Berthelot, 478, fig. 30 *bis*) involves an axle turned by an endless sprocket chain geared up to another axle whose rope has two buckets. The double-acting principle is employed here, and the sprocket chain is the first to be illustrated.

[46] Ramelli, fig. 2, for example.

[47] Petrus Alphonsus, *De disciplina clericalis,* (early twelfth century), relates the tale of the fox's luring the wolf into a well by first descending into it himself with one bucket. When the wolf leaped into the second bucket to follow the fox, his greater weight lifted the fox to safety. See Angel Gonzales Palencia (ed.), (Madrid: 1948), 60–62. Essentially the same tale appears in a late eleventh-century Talmud commentary done by the French rabbi, Raschi. See Johannis de Capua, *Directorium vitae humanae,* ed. Joseph Derenbourg (Paris: 1887), viii, note 1. I thank Mr. Bruce Knapp of the Department of English, University of California, Los Angeles for pointing out these sources to me.

[48] Lillian M. C. Randall, *Images in the Margins of Gothic Manuscripts* (Berkeley and Los Angeles: 1966), fig. 598. This is taken from fol. 138v of Cambridge, Fitzwilliam Museum, Ms. 298, the »Metz Pontifical,« done 1302–16.

CHAPTER IV

SIEGE MACHINERY

Siege machinery was the main business of the medieval military engineer. From the time of Guido da Vigevano's treatise, illustrations of siege machinery were the dominant feature of many of the technological works, especially those by German engineers. The reasons for this concentration of interest are easy to understand. Fortification techniques were such that the defenders of a well-built castle or town had a great advantage over attacking forces. Towns and castles were difficult to take by direct storm, and it was more usual for the besieging army to try to starve the defenders into surrender. The responses to the advantage held by the defense were many and interesting. For example, the extremely rapid development of large-caliber cannon was caused by the role such weapons could play in sieges. On a less dramatic level, one notices in almost all the technological treatises how much attention was given to devices and methods of taking a castle and how little to defending it. The *Bellifortis*, for example, contains only three folios devoted to defensive operations.[1]

Many devices were used for attacking fortified positions, including those that also played a role in field operations. For our purposes we shall consider as »siege machinery« only devices whose primary or exclusive purpose was to assist in sieges. This chapter will discuss four groups of machines: 1. mobile shields and shelters; 2. towers for attack and for observation; 3. ladders; and 4. trebuchets. Other weapons that were used in all types of combat, e.g., crossbows, will be discussed in the subsequent chapter.

Mobile Shields and Shelters

Protective shelters for attack squadrons are very old. The earliest examples are Babylonian or Assyrian and date from the eighth and seventh centuries B.C. In appearance they resemble tortoises, having rounded covers to protect the soldiers from enemy missiles and battering rams on their front end.[2] The evolution of such shelters was slow, and improvements were

---

[1] Quarg, *Kyeser,* I:82r and 83v–84r (one drawing); II:52. The former shows a Nuremberg shears that can grip and shove away the ladders of an attacking force and the latter depicts soldiers rolling rock-laden wagons downhill onto a group of attackers.

[2] A. Rupert Hall, article »Military Technology,« Singer II,715, fig. 646 (hereafter Hall, Singer II), cites a seventh-century B.C. relief from Nineveh. Wilhelm Treue, *Achse, Rad, Wagen* (München: 1965), 44 (hereafter Treue *ARW*), cites a relief from Kalach dated 730 B.C. August Demmim, *Die Kriegswaffen in ihren*

generally intended to increase the strength of the shelter or to permit easier entry and egress by the soldiers who manned the tortoise. The second-century military engineer Apollodorus of Damascus, for example, sketched designs for an assault shelter which included a ladder projecting from two uprights extending above the shielding so that the attacking soldiers would have the means to climb over walls or parapets.[3] Such designs became part of the military engineer's craft and were repeated with many variations throughout the centuries to follow. The invention of cannon changed matters slightly as we can see from the *Bellifortis*, but it did not render such shelters useless. Kyeser's tortoises are faced with heavy wooden armor and are usually designed to present an oblique surface to enemy shots. At times Kyeser shows shelters that carry cannon and could return an enemy's fire (an interesting forerunner of the modern tank). Otherwise his designs resemble their ancient forerunners quite closely.[4]

Section A is exceptional among fifteenth-century illustrated technological manuscripts in that it shows very few mobile siege shelters. It includes only two sketches of such machines and A clearly did not lavish much attention on either. Fol. 15v shows an obviously incomplete drawing of a mantlet with what appears to be a pivoting shield. The drawing is so poorly done that further identification is impossible. Author A's other illustration (fol. 16v), while scarcely better than fol. 15v in artistic skill, occupies a central place in discussions of the manuscript because of its accompanying text. The drawing shows a crude, cannon-bearing, triangular siege vehicle similar to some shown in the *Bellifortis*. The text, however, claims much more:

> Herr Arkinger had this shield before Žatec. A hundred men can easily go safely behind it. The winch is inside and when one arrives at the city, he raises the shield and shoots and then lets it fall shut again. Then he turns the winch inside again so that the shield departs from the place and the soldiers stand behind it without harm.

Some of the historical details contained in the passage are verifiable. Erkinger von Seinsheim,[5] to use the modern form of his name (1362–*post* 1429), was the progenitor of a noble family, the »Freiherrn und Bannherrn zu Schwarzenberg und Hohenlandsberg.« Seinsheim is near the family's original castle at Schwarzenberg.[6] Erkinger received his title partially in

---

geschichtlichen Entwicklungen (4th ed.; Leipzug: 1893), 160, cites another Assyrian specimen from the seventh century B.C.

[3] Gille, *Engineers*, 18–19, reproduces this and other designs by Apollodorus.

[4] The similarity between the designs in the *Bellifortis* and those found in Italian manuscripts suggests that Kyeser depicted siege engines that were standard throughout much of Europe. Even Guido da Vigevano's drawings, done about three-quarters of a century before Kyeser's, are similar (cf. e.g., his fol. 41v). Only rarely do novel designs appear: Valturio drew a siege engine driven by a cranked worm-and-spur drive. See Canestrini, 192, fig. LXVI). Guido designed a windmill-driven one (fol. 14v) and Leonardo sketched a saucer-shaped »tank« with cannon on all sides, *Cod. Atl.*, fol. 4v–a.

[5] Beck, 270; Uccelli, 49 and Feldhaus, *Technik*, 1264, identify him, but give no details.

[6] Seinsheim lies almost on an axis between modern Würzburg and Nuremberg. It is about sixty kilometers from the latter in the Steigerwald district.

recognition of his services in wars against heretics.[7] The siege of Žatec occurred from September 13 to October 2, 1421, and was marked by fierce fighting and heavy use of firearms by the Hussites. The siege failed, not only because of the town's resistance, but because the German troops burned their tents and fled before an advancing relief army headed by Jan Žižka. Because of their heavy losses and their ignominious retreat, the battle became legendary among the Germans.[8] Unfortunately, there are no chronicle sources that place Erkinger at the siege nor any that mention an »automobile« siege engine like the one A describes.[9]

If we take A at his word, Erkinger had a tortoise of impressive dimensions powered by an internally-mounted capstan. This is not what the drawing shows. One would expect that such a machine would have been striking enough to appear in accounts of the siege, but they are silent about it. It is unlikely that A's description can be accepted at face value. He may have been influenced by the attempts of others to design and build »automobile« vehicles. From the time of Roger Bacon's prophetic utterance about »wagons ... which will not be pulled by animals, but which incredible forces will drive,«[10] medieval engineers searched for the means to produce a »horseless carriage.« Guido da Vigevano's windmill-powered wagon represents one early attempt,[11] and Giovanni da Fontana's strange little man-powered cart shows another.[12] The fifteenth-century Italian texts depict a number of vehicles moved by capstans turning the wheels through reduction gearing.[13] Their German contemporaries seem to have been more active in actually making similar devices. One appeared on the streets of Memmingen in 1447, and an ill-fated attempt was made in 1509 to travel in a man-powered cart from Pirna to nearby Dresden.[14] The advantages of a tortoise whose motive force was protected from enemy fire would have been obvious to any soldier, and some attempts were made to design such shelters within which the draft animals could be ac-

---

[7] *Allgemeine Deutsche Biographie*, vol. XXXIII (Leipzig: 1891), 260ff. and 779.

[8] Palacky, III, pt. 2, 252–54. He relates a contemporary German joke about the Catholic forces at the siege, who were so pious that they were unwilling even to set eyes on the loathsome heretics, and so fled the field instead.

[9] Canestrini, 182, in discussing Clm 197, I, fol. 23v, notes, »Di un carro senza cavalli munito di cannoni parlano le cronache a proposito dell'assedio di Saaz. ...« He seems to have confused fol. 16v, which he must have seen during his research, with »le cronache.«

[10] *De secretia operibus*, cap. 4 in J. S. Brewer (ed.), *Fr. Rogeri Bacon Opera quaedam hactenus inedita* (London: 1859), 533.

[11] Guido da Vigevano, fol. 14v.

[12] Giovanni da Fontana, fols. 17v–18r (one drawing).

[13] Cf., e.g., Francesco di Giorgio's ingenious designs (Maltese, fig. 95) and Leonardo's, found in Ms. B, fol. 83v, reproduced in Bern Dibner, *Leonardo da Vinci, Military Engineer* (New York: 1946), 11–12 and fig. 6.

[14] For both instances see Feldhaus, *Technik*, 1264. For later examples see August von Essenwein, »Eigenthümliche Wagen, Schiffe und Schlitten vom 15.–18. Jahrhundert,« *Anzeiger für Kunde der deutschen Vorzeit*, XXX (1883), cols. 57–79.

comodated.[15] The weight of these made them impossible to move by capstans, however, no matter what gearing was employed. Author A lived in a time when the interest in »automobile« vehicles was high. He fell victim to the mania and believed that, in a long-ago battle legendary for its violence, Erkinger had used such a wagon, so large that »one hundred men can easily go safely behind it.«

Author B is much more interested in siege vehicles than A is. He depicts seven examples (see fols. 33v, 39r, 39v, 41v, 43v, 48r and 48v). With the exception of the last-mentioned folio, B's siege engines are all variations on established themes of the sort seen in the *Bellifortis*. But, in contrast to Kyeser's engines, B's are unsuitable for actual use. His siege engines are all too heavy, too over-loaded with iron reinforcements and spikes, and equipped with too many wheels to have been of practical value. Siege engines were always difficult to use. Often they had to have heavy solid wheels to support their weight, and many of them deserved their military nicknames, »swine,« »mother-sow« and the like.[16] Author B's designs accentuate all the problems such machines had without relieving any of them. For example, many of his siege engines have six, eight or more non-steerable wheels. This would help to support their enormous weight, but at the expense of extreme difficulties in maneuvering. Medieval four-wheeled wagons had long before been equipped with pivoting front axles, but these were useless in vehicles with more than four wheels.[17] Author B's engines could have been turned only by main force—literally dragged about in their tracks—but B makes no provision to solve this problem.[18] Indeed, he scarcely seems aware of its existence.

Only one of B's siege engines shows an interesting feature. Fol. 48v depicts a shelter that runs in a U-shaped track or trough. The machine is fantastic in its conception and virtually useless in practice. Who would lay the tracks on which it runs as it approached a city under siege? However, it does show that the principle of guiding vehicles by means of rails was beginning to be widely recognized in the late fifteenth century. Together with the fortification construction machine (fols. 35v–36r) it reveals that author B was interested in this new approach to vehicular movement.

---

[15] Cf., e.g., Taccola (Clm 197,II), fol. 48r [52r].

[16] Treue, *ARW*, 205–06.

[17] For a discussion of steering gear in medieval wagons cf. Marjorie Nice Boyer, »Medieval Pivoted Axles,« *Technology and Culture*, I (1960), 128–38, and A. Rupert Hall, »More on Medieval Pivoted Axles,« *Technology and Culture*, II (1961), 17–22. Their researches date such axles from the late thirteenth century, but White, 66, note 4 cites an example from c. 800. The pivoting front axle would help to turn a cart with six or more wheels, but not very much. Ideally, all the wheels except one pair would have to pivot (but through differential angles) in order for such a wagon to turn without dragging, cf. Alan H. Nelson, »Six-wheeled Carts: An Underview,« *Technology and Culture*, XIII (1972), 391–416.

[18] Some manuscripts cope with the problem a bit more intelligently. For example, Cpg 126, fol. 2r shows a cranked, geared mechanism for turning the rear wheel of a large siege engine. Kyeser seems to avoid the use of more than four wheels wherever possible, and when he was obliged to use six or more, he usually specified that draft animals were to be used.

Siege Towers

The siege tower is a first cousin to the siege shelter. Its general purpose is the same, but it employs its greater height to the advantage of attacking squadrons. Towers appear considerably later than shelters; they are unknown before Hellenistic times. Roman towers are the first for which we have definite measurements. Vitruvius claims impressive dimensions for them: 30–60 meters in height, with bases of 8–10 meters square. As A. R. Hall notes,

> however absurd these dimensions seem, there is plenty of evidence to show that towers were built, and set in motion, capable of overtopping the walls of a first-rank city. The weights suggested for them–100 tons and more–do not seem impossible; doubtless block-tackle was used to haul these monsters into position.[19]

Later authors, e.g., Apollodorus of Damascus, have left us drawings showing that observation towers were as common in the ancient world as assault towers were.[20]

There was no break in the tradition of such towers in the middle ages, nor should we expect one. Byzantine authors, e.g., Hero of Byzantium, replicate the designs of their ancient predecessors,[21] and western sources mention the use of assault towers. Simon de Montfort used a »beffroi« (belfry) at the siege of Toulouse during the Albigensian Crusade.[22] Guido da Vigevano's treatise, devoted to means to insure success against the infidel, gives a prominent place to assault and observation towers.[23] Although one would assume that the advent of firearms would have doomed these vulnerable towers, this was not the case. The shapes of towers were slightly different after Guido's day, but they continued in use until the middle of the seventeenth century, although less and less frequently.[24]

The three examples shown in our manuscript, fols. 7 r, 29 r and 32 r, represent the three main types still in use in the later fifteenth century. Author A's fol. 7 r shows a portable observation tower capable of lifting an observer high enough to look over the walls of a besieged city. Several versions of the same basic device appear in other military technology treatises, particularly Kyeser's and Valturio's.[25] Fol. 29 r shows a fixed »crow's nest« tower that would serve for observation or as an elevated gunner's platform; author B includes a soldier firing an arquebus from it. This was the most popular form of tower in the other treatises, but most authors show them as mobile rather than fixed.[26] Author B's second illustration, fol. 32 r,

---

[19] Hall, Singer II, 716–17.

[20] Gille, *Engineers*, 19.

[21] *Ibid.*, 20.

[22] Robert Coltman Clephan, *The Defensive Armour and the Weapons and Engines of War of Medieval Times and of the »Renaissance«* (London: 1900), 191.

[23] Guido da Vigevano, fols. 9 r and 43 r, for example.

[24] Clephan, 191.

[25] Quarg, *Kyeser* I: 32 v; I: 25; Canestrini, 120, fig. XXXVIII.

[26] Cf., e.g., Canestrini, 66, fig. XI; 100, fig. XXVII; Taccola (Clm 197, II) fols. 73 v [76 v], 83 r [84 r] and 85 r [86 r].

depicts a large wheeled attack tower. Such machines were becoming less useful in the late fifteenth century owing to the increasing efficiency of artillery. Note that this tower moves along its diagonal axis to present oblique sides to hostile fire. Kyeser had shown a similar tower in the *Bellifortis* (but with flat sides presented to fire) and the drawing was frequently reproduced in copies of the work.[27] Some idea of the size of this tower may be gained by comparing it with a similar design in an Italian manuscript. It was probably meant to be about twelve or thirteen meters high and about four wide.[28] Thus it would have been much smaller than those described by Vitruvius and more easily maneuverable.

Ladders

Once an attacking army reached the walls it was attempting to storm, it had to have some means of climbing them. The most ancient device for this purpose was the ladder. The earliest known representation of a ladder used in warfare is an Egyptian tomb painting dating c. 2450 B.C.[29] But carrying and erecting a ladder under hostile fire posed serious problems: a long ladder was extremely awkward to carry and required additional equipment to set it in place. As a result, medieval military engineers devised several different types of ladders that would be more easily portable and faster to erect. Only one of their solutions had an ancient predecessor, the sectional ladder. Ladders constructed in segments that could be put together on the spot as needed appear in the work of Apollodorus of Damascus (second century) and may have been used by the *cohortes vigilium*, Rome's fire brigades.[30]

The ladders shown in our manuscript are a representative sample of contemporary designs (see fols. 5v, 16r, 17r and 34v). The types depicted are: 1. conventional ladders raised by a pole; 2. single-shaft ladders (so-called »Copenhagen ladders«), both sectional and single-piece; 3. rope ladders with hooks; 4. articulated and sectional conventional ladders; and 5. expanding X-frame (»Nuremberg shears«) ladders. All these types appear in a number of fifteenth-century texts.[31] The most interesting of these is the »Nuremberg shears« type (see

[27] Quarg, *Kyeser*, I: 44v and 45v; II: 31.

[28] Venice, Bibliotheca Marciana, Ms. latin VIII 40 (2941) (either Taccola or di Giorgio; reproduced in Canestrini, 102, fig. XXVIII). The wheels of the tower in this drawing are slightly smaller than the height of a human figure drawn next to them, i.e., about 1.5 meters. The tower is almost eight times as tall as its wheels (a proportion that holds true for fol. 32r also), i.e., about twelve or thirteen meters.

[29] Found at Sakkara. See Wolfgang Hornung, »Vom Steigbaum zur Drehleiter,« *Technikgeschichte*, XXXVI (1969), 180 and fig. 3. Hornung's is the only recent survey of this subject, but Feldhaus, *Technik*, 617ff. contains information on medieval and early modern developments.

[30] *Ibid.*, 185–86 argues for their use by the *cohortes* on the basis of their appearance in Apollodorus of Damascus.

[31] For ladders lifted with poles: cf. Quarg, *Kyeser*, I: 70r; II: 45. For single-piece Copenhagen ladders: cf., *ibid.*, I: 70r; II: 45; Taccola (Clm 197, II), fols. 11v [12v], 17r [18r]; Paris, B.N., Ms. latin 7295 (c. 1446?), fol. 139r; Bossert and Storck, fig. 63. For rope ladders: cf. Quarg, *Kyeser*, I: 71r; II: 45; Taccola (Clm 197, II), fol. 17r [18r]; Bossert and Storck, fig. 63. For sectional Copenhagen ladders: cf. ÖNB 3069, fols. 45v, 59r–

fols. 16r, 17r and 45v). The origins of this device are unknown, and the reason for its name is unclear. They appear in Europe first in the *Bellifortis*, where they are referred to as »serpens« or »vippere.«[32] Their chief military uses were as ladders or grappling hooks, since they could be carried collapsed in compact form, but could be extended into a long unit. They could, however, also be applied to machines to transmit motion in a unique fashion. The Italian engineers were quick to recognize and utilize their interesting possibilities and in the sixteenth century they were used by Jacques Besson and Juanelo Turiano, among others.[33] In our manuscript Nuremberg shears serve as ladders, although B shows one carrying a block of material (fol. 45v).

One other point of interest concerns how high ladders like those illustrated in our manuscript might reach. Author A states in the text on fol. 16r that one ladder is to be made in sections seven meters long and that three or more be joined together to make a completed ladder. This would indicate that heights of twenty-one meters and up were considered within the normal reach of such ladders. If this is so, it clarifies why all A's ladders have iron hooks at the upper end. Not only would such hooks help prevent an enemy's dislodging the ladder from the wall, but they would tend to relieve some of the stress on the joints of an articulated or sectional ladder. Note that A also reinforces the joints of many of his sectional ladders with iron. Without such hooks and joints a ladder twenty meters or more in height would be in grave danger of collapsing from excessive lateral stress.

Trebuchets

Counterweight-powered artillery, »trebuchets,« were the last form of missile-hurling device to appear before the invention of cannon.[34] They were extensively used in sieges even after cannon became common, and it is not clear when they were completely replaced by firearms.[35] Certainly they were in use in the fifteenth century, and they appear in treatises on

---

59v; ÖNB 2952, fol. 91r; Paris, B.N. Ms. lat. 7295, fol. 139r; Canestrini, 110, figs. XXXI–III. For articulated and sectional conventional ladders: cf. Quarg, *Kyeser*, I: 69r; II: 44–45; Taccola (Clm 197, II), fol. 17v [18v]; Canestrini, 110 and 116, figs. XXXI and XXXVI; Bossert and Storck, fig. 63.

[32] Quarg, *Kyeser*, I: 70v and 82r; II: 45 and 52; cf. Feldhaus, *Technik*, 617ff. who notes their use by a nomadic tribe in south Russia as tent supports (from what date?).

[33] For example, Paris, B.N. Ms. latin 7295, fol. 139r; Maltese, figs. 115ff.; Canestrini, 118, fig. XXXVII (from Valturio); Besson, figs. 10 and 14. For Turriano, cf. Beck, 365–90; Wolf, 526–31; and most recently Ladislao Reti, »The Codex of Juanelo Turriano (1500–1585),« *Technology and Culture*, VIII (1967), 53–66.

[34] The oldest known version of this is Chinese and uses a crew of men to pull down a pivoted beam. It first appears in 1004 A.D. The first reference to a counterweighted version is Syrian, 1187–92. This was known in Europe under the name »trebuchet« by 1199; cf. White, 102. New studies of the trebuchet in East Asia and Islam include Joseph Needham, »China's Trebuchets, Manned and Counterweighted,« Hall and West, 107–45 and Donald R. Hill, »Trebuchets,« *Viator*, IV (1973), 99–114.

[35] Eric Brockman, *The Two Seiges of Rhodes, 1480–1522* (London: 1969), 45 and 83 claims that their last recorded use by a western army was during the first siege of Rhodes. Even there, he claims, they were em-

military technology even in the sixteenth century.[36] Unlike cannon, a trebuchet could hurl more than solid or exploding shot. It could, for example, cast rubble or bales of hay into the moat of a besieged castle to make a pathway for attacking troops. Or, as our text illustrates, it could throw carcasses of dead animals or humans or offal into a besieged area in a primitive form of bacteriological and psychological warfare.[37] That even the corpses of large bovine or equine animals would not be beyond the capacity of such engines is demonstrated by modern experiments, which show that they can loft 100–150 kilogram loads over about 300 meters.[38]

The trebuchet is a very simple instrument having an unequal lever arm with a very heavy weight on the shorter side. Releasing this weight from a raised position imparts rapid motion to the longer arm hurling the missle. A sling was usually added to the longer arm to give added velocity. The two drawings of trebuchets in our manuscript, fols. 24v and 32v–33r (one drawing) both show this feature. The former uses a simple counterweight made as an integral part of the main arm, while the latter uses a suspended counterweight to give slightly greater mechanical advantage. The latter drawing is especially interesting in that it resembles an illustration in the *Bellifortis* which gives the dimensions of various parts of the machine.[39] On the basis of the strong similarity between the two drawings, we can estimate that B's trebuchet has a counterweight of about two or three thousand kilograms. This would be barely adequate to hurl the dead animal he shows in the trebuchet's sling (cf. fols. 32v–33r Commentary).

---

ployed only as a makeshift defense, constructed on the spot by a Basque sailor. This may be true, but later military engineers did not lose interest in them.

36 Ramelli, for example, illustrated several.

37 Ralph Payne-Gallwey, *The Crossbow* (1903; reprinted New York: n.d. [1958]), 271–72, cites two instances of the use of trebuchets in this way: at the siege of Karlstein in 1422, Corybut is said to have hurled the bodies of the Czech dead as well as 2000 cartloads of manure into the besieged town; Froissart mentions that the body of an emmisary from Auberoche was cast back into the city. Payne-Gallwey's fig. 189 shows a trebuchet loaded with a dead horse copied from Leonardo's *Codex Atlanticus* (no further citation given).

38 White, 103. He also notes Jean Buridan's claim that a trebuchet could throw a ball weighing about 500 kilograms.

39 Quarg, *Kyeser*, I:30r; II:24. B's drawing is not a precise copy of this, but must have been influenced by it.

CHAPTER V

OTHER INSTRUMENTS OF WAR

Sieges, however, were not the only aspect of warfare in which the engineer was expected to exercise his skills. Weapons for use in the field, as well as machines to assist in transporting goods and men on the march, were equally of concern to the army's technical experts. Our authors have sketched several pieces of equipment which fall into these categories. Many of them are simple and self-evident, but others require some clarification. This chapter will consider cannon and cannon mounts, bridges, wagon-fortresses, and crossbows together with crossbow-spanning devices.

Cannon and Mounts

Cannon appeared in Europe in the early part of the fourteenth century.[1] The extent to which they revolutionized the art of warfare cannot be overemphasized, but the pace at which this revolution took place is sometimes over-estimated. Cannon and their offspring, handguns, did not make all other forms of missile weapons obsolete overnight. Nevertheless, cannon quickly became a standard part of the equipment of an army, and much work was devoted to their improvement and adaptation to a variety of combat situations. Cannon, in other words, became a subject for intensive and rapid development, and the fertility of efforts in this direction has been characterized by A. R. Hall as »astonishing.«[2]

The very earliest forms of cannon were shaped like pears, with the thickest part of the barrel concentrated near the breech. These were soon replaced by pieces shaped more like megaphones or cones, having the largest diameter at the muzzle and the smallest at the breech.[3] The larger the piece, the more conical the shape–to bear the increased force of larger shot. Conversely, smaller pieces tended to be more nearly cylindrical. These shapes continued to be favored for the better part of a century, until about the mid 1400's, when advances in

---

[1] For a discussion of the advent of cannon, see White, 99 ff. The first picture of a cannon in Europe dates from 1327 and at least one extant artifact in Stockholm resembles the drawing quite closely. See Carlo M. Cipolla, *Guns, Sails and Empires* (New York: 1965), 21.

[2] Hall, Singer II, 727.

[3] One of the earliest illustrations of this type occurs in a fresco, »The Loggia of Hell,« done in 1340 by Paolo del Maestro Neri at the monastery of St. Leonardo in Leccetto; see Oscar Guttmann, *Monumenta pulveris pyrii* (London: 1906), 27–28, figs. 74–75.

metallurgical techniques apparently made them unnecessary.[4] The earliest cannon were mounted quite crudely, usually directly on the ground. By the latter part of the fourteenth century[5] several different types of mounts seem to have emerged, many of them capable of being adjusted for range and bearing. Some of the early illustrations of these, from the opening decades of the fifteenth century, reveal a surprising ingenuity, e.g., mounts built to hold several weapons, designed for »rapid fire.«[6]

The usual form of cannon mount from this period, however, was one in which the cannon was fastened to a long platform which pivoted in a fixed framework. The piece was elevated by raising or lowering a tail fixed to the platform, either by means of a screw or by resting the tail on pegs inserted into a wooden arc at the rear of the fixed frame (see fol. 28v). This system is quite persistent and even was used on wheeled carriages where it was not really necessary. This seems to have been the result of the desire to transfer the shock of recoil to the strongest parts of the frame and to avoid concentrating it on the axles or the pivot.[7] This practice became unnecessary near the end of the fifteenth century when trunnions (projections at right angles to the barrel) were cast on the gun-barrel; these served as pivots on which one could raise or lower the muzzle. The angle of fire was adjusted by inserting a number of wedges between the cannon breech and the carriage.[8]

The problems of transporting cannon were also difficult. Throughout most of the fourteenth century cannon were apparently simply carried into place on wagons, from which they were unloaded and set in position before firing. The earliest wheeled mounts or carriages appear only in the latter part of the century, and even then were not used for large pieces.[9] Because of the difficulties inherent in the recoil of heavy firearms, the early carriages were heavily

---

[4] The pace of change in this matter is uneven. Kyeser, for example, always shows pieces with nearly cylindrical shapes, indicating small bores(?), while Taccola, writing some thirty years later, often illustrates conical or megaphone-shaped pieces. By the 1460's one rarely sees anything but cylindrical pieces.

[5] Dating changes in techniques of using cannon is made more difficult by the fact that a technical literature dealing with firearms did not emerge until the early decades of the fifteenth century. Thus, while one may learn much about the number of cannon owned by a given noble through archival records, it is rare to learn any details of how they were made or mounted. The earliest examples of texts to deal with cannon and gunpowder are the illustrated manuscripts Cgm 600 and Nürnberg, Hs.25,801. The *Feuerwerkbuch*, the most widely-copied artillerist's manual, does not seem to antedate the early 1420's, although a number of undated copies remain to be examined. Kyeser's *Bellifortis*, however, shows a number of cannon mounts which are too sophisticated to have been all Kyeser's own novelties. They must have been in use for some years before.

[6] Kyeser shows, e.g., a revolving and a rotating mount (Quarg, *Kyeser*, I:109r; II:83) and a triplex mount (I:108v; II:82).

[7] O. Barrmann, »Die Entwicklung der Geschützlafette bis zum Beginn des 16. Jahrhunderts und ihre Beziehungen zu der des Gewehrschaftes,« *Beiträge zur Geschichte der Handfeuerwaffen: Festschrift zum achtzigsten Geburtstag von Moritz Theirbach* (Dresden: 1905), 70 and 78–79.

[8] Hall, Singer III, 362.

[9] The cannon hoists discussed earlier were doubtless used to lift moderately heavy pieces onto their carriages, but were also intended to lift heavier pieces onto wagons for transport.

constructed, and, as noted above, were equipped with suspended platforms to bear the thrust of the weapon. These carriages were normally two-wheeled, with the axle fitted under the muzzle or, at times, the midpoint of the cannon.[10] There were some attempts near the end of the fifteenth century to make carriages which could be linked together and be hauled in a train (see fols. 43r and 46r). This practice never became widespread, probably because of the difficulties of drawing such a burdensome load along poor roads.[11]

The most severe problems in using cannon were those of loading and firing them. The necessity of cleaning the piece after each firing, reloading and tamping powder and placing a new shot (which could not be done until the piece had cooled) caused most medieval cannon to have a very slow rate of fire. The desire to produce a more rapidly-firing weapon, combined with difficulties in casting or forging a tube closed at one end, led to early attempts to make a serviceable breech-loader. The technical problems were formidable. Obturation had to be sufficient to prevent too much of the force of the expanding gases from leaking through the breech, yet the breech had to be able to be removed quickly and easily. A breech which was screwed into the barrel, like those found in such pieces as »Mons Meg« at Edinburgh and the Turkish bombards used at Constantinople in 1453,[12] resulted in sufficient obturation, but the heat of firing expanded the screw threads to such an extent that several hours cooling time was required before the breech could be removed.[13] A more popular solution sacrificed obturation for speed of loading. In this system a powder chamber resembling a German beer-stein was wedged into the cutout breech of a cannon. The chamber could be removed and reloaded in a short span of time, even when the weapon itself was quite hot, but the seal against escaping gases was probably not very good. Examples of this type appear in the early fifteenth century and persist well into the early sixteenth.[14] Generally, however, this seems

[10] Barrmann, 70; 929, fig. 28 *bis* shows a photograph of one of the early carriages, dated c. 1420.

[11] One of the best designed of such systems is found in *Das mittelalterliche Hausbuch*, Bossert and Storck, fig. 58, where a six-horse team pushes rather than pulls the train.

[12] Hall, Singer III, 361.

[13] Cippola, 24, n. 1.

[14] Robert Held, *The Age of Firearms* (New York: 1957), 20 dates this type from 1380, but his dating seems to be based on Cgm 6000 or Nürnberg, H. 25,801, and should probably be c. 1410–20. O. Mahr, »Zeittafel zur Geschichte des Geschützwesens bis zum Weltkrieg,« *Technikgeschichte*, XXVII (1938) p. 108 notes that Frankfurt had copper pieces of this type firing stone shot in 1412. Durdík, *HH*, 106–07 thinks that this type might have been used at the siege of Karlstein in 1422, but nowhere else during the Hussite War. Charles Ffoulkes, *Arms and Armament* (London: 1945), 93 and 95, figs. 101 and 104 illustrates two such pieces, one from 1424, the other undated. Demmin, 928, figs. 20–23, shows four examples, but dates them only as fifteenth century.
Regarding larger pieces, see August von Essenwein, *Quellen zur Geschichte der Feuerwaffen* (2 vols., 1877; reprinted Graz: 1969), Vol. A. plt. LXVI, where he reproduces an engraving by Israel of Meckenen (done after 1480) which shows four siege pieces of this type in place; also Cod. icon. mon. 222, fols. 21r, 22r, 22v, 23v, 62r, 63r, 63v, 67r, 68r, 68v and 69v.
Later examples include Seselschreiber (Cgm 973) fol. 46r and Clephan, 210, fig. 50, dated early sixteenth century; both illustrate small pieces.

to have been used only for smaller weapons until the latter third of the fifteenth century.[15] Technically, a system of this type would be difficult to adapt to larger caliber weapons.[16] The inefficiency of breech-loaders of all types, together with improved methods of casting a closed-breech (muzzle-loading) cannon, doomed the breech-loader and made the muzzle-loader the standard cannon from the early sixteenth century until the later nineteenth.

The cannon in our manuscript reflect some of the developments outlined above, but often in curious ways. Section A in particular exhibits much less interest in cannon than do other contemporary treatises; it illustrates them only five times. Three of author A's illustrations (fols. 2r, 5r and 9r) show cannon being lifted by hoists; the emphasis is clearly on the hoist, not the cannon. Similarly fol. 16r depicts a cannon on the siege engine used at Žatec, but the emphasis is on the engine, not the cannon. In only one instance, fol. 28v, does A concentrate exclusively on cannon and mounts. Moreover, as Max Jähns points out, all A's cannon have a »surprisingly old-fashioned character,«[17] i. e., they are all conical in shape, a form virtually obsolescent by the 1470's. Author A clearly is not interested in cannon *per se*, but only in ancillary devices such as hoists and mounts; he may have drawn old-fashioned models deliberately so that they would be recognizable by any reader. In this connection it is also interesting to note that the mounts and carriage on fol. 28v are also of a type which would have been familiar many decades before A's writing. This is not so striking as are his old-fashioned cannon, however, since other contemporary manuscripts illustrate mounts similar to A's.[18]

Section B illustrates cannon more frequently than A; nine folios contain them (see fols. 33v, 38v, 41r, 43r, 43v, 44v, 47r, 48r). B also shows breech-loaders, apparently designed to use the »beer-stein« powder chambers described above. His cannon are all identical to one another, however, and he seems to have little interest in depicting fine details of their construction. No attempt is made to show the powder-chamber in any of B's drawings, nor is it made clear how the chamber was fitted into the breech. B's interest lies not in cannon themselves, but in the various types of emplacements for them. His drawing of a boat carrying a cannon mounted on its bow, for example, (fol. 44v) is not an illustration of a real boat or an actual cannon, but rather an indication of a possible means of using cannon on board a

---

[15] Karl Ritter, »Die mittelalterliche Steinbüchse aus Schmiedeisen,« *Technikgeschichte*, XXVII (1938), 23, also plt. 6, fig. 9, claims that large pieces of this type originated in Spain in the latter part of the fifteenth century.

[16] The reason for this is that the propulsive force of gunpowder as well as the resistance of the shot increases as a function of volume, while the size of the sealing surface increases as a function of area. Doubling the size of the cannon's bore will yield eight times the propulsive force, but only four times the sealing surface. Thus a larger piece will produce higher pressures per unit of sealing surface, resulting in more leaking and lower muzzle velocity. For small pieces the leakage is probably negligible, while for large ones it is considerable.

[17] Jähns, 259.

[18] See Canestrini, 224, fig. LXXVII (Valturio); 228, fig. LXXIX (di Giorgio); and 230, fig. LXXX (Taccola).

ship. This was not a new idea in B's time,[19] but might have been somewhat novel to inland dwellers. In the same fashion, his drawings of cannon carriages linked together to be drawn by a single team (fols. 43r and 46r) are not very detailed technically (all the carriages lack elevating devices), but seem intended rather to capture the essential features of such a system. Other illustrations of cannon follow a similar pattern. They do not give much detailed information about the cannon themselves, but show their use in a variety of different situations, e.g., on siege engines (fol. 48r for example) or in stationary fortified emplacements (fol. 41r).

Bridges and Other Devices for Transport

Among the tasks of the military engineer were those relating to problems of supporting an army on the march. As a rule devices which served this category of duties make only a few fleeting appearances in technical treatises in our period. Most such machines were either so well understood or so commonplace that the engineers found them uninteresting and chose to ignore them. The exceptions to this rule are cannon carriages and portable bridges. Both were areas in which severe problems had to be solved. We have already discussed cannon carriages, but the subject of bridges remains to be considered.

The most common types of bridges that one encounters in fifteenth-century manuscripts can be divided into three classes: 1. pontoon bridges; 2. bridges made of folded sections to be assembled as needed; 3. amphibious wagons with extension leaves to be linked together if necessary to form a bridge. The first two classes were known and used in antiquity, although pontoon bridges were somewhat more usual than sectional ones.[20] From the earliest of the late medieval illustrated treatises we find these same two types discussed, e.g., by Guido da Vigevano and Kyeser.[21] The amphibious wagon-bridge seems to have been a novelty among engineers of the fifteenth century, however. I can find no examples of it earlier than the *Bellifortis*.[22] It is a device which captured the imagination of several German engineers, including author A, but I can find no indication that it was depicted south of the Alps until the late sixteenth century.[23] It was a device which possessed a certain elegance, since as a self-contained unit it made it possible to build a bridge without having to carry extra materials with the army on the march. Indeed, the wagon could even carry material or supplies

[19] Cannon were mounted on board ship from 1338, cf. Cipolla, 75; Feldhaus, *Technik*, 414. Similar drawing of boat hulls with cannon mounted in various positions occur in other manuscripts from our period, e.g., Waffensammlung, Ms. P 5135, fol. 20v.

[20] Cf. E. A. Thompson, *A Roman Reformer and Inventor: Being a New Text of the Treatise* **De rebus bellicis** (Oxford: 1952), 58ff. for a discussion of ancient portable bridges.

[21] Vigevano, fol. 48v; Quarg, *Kyeser*, I: 60r; II: 39 (pontoon bridges); I: 61r; II: 39 (sectional bridges).

[22] *Ibid.*, I: 59r; II: 38.

[23] When some were pictured by Ramelli, plts. 150–51.

for other purposes and could be used as a boat as well. The difficulties in actually making such a craft, e.g., designing the wheels and axles so that the hull would be water-tight while still allowing the wagon to turn, were so severe that few, if any, were probably ever built.

The two authors of our manuscript show a wide variance in the attention they give to transportation. Author A depicts seven devices in this category, while B shows no interest in it whatsoever. All but two of A's drawings illustrate bridges; see fols. 5r, 11v, 12r, and 26v–27r. Most of these are simple practical bridges in the same vein as those drawn by Kyeser and other German engineers of the period, e.g., a bridge made of planks fitted onto trestles (fol. 5r), a folding bridge (fol. 11v) and an amphibious wagon-bridge (fol. 27r). Two of A's bridges are worthy of comment. His pontoon bridge (fol. 12r), in contrast to most such bridges, floats on inflated leather bladders rather than wooden barrels. This was unusual in his day,[24] probably because of the danger of leakage, but his use of leather here is consistent with his interest in inflatable leather flotation collars.[25] Also, A's schematic diagram for a suspended bridge (fol. 27r) is unusual in that it does not appear to be a pontoon bridge, as we might expect. Neither the drawing nor the accompanying text (fol. 26v *supra*) is particularly clear on details of its operation, but this provides us with an interesting early example of the idea of a suspension bridge.[26] A's other transportation machines (fols 7v and 11r) are merely simple wheeled devices to transport material in and around an army camp.

Wagon-Fortresses

One of the more interesting machines of war to be depicted in the manuscript is the wagon-fortress *(Wagenburg)* on fol. 23v. We have already presented some discussion of this device in Chapter II to show how widely it was used in the fifteenth and sixteenth centuries. Here we shall confine the discussion to a consideration of the wagon-fortress as a piece of equipment and the ways in which it was employed.

A wagon-fortress is essentially a group of armored vehicles which can be drawn up into a closed formation. As noted above, such fortresses had been used from time to time long before the Hussite Wars of 1419–34, and there are indications that the early fifteenth century witnessed a rebirth of interest in their use.[27] It was the Hussites under their famous general, Jan Žižka, however, who brought the wagon-fortress to a new peak of effectiveness as a tactical weapon.[28] To some extent, their use of such fortresses was born of necessity,

---

[24] Although di Giorgio used this system; see Maltese, fig. 113.

[25] Cf. fols. 13r–v.

[26] For later examples see Feldhaus, *Technik*, 152ff.

[27] Cf. Chapter II, note 7; von Wulf, 23.

[28] Charles Oman, *The Art of War in the Middle Ages* (2nd ed.; Ithaca, New York: 1953), 153–54 speculates that Žižka may have based his fortresses on a Russian weapon, the *goliaigorod* or »moving fortress,« used around the end of the fourteenth century. However, von Wulf, 23 points out the extremely close agreement

since the Hussite movement by its very nature produced an army largely composed of devoted but ill-trained peasants. Such an army could not stand up to the feudal levies of the Emperor Sigismund without radically new tactics. To maximize the effectiveness of the forces at his command, Žižka used wagon-fortresses as a defensive formation in which his troops were concentrated.[29] In its fully-developed form, each wagon carried about twenty men armed with flails, pikes, crossbows and firearms.[30] Two key elements in Žižka's successful use of such wagons were the employment of firearms as an integral part of infantry tactics and the high level of command control Hussite commanders maintained over the movement of the wagons. Upon making contact with Imperial forces a Hussite army would arrange the wagons in a suitable closed formation in a convenient location. The emphasis was on speed and order, for the formation had to be secured before hostile forces attacked it. Once the formation was set the Hussites waited to be attacked by heavy cavalry. Cavalry armed with lances could not break the wagon formation, however, and it was exposed to heavy close-range fire as it tried. When the cavalry had been decimated, the Hussite troops, cavalry and infantry, would sally forth to engage the remaining Imperial troops.

To be sure, the effectiveness of such tactics was limited and the general success of the Hussites depended on many factors other than just the wagon-fortress, e.g., their religious and national fervor. Nevertheless, the wagon fortress was an impressive weapon to contemporaries and exaggerated tales of its efficacy found their way into many chronicles of the period.[31] This reputation, together with the increasing use of light, hand-held firearms and light cannon, contributed to its popularity in successive decades. As we discussed in Chapter II, wagon-fortresses were employed throughout central and eastern Europe in the fifteenth and early sixteenth centuries.[32] Older scholarship did not recognize this fact, contending that such fortresses were limited to a defensive kind of tactics and unsuitable for widespread use and offensive campaigns. Indeed, it is difficult to imagine how trains of such wagons could have been used offensively, but several military writers from the later fifteenth and sixteenth century give us directions as to how this could be accomplished.[33] The idea that the wagon-fortress was born with and died with the Hussite Wars of 1419–34 is a myth that should be laid to rest.

---

between the wagon-fortresses described in the *Kriegsordnung* of Hajek von Hodjetin (1413) and later Hussite wagons. This would indicate that Žižka did not necessarily reach into foreign experience for his wagon-fortresses.

[29] Von Wulf, 23–24 describes the early Hussite wagon-fortresses and also notes that they were used very sparingly at first; only by about 1423 were large numbers of wagons employed.

[30] Durdík, *HH*, 69; the description of tactics which follows is based on Durdík's.

[31] Oman, 156 complains that these chronicle tales have biased our view of the wagon-fortress. One of the most noted exaggerators is Aeneas Silvius Piccolomini (Pius II); cf. von Wulf, 12.

[32] Cf. Chapter II, note 7.

[33] Von Schwendi, for example, describes their use in advances and retreats and includes some complex turning maneuvers to flank an enemy's wagon-fortress. His discussion does not sound like that of an impractical dreamer. See von Frauenholz, 217–19.

The individual wagons that composed the wagon-fortress during the Hussite War differed from the ordinary farm wagons of the period, but were probably originally adapted from them.[34] The wagon was a four-wheeled one, equipped with a flat top hinged on one side. On the march the wagons did not carry the ordinary supplies of the army–special supply wagons were used. In formation the top was folded down toward the outside and left to hang there, forming a palisade covering the wheels down to the hub. The inner wall had a section hinged at the bottom that dropped down to form a gang-plank. A board running the length of the wagon was hinged underneath it along the center-line; this too was dropped down to form additional shielding behind the wheels. Within the wagon was a trough which could be weighted with stones to make the wagon heavier and more resistant to shock tactics.[35] Additional palisades of staves or a shield *(pawež)* braced by poles was erected between certain wagons.[36] Every fifth or sixth wagon carried a light cannon, the *tarasnice* (or simply *taras*). When drawn together in formation, the wagons were arranged diagonally to the line of the fortress–rather like autos in some modern parking lots–with the right front wheel of each wagon chained to the left rear of its nearest neighbor. Gaps were left every fifth or sixth wagon, protected by staves or shields and a *taras*.[37] Squares or rings of such wagons linked together, at times in ranks of two or three, made up the completed fortress.

The drawing in our manuscript does not correspond to the description given above in particulars, but the salient feature, the palisades along the sides of the wagons, is clearly shown. Elements missing from our drawing include the additional shielding and the diagonal arrangement of wagons in a line. It is possible that the discrepancies are the result of failures of modern scholarship in reconstructing a composite picture of what a wagon-fortress looked like,[38] but it seems more likely that A simply had never been able to study a Hussite wagon-fortress closely and that he drew his illustration on the basis of inaccurate descriptions or later versions of such wagons. As far as I can determine, A's drawing does not correspond with any other illustration of a wagon-fortress now extant, and so we must conclude that it is inaccurate.

[34] Von Wulf, 24. The description which follows is necessarily of a wagon-fortress of the Hussite period because that is where Czech historiography has focused its efforts. To gain a composite picture of later wagon-fortresses, one would have to collate the later *Wagenburgordnungen*, and this task has not yet been done. The most recent works dealing with Hussite wagon-fortresses are Durdik, *HH* and Eduard Wagner, Zoroslava Drobna and Jan Durdik, *Medieval Costumes, Armour and Weapons* (London: n.d. [1959 ?]), part X. Von Wulf, 25–26 and 41–44, also describes the Hussite wagons in some detail. The only drawing of a complete Hussite wagon-fortress in formation from a contemporary source is ÖNB 3062, fols. 147v–148r. A model reconstruction of a single wagon is now in the National Museum, Prague; Howard Kaminsky, *A History of the Hussite Revolution* (Berkeley and Los Angeles: 1967), fig. 5 reproduces a photograph of it. Cf. Bossert and Storck, figs. 60–61 for a later wagon-fortress in formation.

[35] Wagner *et al.*, part X plts. 4–6.

[36] Von Wulf, 41.

[37] Durdík, *HH*, 163 ff.; von Wulf, 43–44 thinks that the *tarasnice* was little used by the Hussites.

[38] Or that this scholarship reflects only the practices of the early period (1419–34). As noted above, much work remains to be done to determine exactly what a later wagon-fortress looked like.

Crossbows

Both author A and author B devote some attention to crossbows and to devices for helping span them (see fols. 18r, 29v, 30r, 40r and 40v). The historical development of such weapons is tortuous and very difficult to describe briefly. The literature on this subject is extensive, but contradictory on many major points.[39] To avoid adding by these comments to the total confusion that prevails in this area, we shall restrict the discussion to the main outlines of crossbow development. The nomenclature of crossbows and related devices is more than a little misleading (indeed, terminological confusion has fueled many of the fires of scholarly debate on this topic), and for the sake of clarity we shall introduce some purely arbitrary conventions at the beginning. A crossbow is a tension-powered (i.e. having a bow) machine with a stock and a trigger mechanism; it is small enough to be carried, loaded, spanned and fired by one man (equipped, if necessary, with an ancillary spanning device). The missile hurled by such a machine may be either an arrow, a bolt, or a stone: for our purposes the nature of the missle is unimportant.[40] Crossbows large enough to require a structure to support their weight will be designated »siege crossbows,« since if such machines were ever in fact built, their bulk would have limited them largely to siege warfare.[41]

A related device that first appeared in the Hellenistic era was similar in appearance to a siege crossbow, but mechanically quite different. In place of the large resiliant bow as power source, these machines employed two rigid arms which were moved in arcs by twisted bundles of sinew rope. We shall call such torsion-powered engines »torsion ballistae,« again without further distinction as to the type of missile hurled.[42] Since we are concerned only with crossbows and torsion ballistae, we need only note in passing that one-armed torsion-powered devices existed in antiquity and the middle ages under a variety of names, the most common being mangonel and onager; the latter is an Asiatic wild ass which reputedly kicks rocks at pursuers with its hind legs.

---

[39] The major studies in this field are: Eric M. Marsden, *Greek and Roman Artillery*, Vol. I *Historical Development* (Oxford: 1969) and Vol. II *Technical Treatises* (Oxford: 1971); Kalvero Huuri, *Zur Geschichte des mittelalterlichen Geschützwesens aus orientalischen Quellen* (Helsinki: 1941); Ralph W. F. Payne-Gallwey, *The Crossbow, Medieval and Modern, Military and Sporting* (London: 1903; reprint New York, n.d. [1958]); Napoleon III, Emperor of the French and Ildefonce Favé, *Etudes sur le passé et l'avenir de l'artillerie* (Paris: 1846–71); G. Köhler, *Die Entwicklung des Kriegswesens und der Kriegsführung in der Ritterzeit* (Breslau: 1967).

[40] The historical terminology, on the other hand, originally indicated precisely this distinction. Technically, an arrow-shooting engine is a *catapult* (Greek literally »through the shield«) while a stone-thrower was usually called *ballista*. Needless to say, even this fundamental distinction was blurred beyond recognition by later antiquity. See Marsden, *Hist. Dev.*, 2 and 188ff.

[41] This term is preferable to the Latin *arcuballista* (cf. *infra* note 52) and the German *Standarmbrust*, both of which are now often used to designate *any* two-armed, stand mounted piece of artillery, no matter what its mechanical design.

[42] For a fuller description of such machines, see Marsden, *Hist. Dev.*, 16ff.

The historical origin of the crossbow is uncertain, but the most reasonable hypothesis is that it evolved from self-acting bow-and-arrow traps designed to kill animals who tripped a trigger.[43] The earliest use of true crossbows in historical times took place in China, where they were quite common by the early Han and may date back to the Chou or even earlier.[44] Crossbows in China were developed significantly in subsequent generations, acquiring sophisticated trigger-mechanisms and evolving into large, multiple-bowed, wagon-mounted weapons of considerable power.[45] Curiously, the Chinese do not seem to have developed torsion ballistae at all; the crossbow remained the standard weapon, even down to the last decades of the nineteenth century.

The history of the western crossbow is far more complex and is subject to a great deal of confusion over the technical terms used in the sources to describe all types of missile-hurling machines. If we restrict ourselves to the least ambiguous sources, the following picture emerges. Heron of Alexandria's *Belopoiika* is the best early text still extant to discuss crossbows. Although it was written in the first century, it probably reproduces in substantial measure a now-lost treatise by the Alexandrian engineer Ktesibios from the third century B.C.[46] Heron describes briefly what he calls a *gastraphetes* or »belly bow.« This was a portable machine with a hand-held stock and a large bow mounted on its front. Heron explains that the stock was developed originally as a means of spanning bows that had grown too large to be bent by unaided human muscle power. The stock, with its ingenious ratchet slider and trigger mechanism, gave mechanical assistance to the archer, allowing still larger and more powerful bows to be developed. To span the *gastraphetes,* its operator used the weight of his body to compress the slider-stock apparatus, thereby retracting the bowstring–hence the name »belly bow.«[47]

Heron (Ktesibios) gives us no historical dates for the invention of the *gastraphetes*: indeed, he treats the entire matter only as a prelude to a discussion of more advanced types of torsion-powered artillery. It can be plausibly argued on the basis of a large number of nontechnical references that the *gastraphetes,* or something similar, first appeared in Syracuse c.399 B.C. and was in common use in the Mediterranean by the middle third of the fourth century B.C.[48] The problem is to determine what became of it thereafter. Technically, such devices were superseded in every respect by the torsion ballista, which makes its appearance

---

43 C. Martin Wilbur, »The History of the Crossbow,« *Annual Report of the Smithonian Institution* (1936), 428.

44 *Ibid.,* 428 ff.; White, 151–52. Unfortunately Needham has deferred a discussion of Chinese crossbows to a later volume. However, an oblique reference in Vol. IV. pt. 2, p. 126 suggests the later Chou as the period when they originated in China.

45 Huuri, 197 ff.

46 See Marsden, *Treatises,* 18–60 for Heron's text. On the relationship between Heron's work and Ktesibios, see *ibid.,* 1–2.

47 *Ibid.,* 21–23.

48 Marsden, *Hist. Dev.,* 48 ff. reviews in great detail the available evidence.

under Alexander the Great or his father.[49] Of all the other poliorcetic writers of antiquity, only one mentions the *gastraphetes* again. This is the shadowy Biton, about whom virtually nothing is known. His treatise on war machines (date and place of composition unknown; probably Pergamon, c. 250 B.C.) deals with bow-engines in the form of siege crossbows, spanned with winches. Despite the fact that machines, not bellies, span Biton's bows, he nevertheless still calls them *gastraphetes*.[50] The numerous difficulties in Biton's text make it dangerous to assume that he was describing machines in common use in his day, and, in any case, there is no subsequent source, technical or non-technical, which refers to them again. Thus it is most reasonable to assume that the military crossbow was invented in antiquity c. 399 B.C. but that it was quickly rendered obsolete by the development of the torsion ballista and that it disappeared from use by c. 250 B.C., if not earlier.

Outside the field of war, the crossbow seems to have fared somewhat better. It probably remained in use as a hunting weapon throughout the Roman period. We may assume this from its appearance in two reliefs from Le Puy dating from the first and second centuries. An additional second-century relief showing the crossbow completes the scanty repertory of evidence for its continued existence in antiquity.[51] We should note here, however, that even if the hunting crossbow remained in use, we can derive no conclusions thereby about the use of military crossbows. Hunting weapons are generally of inferior strength when compared with military weapons; certainly the crossbows depicted on the various reliefs would have made poor weapons in combat, since they appear to have been intended mainly for small birds.

The most difficult periods in which to discuss any type of missile engines are late antiquity and the early middle ages. The historian is faced here not only with a dearth of any comprehensible technical descriptions, but also with a level of terminological confusion that surpasses anything previously encountered. The only honest and simple claim that he can make is that some form of missile-hurling machinery remained in use for military purposes throughout both periods, but we cannot know any of the technical details of these machines with certainty. I have argued at length elsewhere that torsion-powered engines seem to have remained the standard forms among Roman legions and that all attempts to interpret ambiguous references in late ancient texts as referring to crossbows are erroneous.[52] While we

[49] On the advent of torsion artillery, see *ibid.*, 56ff. On its technical superiority, cf. 15 and 86ff.

[50] Marsden, *Treatises*, 106ff. gives the text of Biton.

[51] See Feldhaus, *Technik*, 33ff. and White, 152 for the Le Puy reliefs. The *Illustrated London News*, April 6, 1953, 501, fig. 7 shows the most recently-discovered relief.

[52] This opinion follows and extends the arguments of Marsden; cf. my review of his works, »Crossbows and Crosswords,« *Isis*, LXIV (1973), 527–33. The most significant element of the argument derives from Marsden's reconstruction of the machine described in Heron's *Cheiroballistra (sic: ballistra* is a late form of *ballista*) as a torsion-powered, stand-mounted device with a metal frame and a conspicuous sighting-arch or *kamaron* (see *Treatises*, 206ff.). Following this, one can interpret most references to »arches« in later artillery as referring to a sighting-arch rather than a bow. Vegetius' *arcuballista*, for example, is an »arch-ballista,« not a »bow-ballista.« So simple and compelling an explanation renders obsolete the tortuous interpretation of such tech-

know very little regarding early medieval artillery, I am inclined to believe that true crossbows as military weapons were absent or very rare before the eleventh century.[53] What can be stated with certainty is that crossbows were in common use among western armies by the time of the First Crusade,[54] and that both Byzantines[55] and Muslims[56] regarded them as new, western weapons. How and when the crossbow emerged as a significant military weapon in the middle ages is entirely unknown. Possibly it was re-invented then, or learned from the Chinese, or adapted to military purposes from existing hunting crossbows (if indeed hunting crossbows themselves survived).[57] In any case the crossbow became a weapon

---

nical puzzles as the »arcus fereus« in the *ballista fulminalis* described in *De rebus bellicis*, xviii (1), which Thompson, 61–68, seeks to explain as a steel spring powering what is essentially a bow-engine. This view, which contravenes everything we know about ancient metallurgy, was reinforced with more philological than technological rigor by Revilo P. Oliver, »A Note on *De rebus bellicis*,« *Classical Philology*, L (1955), 113–117, and still has surprisingly wide currency. Marsden's treatment of Heron's text (*Treatises*, 240–46) invalidates all previous attempts at interpretation.

[53] This statement deliberately glosses over one of the more acrimonious (and less enlightening) scholarly debates of the late nineteenth century: did the early middle ages continue to use the two-armed torsion ballista of the ancients? Since all participants to the controversy lacked adequate information from the texts, each question turned on conflicting interpretations of ambiguous terms and dubious inferrences from questionable texts. For a balanced, if now slightly outdated, summary and verdict on this matter, see Huuri, *passim* and especially 47–50.

[54] White, 35; Hall, Singer II, 723; Huuri, 44–45.

[55] Huuri, 72ff. notes that Byzantine sources refer to several types of artillery, but that the most frequently used term is *toxoballistra*. He considers this the equivalent of Vegetius' *arcuballista* and thinks that it meant either a siege crossbow or a torsion ballista. If we follow the interpretation suggested in note 52 *supra*, then the latter seems more likely. Huuri assumes that a rare compound, *cheirotoxoballistra*, means a crossbow, but his grounds for this are extremely weak (cf. 73–74). More significant is the shift in Byzantine terminology from *toxoballistra* to *tzaggra* that occurs suddenly c. 1100, or just about the time of the First Crusade. Since *tzaggrai* are often mentioned in extremely large numbers, e. g. as equipment for an entire battallion (76), it seems possible that the new word referred to an imitation of the Frankish crossbow.

[56] Huuri, 94ff. The Islamic situation seems even more confusing than that in the west. In essence, two types of crossbow were in use in Muslim lands by the fourteenth century, a western type (possibly related to Frankish models) and an eastern type (which may have derived from China via the Mongols). Neither description is clear enough, however, to allow more than speculation on this point. Significantly, Muslim writers fail to mention the crossbow prior to the period of the early Crusades. The only exception to this rule is Mohammed ibn Garir al-Tabari (838–923), whose description of the revolt of Ethiopian and Nubian slaves in Iraq c. 880 contains the idiom later employed for the crossbow, *qaus al-rigl* (lit. »foot-bow«). Thereafter there is no further mention of this device in Muslim sources until the twelfth century. Ignoring the obvious questions of emendation or interpolation by later copyists, it still seems reasonable to ask whether *quas al-rigl* meant »crossbow« to al-Tabari. Perhaps his reference is to powerful hand bows, which were later often strung by being held with the foot or ankle. In any event, extensive use of the crossbow in Islam does not take place before the period of Frankish (and later Mongol) influence.

[57] As White, 152 notes, the only evidence for the hunting crossbow in western Europe during the early middle ages is one possible interpretation of an ambiguous Anglo-Saxon riddle.

closely identified with western peoples, the »Franks,« in the minds of others. Anna Comnena, for example, was astonished when she viewed crossbows in the hands of western troops during the First Crusade.[58] From this time on the Latin West remained the center of developments in and diffusion of crossbows.[59] They were a specialized weapon (though not so demanding of skill as the English longbow) and the men who used them were often organized into special companies. Later certain areas, e.g., Genoa, were to become noted for their mercenary crossbowmen who participated in wars throughout Europe.

The technical development of the crossbow from the eleventh century on is fairly clear, although precise dates for the introduction of various improvements and new components are difficult to determine. In general improvements were directed toward making the bow more and more powerful, even at the expense of increased difficulties in spanning it. Before the fourteenth century bows were made of tempered wood, or of organic animal material such as horn, bone or sinew; resilience was achieved by laminating these and similar materials together in differing manners and amounts.[60] Sometime in the latter half of the fourteenth century bows of tempered resilient steel began to appear.[61] These were, naturally, considerably more powerful than bows of organic material, and they demanded more sophisticated mechanical aids in order to span them. Early crossbows apparently were spanned by manual power alone. With the introduction of composite, laminated bows, it became necessary to equip the bow with a stirrup at its front end into which the bowman could step, holding the bow inverted against the ground. He would then bend over and grasp the bowstring with both hands pulling up on it to span the bow. As bows became more powerful, this method was replaced by one in which the bowman used a stout cord fixed to his belt and hooked to the bowstring in place of his arms (cf. fol. 30r). This method enabled him to use his leg and back muscles more effectively than the earlier method.[62] Ultimately a metal hook replaced the cord and the bowman spanned his weapon from a more upright position, using his leg to thrust downward in the stirrup.[63]

As bows became even stronger, however, the limits of the force which human muscles could produce were reached. More powerful mechanical apparatus had to be employed. The fourteenth century saw the introduction of the goat's-foot lever (an offset lever with a hook) as a means of exerting even more force while spanning a bow. Other devices for this purpose include windlasses which fit onto the stocks of the crossbows and span them by means of pulleys and the *cranequin*, a cranked reduction gear which also fitted the bow's stock; both

---

[58] Anna Comnena, *The Alexiad*, trans. Elizabeth A. S. Dawes (London: 1928), 255.

[59] See White, 151ff.

[60] Payne-Gallwey, 57ff. dates the composite bow from the early thirteenth century.

[61] *Ibid.*, 90ff. dates the introduction of the steel bow at c. 1350–1370; elsewhere he seems to prefer the latter date.

[62] *Ibid.*, 73–75. This was sometimes used with a pulley. Fol. 30r shows a variant of this method.

[63] *Ibid.*, 77–80.

date from the fifteenth century.[64] There has been no discussion, to the best of my knowledge, of an alternative system which appears in several German manuscripts from the fifteenth century (cf. fols. 18r and 29v). In this a larger windlass or a lever was mounted on a small cart and was apparently shared by a group of bowmen in turn.[65] This allowed the construction of a fairly large machine, capable of more force than the smaller devices which attached to an individual bow, and at less cost than alternative systems. It may be that the small individual bow windlass developed out of the larger ones for several bows, since the cart-mounted crossbow windlasses and levers antedate the smaller ones by at least a few years.

The vast majority of medieval crossbows were personal weapons small enough to be carried by one man. A few, however, were apparently much larger and would seem to have demanded some sort of mounting stand. It is difficult to determine, however, if such larger machines, true siege crossbows, were in fact ever used in the later middle ages. Mounting stands were common in antiquity for use with torsion ballistae (which were heavier and often employed a higher angle of fire than crossbows). Indeed, if Biton's descriptions are to be trusted, then stands were used with the heavy *gastraphetes*.[66] It would seem reasonable, therefore, that the medieval development of crossbows led ultimately to attempts to mount them on stands or carriages. (Alternatively, since carriage-mounted crossbows were frequently used in China, we might expect the Mongols to have brought such devices to the attention of Europeans.[67]) Unfortunately, even though we encounter mention of very large crossbows from the thirteenth century onwards,[68] we do not have any descriptions of mounting stands nor any illustrations depicting siege crossbows before the fifteenth century.[69] Two types appear then. In one, the crossbow is mounted on a tripodal or circular base with a pivot joint.[70] In the other, crossbows are shown mounted on a flat, table-like trestle having four

[64] *Ibid.*, 84–88. The windlass is credited to the fourteenth century by *ibid.*, 71. White, 111 notes that there is no firm evidence for its use before Kyeser. Payne-Gallwey correctly dates the *cranequin* as c. 1450. Hall, Singer II, 722 apparently errs in implying that it is from the fourteenth century.

[65] A similar device appears in *Bellifortis*: Quarg, *Kyeser*, I: 78r; II: 49–50. This is the earliest illustration I have seen of this type. An example of a windlass used for the same purpose is found in Waffensammlung Ms. P 5014, fol. 14v; it is capable of spanning two crossbows at once.

[66] Marsden, *Treatises*, 74–77 and 97ff. The larger of Biton's two belly-bows appears to have a bowstring about 2.5 meters long (8.15 ft.).

[67] Huuri, 51 and 197ff.

[68] *Ibid.*, 50 cites a description of c. 1275; »Vidi in castro Friburc balistam, cuius arcus de cornu nobili longitudo 13 pedes habebat.«

[69] Payne-Gallwey, 301 notes that medieval authors often describe ancient ballistae »as if they had a gigantic bow in one piece.« He asserts, however, that he found no definite evidence that such siege crossbows were ever constructed.

[70] The best example I have seen of this type occurs in Waffensammlung, P5014, fol. 51v, which clearly depicts a swivel pivot. Ramelli includes a similar design with three bows instead of one; see Payne-Gallwey, 302.

legs. Since it would be very difficult to elevate the latter types, they would seem to have been of limited utility in combat. Nevertheless, such mountings are depicted in several of the Germanic treatises.[71] The pivoting style, which is the only practical one, appears much less frequently. The presence of illustrations of the clumsy trestle-mounting is *prima facie* evidence that what is being depicted is an engineering fantasy, and that siege crossbows were not in widespread use among late medieval armies. It surely would not have been beyond the technical capabilities of late medieval engineers to have made siege crossbows but they apparently did not choose to do so. They only visualized them as possibilities, usually in crude form. This strange situation may have resulted from the nearly simultaneous advances in metallurgy and artillery made in the thirteenth century. The capacity to make resilient steel bows gave the crossbow for the first time the ability to compete successfully with older, torsion-powered, two-armed ballistae in range and weight of missile. Simultaneously, however, ballistae were rendered obsolete first by counterweight devices (trebuchets) and shortly thereafter by cannon. Thus we find the speculative writers of treatises repeatedly bemused by the possibility of designing a workable siege crossbow, while the practical military specialists in the field apparently preferred to concentrate on more promising newer types of artillery, especially cannon.

[71] Fol. 40r shows a siege crossbow mounted in this way. It is difficult to imagine how this could be aimed. Similar examples appear in Cgm 600, fol. 18v; ÖNB 3069, fol. 22v; Waffensammlung, Ms. P 5014, fol. 16r and 74r (the former has two bows). Leonardo designed a bow of enormous dimensions; *Cod. Atl.*, fol. 53v; cf. Payne-Gallwey, 262. It is interesting to note that the illustrations to *De rebus bellicis*, vii and xviii depicting artillery also show pieces mounted on flat-bedded carts without any adjustment capacity, despite the fact that elevating screws are specified in the text (see Thompson figs. III and XII). The extant illustrations to this work are the products of the fifteenth and sixteenth centuries, and they are stylistically reminiscent of the Germanic tradition, especially Kyeser. I suspect that the illustrations tell us more about fifteenth century than about ancient technology.

CHAPTER VI

AQUATIC DEVICES

As engineers both author A and author B concerned themselves with devices that relate in some way to water. Author A has left us several interesting illustrations of devices with which to move about, on or under water, and both authors include machines used in municipal water systems. This chapter will consider: 1. flotation collars; 2. diving apparatus; 3. paddle wheel boats; and 4. borers and water systems.

Flotation Collars

Inflated skins are one of the oldest means of keeping a man afloat in water. The earliest examples of this are found on an Assyrian relief from the ninth century B.C. and similar pieces of apparatus are known from Roman times.[1] For the most part these were simply leather bags, resembling pillows, inflated by the would-be swimmer who held on to them with his hands. There are no records telling us that similar devices were used in the early middle ages, but it is probable that they were. The treatise of Guido da Vigevano shows inflated bags that resemble the ancient examples closely.[2]

The flotation collars that appear in the *Bellifortis* are much more advanced. They are tailored to the human anatomy, designed to be buckled about the waist or torso. They look as if they could keep a heavily-armed man afloat.[3] Moreover, Kyeser's comments on how they were to be worn have the ring of practical experience about them, indicating that such collars were in actual use in his day.[4] The drawings of flotation collars in Kyeser's work were popular among those who copied the *Bellifortis*,[5] and they influenced the authors of non-Kyeserian

---

[1] Feldhaus, *Technik*, 1010 ff. Some Roman examples apparently used floats made of cork.

[2] See Gille, *Engineers*, 29.

[3] Quarg, *Kyeser*, I: 55 r, 63 v, 66 r; II: 37, 41, 43 respectively. Not only are his collars well-shaped, but one of them uses a mouth-piece with a stopper on its inflating tube.

[4] Also one was apparently used to attempt to rescue King Maximilian from the castle of Brügge in 1488; cf. Feldhaus, *Technik*, 1012.

[5] ÖNB 3069, for example, includes seven different examples, including (fol. 61 v) inflated boots with spiked soles and heels (for walking on water?).

treatises as well.[6] Author A's collars (fols. 13r and 13v) are clearly part of the tradition stemming from the *Bellifortis*, although they are in no way copies of Kyeser's drawings. Indeed, A's drawings indicate the way in which later generations were finding new applications for devices Kyeser had shown. Fol. 13v shows a flotation collar intended for a horse, not a man, which would (at least in theory) give a cavalryman the same advantage as a foot-soldier when faced with a river to cross.

The Italian manuscripts give less attention to flotation collars, and the designs they show are generally inferior. Taccola's, the earliest I have seen, show rigid pontoons strapped about the upper torso.[7] It is only with Leonardo da Vinci that the sophistication of Italian design in this matter approached that of the German texts.[8] Interest in these collars never died out. They continued to be sketched, in ever more varied forms, in the sixteenth and seventeenth centuries.[9] One of the men who included an assortment of flotation collars in his treatise was Christof Seselschreiber, and he copied some of them from our manuscript. His comments, however, indicate that he considered them representative examples of old-fashioned designs.[10]

Diving Apparatus

If men could walk on the water's surface, could they not also walk beneath it? This seemed reasonable enough to author A, who drew a suited diver hoisting sunken goods to the surface on fol. 14r. Perhaps unwittingly he was following a tradition of speculation on the topic that had a long and distinguished past. The earliest reference to any apparatus for sustaining life underwater occurs in the pseudo-Aristotelian *Mechanical Problems*, where an elephant's trunk is compared with the tubes used by divers and where air reservoirs carried by divers are mentioned.[11] The real source of the idea of underwater exploration came to the middle ages in the form of fantasy, not fact. In the romances concerning Alexander the Great there is the tale of his descent into the bottomless deep in a glass bathysphere. There he meets a number of frightful creatures and has several hair-raising adventures, finally returning to

---

[6] For example, Cgm 356, fol. 159r; Cgm 599, fol. 46v; Weimar, Hs. fol. 328, fol. 134r (cf. E. Marx, 319, fig. 9).

[7] Clm 197, II, fol. 90r [91r]. Di Giorgio's designs are slightly improved, but he also includes rigid pontoons; see Maltese, figs. CXII–III.

[8] Cf. *Cod. Atl.*, fol. 276v–a and Ms. B, fol. 81v.

[9] For example in Veranzio's *Machinae novae* (Venice: 1609), plt. 39. Cf. Feldhaus, *Technik*, 1013. Even quite late the German influence on collar design remained strong. Paris, B. N., Ms. franc. 20006 (early seventeenth century), fols. 108v–108r, shows examples quite similar to our fol. 13r–13v.

[10] Cgm 973, fols. 139v–140r copy Clm 197, I, fols. 13r–13v respectively. The caption to the latter of Seselschreiber's drawings calls it »ain alter Baszer [i. e., wasser] Zeug.«

[11] Feldhaus, *Technik*, 921 and 1124ff.

dry land in safety. This tale was popular in the west from the tenth century onward.[12] Alexander had at least one German imitator. The late twelfth-century vernacular poem *Salman und Morolf* describes Morolf's descent into a sunken ship; air was supplied to him through a leather tube.[13] Morolf was better off than the fabled conqueror, who had to make do with the air contained in his glass vessel.[14]

If these fanciful tales did not contribute materially to the early development of underwater apparatus, at least they kept the idea of submarine activity alive. With the fifteenth century we find a number of engineers in Germany and Italy beginning to explore methods of supporting human life underwater. Kyeser's illustrations show a confused groping for a suitable solution. He depicts, for example, two divers in combat, each of whom breathes from a small reservoir of air carried with him in a bladder. The divers both wear leather suits.[15] Taccola's drawings, done about three decades after Kyeser's, show a distinct advance. His divers breathe through leather tubes held above the surface.[16] In Taccola's earlier drawings the breathing tube enters the diver's mouth. In his later ones it is connected to the top of the diver's helmet and is supported on the water's surface by a float. This is essentially the same system used by author A and the one favored by most later engineers. There seems to have been no recognition of the fact that pumping air down to the diver was superior to forcing him to inhale and exhale through a lengthy tube. Most of the later drawings of such divers make no provision for air pumps of any kind.

The later fifteenth and the sixteenth centuries saw little improvement on the basic designs of Kyeser or Taccola. Only Leonardo's notebooks show a pump to supply the diver with air.[17] It is surprising, therefore, that Seselschreiber, who copied author A's drawing of a diver, called it the »old method.«[18] There were no »new methods« (save for Leonardo's) nor were there to be any until modern times. The sixteenth century manuscripts simply repeat

[12] The submarine episode first appears in a sixth-century wonderbook version of the biography of Alexander by the pseudo-Callisthenes. This came to the west in a Latin translation by the archpriest Leo of Naples after 952. The first German versions of the tale date from 1080–1120 and become fully developed by the mid-thirteenth century; cf. David J. A. Ross, *Alexander and the Faithless Lady: A Submarine Adventure* (London: 1967), 5–6, 8–9 and 12–13.

[13] Feldhaus, *Technik*, 1119; the relevant lines read: »Eyn rore in daz schiffelin ging da mit Morolf den atem ving.« Uccelli, 232, fig. 23 cites a ninth-century representation of a mer-man who appears to be breathing through a tube.

[14] Although by the late thirteenth century at least one version of the tale provided Alexander's bathyscaphe with a long glass tube that jutted above the water's surface; cf. Ross, 17 and fig. 6.

[15] Quarg, *Kyeser*, I: 62 r; II: 40.

[16] Clm 197, II, fols. 11 r [12 r] and 54 r [57 r]; cf. also the drawing (later copied by di Giorgio) reproduced in James Beck, »The Historical ›Taccola‹ and the Emperor Sigismund in Siena,« *The Art Bulletin*, L (1968), 318, figs. 24, 25.

[17] *Cod. Arundel*, fol. 24 v, cited by Uccelli, 551, fig. 109. Most of da Vinci's drawings of divers do not show this feature, however; cf. *Cod. Atl.*, fols. 7 r–a and 368 r–b; Ms. B, fol. 18 r.

[18] Cgm 973, fol. 135 v.

older solutions without adding anything new or better.[19] The immediate future of submarine activity lay in a different direction, one suggested by the Alexander romances. The only successful devices put into practice were bathyscaphes capable of working at shallows depths. There were many such devices made in the late sixteenth and seventeenth centuries.[20]

Paddle Wheel Boats

Author A's drawing of a boat moved by paddle wheels (fol. 17v) has a long history. The earliest record we have of such a boat is the anonymous fourth-century treatise *De rebus bellicis*, where one is depicted under the name *liburna*. It is shown with two sets of paddle wheels turned by oxen.[21] There is no indication, however, that any boat of this type was ever constructed in Roman times in the west. In China the story was different. There man-powered paddle wheel boats were invented between 494–97 by an engineer and mathematician named Tsu Chhung-Chih, and remained in use until modern times.[22] In Europe the idea of such boats lay dormant[23] until the fourteenth century, when it was revived by Guido da Vigevano.[24] His boat, like the one in our manuscript, was powered by men turning crankshafts to which the paddles were directly attached. (This is also the earliest known depiction of a crankshaft.) Guido's illustration marked the beginning of renewed interest in boats of this type. Kyeser illustrated one in the *Bellifortis* which resembled the *liburna* more than it did Guido's.[25] The works of Taccola depict a different sort of paddle wheel boat,

---

[19] For an example of a sixteenth-century text that reproduces old solutions, see Joseph Werlin, »Ein Feuerwerkstext aus dem 16. Jahrhundert,« *Centaurus* IX (1963), 283–85.

[20] The earliest of these was a successful demonstration of a diving bell by two Greeks before Charles V at Toledo in 1538; for this and other examples from the sixteenth and seventeenth centuries cf. Uccelli, 551–53.

[21] Thompson, 50ff., 102, 119–20 and fig. XI. Thompson concludes that the *liburna* was an original invention by the anonymous author and that it was probably inspired by the undershot water wheel. For a survey of paddle wheel ships see Hugo Th. Horwitz, »Zur Geschichte des Schaufelradantriebes,« *Zeitschrift des österreichischen Ingenieur- und Architekten-Vereins*, LXXXII (1930), 309–13 and 356–60.

[22] Needham, 416; he also suggests that they may have been invented even earlier, in 418; the period of extensive use of such vessels began in the twelfth century. Europeans have almost always considered these man-powered boats to be crude imitations of western steamers.

[23] Save for Roger Bacon's vague description of boats without oars. Some scholars feel that Bacon had seen *De rebus bellicis*: cf. Needham, 413.

[24] Vigevano, fol. 49r; the sources of Vigevano's idea remain unknown. Thompson's assertion (p. 53) that subsequent drawings of paddle wheel boats were »inspired directly« by *De rebus bellicis* does not hold true in Vigevano's case. His boat differs greatly from that in the Roman treatise. Since such boats were invented independently in China and Europe, surely it is credible that they could be reinvented independently in the fourteenth century.

[25] Quarg, *Kyeser*, I: 54r; II: 36. The drawing appears in many of the *Bellifortis* copies, and may have inspired an illustration of a paddle wheel boat in Vienna, Waffensammlung, Ms. P 5014, fol. 10v.

one in which the paddle wheel turns with the current and winds in a rope attached to its shaft. With the rope anchored upstream, the boat pulls itself against the river's flow.[26] From this time on illustrations of paddle wheel boats become fairly common in technical treatises.[27]

The earliest reported use of such boats was in Nuremberg in 1497,[28] but the most famous attempt to make one came in 1543. In that year Blasco de Garay fitted a ship of some two hundred tons, the *Trinidad,* with paddle wheels worked by forty men on capstans (or treadmills) and demonstrated it in Barcelona before Charles V. Although nothing concrete came of de Garay's work, there were a number of later attempts to put similar boats into regular service.[29] It was not until the advent of steam power that paddle wheel boats could be used on long voyages, but there were some animal-powered examples that saw regular service on short runs. As late as 1818, for example, a horse-powered paddle wheel packet operated at Great Yarmouth, Norfolk, England.[30]

The paddle wheel boat drawn and discussed by author A resembles Guido's more than it does any of the other German illustrations of such boats. Perhaps A knews of this type from Valturio's *De re militari* or perhaps he attempted to adapt Kyeser's capstan-powered design to use a crankshaft. Author A was obviously very interested in crankshafts and skilled in their application,[31] and some of the *Bellifortis* copies modified the master's boat by equipping it with a crank.[32] It would have been easy for him to envision a crankshaft driving the paddle wheels in place of Kyeser's bulky capstan and gear wheels.

An interesting problem concerns A's attribution of his boat to the Catalonians in the text accompanying the drawing: »The people of Catalonia use this type to prevail over all other ships.« There is no record of Catalonian sailors' having used paddle wheel craft for any pur-

---

[26] Clm 197, II, fol. 49r [1r]; cf. also James Beck, 318, figs. 23 and 26.

[27] Examples include: Canestrini, 350, fig. XCCII (di Giorgio); Valturio, fol. 215r; London, B.M. Ms. Add. 197621, fols. 43r–43v; da Vinci, *Cod. Atl.*, fol. 844r; Ramelli, fig. 152, to name but a few.

[28] One is mentioned in the *Nuremberg Chronicle* for that year and is cited in Thomas Ewbank, *A Descriptive and Historical Account of Hydraulic and Other Machines for Raising Water, Ancient and Modern* (3rd ed.; New York, Greeley and McElrath, 1849), 406.

[29] The accounts of this have been confused by the claim in Martin Fernandez de Navarrete, *Coleccion de los viages y descubrimentos que hicieron por mar los Espanoles desde fines del siglo XVI,* Vol. I (1826; reprinted Buenos Aires: 1945), 122–23, that de Garay's boat was steam-powered. The distinguished Spanish historian, Modesto Lafeunt y Zamalloa, disproved Fernandez de Navarrete's claim in his *Historia general de España,* Vol. X (Barcelona: 1888), 357–62. Cf. also John MacGregor, »On the Paddle-Wheel and Screw Propellor from the Earliest Times,« *Journal of the Society of Arts,* VI, 282 (April 16, 1858), 335–42 and Feldhaus, *Technik,* 938. Needham, 414, note »a« states that Fernandez de Navarrete was misled by forged interpolations in some archival documents. Cf. Feldhaus, *Technik,* 938ff. for subsequent attempts to use paddle wheel boats.

[30] Cited in Thompson, 54.

[31] See fols. 18r, 20r, 21r and 22v.

[32] See Feldhaus, *Technik,* 938.

pose in our period.[33] While de Garay's experiments took place in Spain, there is no reason to assume that they reflect a local tradition of such ships.[34] There were many designs on paper for paddle wheel boats by 1543, and de Garay could have been inspired by any one of them. H.T. Horwitz speculated in respect to fol. 17v that there may have been attempts in Spain in the fifteenth century to use such craft as ferryboats, but he cited no evidence for this view.[35] It is more likely that author A was misinformed, as he was in his ascription of an »automobile« siege engine to Erkinger von Seinsheim (fol. 16v).

Finally, we should also note that A's text describes a vessel different from the one he draws. He describes a boat large enough to carry twenty armed men and equipped with shielding to hide the soldiers; the drawing depicts a very small boat with no shielding. Author A states that four men power the boat; the drawing shows only two. We should not assume that the text describes some other boat,[36] but rather that A wanted to make clear the salient feature of his machine, its crankshafts. These would be completely hidden if the vessel were covered with shielding or filled with soldiers.

Borers and Water Systems

Both authors depict devices that were part of the water-distribution system in many medieval cities (see fols. 15r, 35r and 35v–36r).[37] A concern with such devices was not unusual for an engineer in the fifteenth century. He was expected to earn his living in peacetime by assisting in building projects, including canals and fountains,[38] and many treatises on military

---

[33] Alexander Keller, »Mechanical Linkages,« *Chartered Mechanical Engineer*, XIV (1967), 323 claims that Catalonian pirates used paddle wheel ships in the fifteenth century but he later admitted this was based on our fol. 17v and no other source (letter of April 1, 1971 in response to my inquiry).

[34] Canestrini's suggestion (p.342) that this type of boat might have been imported from Spain, perhaps by the Romans, lacks foundation. He may have had *De rebus bellicis* in mind.

[35] Hugo Th. Horwitz, »Zur Geschichte des Schaufelradantriebes,« *Zeitschrift des österreichischen Ingenieur- und Architekten-Vereins*, LXXXII (1930), 312.

[36] *Ibid.*, suggests that the description applies to Valturio's drawing. I disagree. It may have been influenced by Valturio's illustration, which shows a covered boat, but it could not be a description of that drawing.

[37] For general surveys of the history of water supply, see: Ewbank as cited in note 28 *supra*; Bernard Buffet and René Evrard, *L'Eau potable à traverse les âges* (Liège: 1951); F.W. Robins, *The Story of Water Supply* (London: 1946); R.J. Forbes, article »Hydraulic Engineering and Sanitation,« Singer II, 663–94. Detailed information on the system of Nuremberg is given by Karl H. Fischer and Leo Walther, *Die Wasserversorgung der Stadt Nürnberg von der reichstädtischen Zeit bis zur Gegenwart: Festschrift zur Eröffnung der Wasserleitung Rama* (Nürnberg: 1912).

[38] Leonardo da Vinci's famous letter to Ludovico Sforza in 1482 outlining his engineering skills lists a number of military machines he can build and then concludes:

> In time of peace I believe I can give perfect satisfaction ... in architecture and the composition of buildings public and private, and in guiding water from one place to another.

Quoted in Gille, *Engineers*, 125–26.

technology also devote attention to reservoirs, pipe-systems or pumps.[39] By the late fifteenth century most large cities had some sort of pipe network to bring water to public fountains for the use of their citizens.[40] If local geography necessitated it, the water would be raised into an elevated reservoir so that it would flow to most parts of the city. In Germany, the favorite machine to lift the water was the *noria*, a peripheral-pot wheel turned by the current (see fol. 35 r).[41] In most cities in northern Europe the pipe system was made of bored-out logs joined together and buried in a shallow trench. Wooden pipes were generally preferred to lead ones because the latter were more expensive (although longer-lived) and had to be made with smaller diameters than the former.[42] In a large, technologically advanced city like Nuremberg, there were several networks of pipes, each drawing from a different source and supplying a different quarter of the city, all maintained at public expense.[43]

Fol. 15 r shows a machine for boring out the centers of logs to make pipes. It is a simple device that allows a log to be pressed against a rotating bit. The descriptive text is very interesting. It states that the borer was in use in Nuremberg and could manufacture fifteen pipes per day, each eighteen *schuch* long (about six meters). This dimension agrees reasonably well with the known length of such pipes in Nuremberg, which varied between twenty and twenty-two

[39] For example, Quarg, *Kyeser*, I:54r; II:36 (an inverted siphon leading to a fountain); I:61v; II:39 (a siphon carrying water over a hill); I:63r; II:40–41 (an Archimedes screw). Francesco di Giorgio sketched siphons and pipe joints. See Maltese, (figs. 48–49 and 78). Taccola (Clm 197, II), fol. 70v [73v] shows similar devices.

[40] Especially important citizens or corporations might be privileged to tap public mains to supply water directly to their quarters. Often this led to so many abuses that the practice was prohibited (usually ineffectively); cf. Forbes, Singer II, 691.

[41] Feldhaus, *DM*, 218 cites one such wheel at Breslau in 1386 and another at Bremen in 1394. The *noria* itself dates from classical times, both in China and in the west. See Needham, 356. Piston pumps began to replace the *noria* in Germany in the late fifteenth century. See Forbes, Singer II, 691.

[42] Fischer and Walther, 86–88 note, on the basis of sixteenth and seventeenth-century records, that lead pipes cost up to seven times as much as wooden ones and that they could be made only 57–63 millimeters in interior diameter, while wooden pipes were made 89–101 millimeters in interior diameter. The average life expectancy of a wooden network was about forty years, with fifty to sixty years as a maximal figure. Some cities installed lead pipes after the premature failure of wooden lines, e.g., Dijon. Cf. William B. Parsons, *Engineers and Engineering in the Renaissance* (Baltimore: 1939; reprint Cambridge, Mass.: 1968), 244. Wood was nevertheless the most frequently-used material until the introduction of cast-iron pipes in the late eighteenth century. In 1745, the London Bridge Waterworks had about 48,000 meters of wooden pipe in operation and only about 3400 meters of lead. Dublin replaced lead pipe with wood as late as 1777. See Robins, 167–68.

[43] An official »Master of the Pipes« *(Röhrenmeister)* was in charge of Nuremberg's system. This officer's accounts appear prominently in Fischer and Walther, *passim*. Augsburg likewise had a municipal system from 1416 and in the sixteenth century the famous Red Tower with its seven Archimedes screws was erected. See Kellermann and Treue, 30. The actual work of constructing a water system was often contracted out to local or foreign artisans. For example, Hans Selber (or Felber) of Ulm built Augsburg's original line in 1416. See Paul von Stettin (der Jünger), *Kunst, Gewerb, und Handwerks Geschichte der Reichs-Stadt Augsburg*, Vol. I (Augsburg, 1779), 145.

statschuch (about 6.5–7 meters).[44] Also the daily rate of production given, which is probably a maximal figure, would be adequate for the needs of a city the size of Nuremberg. The city's »Master of Public Works« recommended in 1471 that one hundred to one hundred-fifty pipes be kept on hand in storage at all times. This would equal one or two week's production for A's machine.[45]

The drawing is not clear as to whether the machine is water- or manually-driven. Despite this difficulty, there is circumstantial evidence to suggest the former as a strong possibility. Water-powered borers are first illustrated in a German technological manuscript that antedates section A by about 35 years;[46] therefore, the possibility of a borer turned by water may not have been utterly novel to author A's generation. The large diameter of the wheel to the right of the drawing, as well as the square-shaped extensions on its periphery, are both typical of water wheels. The only other appearance in our manuscript of a similar configuration of elements (fol. 17v) shows carefully rounded extensions indicating hand spokes. Further evidence comes from the history of Nuremberg itself, where machinery to make water pipes was used in the fifteenth century. Endres Tucher, the »Master of Public Works« *(Baumeister)*, tells of making a contract for a trial run of a pipe boring device in 1462. It worked successfully and was installed in the municipal factory, where, he tells us, it may still be seen (1471).[47] Prior to the installation of this machine, pipes were made »by hand,« a

---

[44] Fischer and Walther, 88. These figures are from late sixteenth and early seventeenth-century documents respectively; no earlier figures are available.

[45] Tucher, 197–98 [131b–132a]. He further states that at least one pipe was needed every day for repairs to the city's lines and that the sale of municipal pipes outside the city was forbidden without the consent of the city council. Nuremberg apparently took some pains to insure a supply of pipes.

[46] Vienna, Waffensammlung, Ms. P 5014, fol. 11r, showing a horizontal paddle wheel of the Alpine *floder* type turning a vertical shaft-bit, against which a cylinder (presumably a log) descends by the force of gravity. The later development of such devices is very obscure. Vertical borers for cannon are known from the fifteenth century (cf. note 51 below), but they are always shown being driven by human or animals, never by water. Feldhaus' often-repeated assertion (*Technik*, 125) that Biringuccio's cannon-borer (1540) is water-powered is incorrect. Biringuccio recommends water power for almost all other metallurgical operations, e.g. for bellows, but he specifically states that his cannon-borer is to be turned manually. See Smith and Gnudi, 309. Agricola's only illustration of pipe-boring shows a log strapped to a frame being drilled by a T-handled auger turned by a single man; see Hoover (ed.), 177. Water-powered borers only begin to develop more fully in the seventeenth century, beginning with Salomon de Caus, *Raison des forces* (Frankfurt: 1615), fig. 26; cf. also Feldhaus, *Technik*, 125. It may be the case that difficulties in controlling the material or the bit's speed retarded the widespread use of water-powered borers. Even in the eighteenth century, most borers were powered by men or animals.

[47] Tucher, 197 [131b]. He made the contract on August 10, 1462 with a miller from Herzogaurach named Hanns Tölzer. Tölzer was to make the machine *(schragen und gerust)* at his own risk and to demonstrate it to Tucher's satisfaction. There is every reason to believe that Tölzer's machine became a regular part of the pipe factory. Tucher states that it »... dann jetzunt noch steet unter derselben hutten, doran man sider gesehen auch das etwas gepessert hat.« It was installed in the same building where pipes had formerly been made by hand and it had been improved somewhat by the time of Tucher's writing.

task that made the workmen »quite unhappy.«[48] Tucher does not describe the machine, but quite possibly it was the same as the one illustrated by author A. Tucher informs us of a second, but unsuccessful, attempt in 1462 to make a boring machine. From his remarks one can assume that this was a vertical borer.[49] By contrast, the first machine appears to have been a horizontal one, like the borer A depicts.[50] Moreover, the first machine (with some improvements) was still in operation when Tucher wrote. Therefore it is quite likely that it was still to be seen when author A composed his treatise (1472–85) and that his drawing is based on it. If this identification is correct, and if Tucher's remarks about making pipes »by hand« is meant literally, then we can assume that the new machine mentioned by Tucher and described by author A did employ some non-human power source, most probably water. On this basis, fol. 15r does represent a water-powered boring machine, one of the earliest on record.

The boring device illustrated by author B (fol. 36v) appears at first glance to be unrelated to water supply. It resembles contemporary vertical cannon borers very strongly.[51] The coloration of the drawing reveals that it depicts instead an earth borer. The tubular protrusion into which the bit descends is earth-colored, not metal-colored, as would be necessary if it were a cannon. Several scholars have expressed confusion over the purpose of this machine,[52] but clearly it must be an earth-borer. It was probably used to widen an already-drilled well. The form of the drill bit would have been unsuited to piercing operations, but would have served to scoop out earth from the sides of an existing shaft. Note that it could not remove the earth from the shaft, however; there is no twist to the bit. Drilled wells date from the

---

48 *Ibid.*, 197 [131b] »unter der hent.«

49 Tucher, 197 [131b]. A Jew, Jacob of Ulm, offered to build the borer and was granted permission and a subvention to do so by the city council. His project required the digging of a pit four meters deep and a stone foundation. Jacob died before the project was complete, and his successors could not operate the device properly. Tucher notes that they had problems with the borer's bit binding in the log. All this indicates a vertical borer. Note the probable similarity between this device and that depicted in Vienna, Waffensammlung, Ms. P 5014, fol. 11r.

50 Tölzer's machine apparently fit into the old factory without problems, since Tucher makes no mention of any extra foundation or other alteration. It must have been a horizontal machine.

51 Bernard Rathgen (ed. Otto Johannsen), »Beiträge zur Geschichte des Geschützwesens I: Zur Geschichte der Geschützfabrikation,« *Beiträge zur Geschichte der Technik und Industrie* (now *Technikgeschichte*), XVIII (1928), 41 notes that the earliest types of cannon borers were vertical ones in which a cannon was held muzzle up in a frame and a bit descended into it. Beginning in the mid-fifteenth century this type was superceded by another vertical borer that held a cannon muzzle down and allowed it to descend onto a rotating bit. Rathgen claims that horizontal borers like Biringuccio's were used less than vertical ones. For contemporary examples of vertical borers cf. Vienna, Waffensammlung, Ms. P 5135, fol. 61v; Cpg 126, fol. 16r; the *Wunderbuch* (E. Marx, 318 and 320, figs. 6 and 13); Seselschreiber (Cgm 973), fol. 58r; and Feldhaus, *Technik*, 391–92.

52 Beck, 280; Uccelli, 60; Feldhaus, *Technik*, 120 think that it is probably an earth borer; Gille, »Manuscrit,« 84 and *Engineers*, 64 claims it is either for boring wooden pipes or the earth.

twelfth century in the west, but they are much older in China and Islamic areas.[53] There are few illustrations of large earth borers in the early technical literature, but Taccola includes one, da Vinci two, and there is at least one sixteenth-century German example.[54]
The drawing on fol. 35r is the least complicated of the three. It shows a *noria* raising water into a pipe system which leads to a fountain. It is merely a schematic version of how such a system might operate, and it shows none of the fine details of construction. Author B shows the *noria* operating in an aqueduct rather than a natural stream. There is no outflow shown to carry off excess water from the *noria*. Apparently B expected that it would raise all the inflowing water up to the raised pipe inlet, but this is hydraulically impossible.

---

[53] Gille, »LdV,« 81; Joseph Needham, »L'Unité de la science: l'apport indispensable de l'Asie,« *Archives internationales d'histoire des sciences*, XXVIII (1949), 563–82; cf. also Needham, 52.

[54] Taccola (Clm 197), fol. 93v [94v]. For Leonardo's see Feldhaus, *Technik*, 119ff. Later examples are noted by Parsons, 115.

CHAPTER VII

MILLS

The mills in our manuscript are its most unusual feature. Other than Francesco di Giorgio's, few of the fifteenth century technological treatises concentrate on them because they were of little interest to a military engineer. Clm 197,I is remarkable in that both author A and author B show various types of milling machines. Author A in particular seems to have been fascinated by them, for he shows more detailed views of more mills than any other German manuscript of the period. It is possible that he came from a family of millers and wished to record ways in which a mill might be constructed. He also comments more extensively on his mills than on any other class of machinery. The chapter will consider four different types of mills: 1. those powered by animals or men; 2. windmills; 3. watermills; and 4. stone-polishing mills.

Mills Powered by Animals or Men

Most of the mills in our manuscript are driven by animals or human beings (see fols. 10v, 17v, 18r, 20r, 20v, 21r, 21v, 22r, 22v and 42r). Classical antiquity was familiar with such mills in a variety of forms,[1] but most ancient examples were constructed in ways that limited their efficiency severely.[2] In spite of this, hand, and possibly animal-driven mills were never completely replaced by the more efficient water-powered mills. Hand mills were always needed for domestic use, and animal-driven mills could be used where water supplies were inadequate for water-powered mills. The later middle ages saw the development of such mills into more efficient machines than their ancient predecessors. From the twelfth century

[1] John Storck and Walter D. Teague, *Flour for Man's Bread: A History of Milling* (Minneapolis: 1952) and L. A. Moritz, *Grain Mills and Flour in Classical Antiquity* (Oxford: 1958) both present good surveys of milling and especially of the development of ancient hand- and animal-powered mills. The technical development of mills is treated by Richard Bennett and John Elton, *History of Corn Milling*, Vol. II *Watermills and Windmills* (London: 1899), but the most recent work is Hermann Gleisberg, *Triebwerke in Getreidemühlen* (»Technikgeschichte in Einzeldarstellungen,« XV; Düsseldorf: 1970).

[2] The Roman donkey-mill, for example, used such a heavy millstone, so inefficient a harness and so short a circuit of travel that most of the draft animal's power was wasted. Slaves were preferred to animals on such mills. R. J. Forbes, article »Food and Drink,« Singer II, 104 and figs. 76–77. White, 108 ff. argues that ancient querns (hand mills) had no cranks and were not rotated in a continuous fashion. This too would have limited their efficiency.

we begin to find hand mills turned by means of gearing,[3] and in the fifteenth century there is secure evidence that these types were equipped with cranks.[4] From the thirteenth century we find large mills turned by men or animals through geared gins.[5] Such developments set the trend which our manuscript follows.

Author A seems to have wanted to catalog the various ways in which human or animal power could be applied to milling. He shows, for example, a mill driven by a man on a treadmill (fol. 20v) and two mills turned by horse-gins (fols. 22r–22v). The most interesting mills, however, are the ones employing crankshafts and connecting-rods (fols. 18r, 20r, 21r and 21v). The crankshaft first appears in Guido da Vigevano's treatise, and it was combined with a connecting rod for the first time about a century later.[6] Author A was among the first to depict mills driven in this way, but his knowledge of them is so good that they must have been in use for some time before he composed his work. He is familiar with single and double-throw crankshafts, and he knows that such shafts should be equipped with flywheels to smooth out the »dead spot« in their motion (see fols. 20r and 21r). As a means of generating the continuous rotary motion required in mills, the combination of crankshaft and connecting-rod was close to ideal. It allowed one or two men to drive a large mill with much more force than could be developed with simple cranks. Other engineers recognized the possibilities inherent in such linkages and applied them to mills with increasing frequency in subsequent decades. Fol. 42r shows author B's attempt to depict such a mill, but *Das mittelalterliche Hausbuch* (c. 1480) gives us a better idea of the speed with which crankshafts were being improved.[7] The *Hausbuch*'s mill uses a counter-weighted, double-throw crankshaft moved by a single rod. The counter-weight will compensate for »dead spots« even better than a simple flywheel.

Author A shows a different sort of hand-operated mill as well, the stamping mill (fols. 10v, 17v and 18r). The example on fol. 10r uses vertical pestles lifted by cams on an arbor turned by a crank. Folio 17v, however, shows stamps which are mounted horizontally and which when actuated by their cams, move with a rocking motion about off-center pivots. The appearance of both vertical and horizontal stamps (the latter are more properly called trip hammers or recumbent hammers) in the same manuscript is somewhat frustrating to histo-

---

[3] Forbes, Singer II, 96.

[4] A. T. Nolthenius, »Les moulins à main au moyen âge,« *Techniques et civilisations*, IV (1955), 150, figs. 1 and 2 dating from 1430 and 1475 respectively; cf. Gleisberg, 50–51 and figs. 40–42.

[5] Anonymous (ed.), *Psautier de Saint Louis* (i.e., Paris, B.N. Ms. latin 10525) (Paris: Catala Frères, n.d. [1900]), fig. LXII shows Sampson pushing a circular framework with gearing on its circumference; cf. Usher, 171. Gleisberg, 52–53 and fig. 94 shows another depiction from the thirteenth century of Sampson in a treadwheel turning mullstones.

[6] Vienna, Waffensammlung, Ms. P 5014, fol. 32v, where it turns an Archimedes screw. This is the earliest example I have seen of a crankshaft and connecting rod. For Guido da Vigevano's example, which does not employ a connecting rod, see Paris, B.N. Ms. latin 11015, fols 10v, 14v.

[7] Bossert and Storck, fig. 47.

rians of medieval power technology. Although we know from documentary evidence that the linkage of cam, arbor and stamp dates from the early ninth century, we do not have any solid evidence regarding the type of stamps employed in medieval industrial operations. There is, of course, no lack of speculation on the subject. Joseph Needham, who has acutely observed the exclusive use of horizontal stamps in Chinese technology, insists that Europeans employed only the vertical stamp until sometime in the later middle ages, perhaps the thirteenth century, when the recumbent hammer was taken over from the East. He notes in support of this contention that the recumbent hammer was infrequently illustrated in western sources before the sixteenth and seventeenth centuries.[8] Other writers have assumed that certain processes known to have been practiced in medieval Europe, e.g. fulling or iron forging, could have been performed with water power only if the recumbent hammer were also employed. Those who hold such opinions tend to date the recumbent hammer's appearance with the earliest records of mills engaged in fulling, etc.[9] In fact, as Bradford Blaine admits,[10] there is no way to resolve this question satisfactorily without some sort of pictorial evidence regarding the types of water-powered mills that may have used trip hammers. Such evidence is not known to exist from before the fifteenth century.[11] Hence, the appearance of both types of hammer in the earliest available pictorial sources serves only to perpetuate the argument.

Using the fifteenth-century evidence as a whole, one can construct a weak retrospective argument in favor of the preponderance of the vertical stamp in the preceding centuries. Vertical stamps do appear more frequently in the technological manuscripts than recumbent hammers. (This may, however, be more related to the types of processes shown, chiefly gunpowder preparation, than to the nature of power technology in general.) The treatment of fol. 17v in our manuscript tends to confirm the general preference of Europeans for vertical stamps. In copying this drawing, the unknown author of Weimar, Hs. fol. 328, fol. 271v,

[8] Needham, 390ff. Needham's discussion seems unnecessarily confusing in its distinction between manuscript and printed sources.

[9] For example, Bertrand Gille, »Le moulin à eau: une révolution technique médiéval,« *Techniques et civilisations*, III (1954), 1ff., who dates water powered forge hammers at Issoudon from 1116.

[10] Bradford B. Blaine, »The Enigmatic Water-Mill,« Hall and West, 163–176, provides a thorough analysis and summary of the present state of knowledge regarding water power in medieval Europe. He avoids making any exaggerated claims. I am grateful to Professor Blaine for his kind permission to use his article in typescript.

[11] Fifteenth century examples of cam-actuated vertical stamps may be found in Guttmann, 20, fig. 49 (source unidentified); the *Hausbuch* (Bossert and Storck, figs. 40 and 43); Cpg 126, fols. 3r and 4r. The earliest fifteenth century illustration of recumbent trip hammers I have seen is from Giovanni da Fontana, fol. 8v, which includes cams mounted on a rotating drum. Taccola (Clm 197, II), fol. 51v, 54v shows a recumbent hammer in the form of a pile driver, but of course it is not cam-actuated. Leonardo made use of the horizontal stamp in several different devices, including file-cutting machines and forges; cf. Beck, 434ff. and Needham, 395, fn. »d«. Two printed illustrations of recumbent hammers from the fifteenth and sixteenth centuries exist: Hugo dem Reutlinger's *Flores Musicae* (1492: reproduced in Feldhaus *DM*, 259, fig. 176) and Olaus Magnus, *De gentibus septentrionalibus* (1565); reproduced in Needham, 395, fig. 620.

replaced author A's recumbent hammers with vertical stamps, from which one may infer that such stamps were more familiar to the copyist and that he deliberately modified A's drawing to make it conform more closely to »standard practice.« If we are to judge from this single instance and from the general frequency of various hammer types shown in the late medieval technological literature, then we should assume that author A's drawing represents a fairly novel device in fifteenth-century Germany.

Windmills

There are three illustrations of windmills contained in our manuscript (fols. 19r, 19v and 47r). The earliest European[12] windmill appears in 1185 at Weedley in Yorkshire.[13] The new type of mill spread through northern Europe at an extremely rapid pace and by the early thirteenth century windmills were common in almost all parts of the continent north of the Alps.[14] Illustrations of windmills first appear c. 1290 and they become common from the fourteenth century on.[15] These early illustrations all show »post mills,« i.e., windmills mounted on a pivot so that the entire body of the mill may turn to face the wind.[16] All of the windmills in our manuscript are post mills.

None of the early illustrations gives us any idea what the interior of a windmill looked like. Author A's drawings are the first to do so. His windmills have a crown gear on the main shaft engaging a lantern pinion on a vertical shaft which turns the upper millstone (see fol. 19r). Fol. 19v shows an alternative arrangement that rotates the lower of the two millstones.[17] We have no other evidence, however, indicating that such a system was ever actually employed. It is interesting to note that A's windmills all lack brakes. They could be

[12] The evidence for windmills in other parts of the world points to northern Iran-southern Afganistan and the late tenth century as the place and date where they were first used. Cf. White, 86; Needham, 556–58. These were vertical-axle windmills, however, and are not related to western horizontal-axle windmills. Hero of Alexandria drew a small windmill to drive the pump of an organ. The textual tradition is apparently genuine, but nothing came of the idea. It remained »a suggestion on paper.« Drachmann, *MT*, 206.

[13] White, 87. There are less certain references to what may be windmills from 1170 on.

[14] *Ibid.*, 88. Their diffusion south was slower, but in Italy, at least, they were familiar enough by the early fourteenth century for Dante to use one in a simile.

[15] The first illustration is in the *Windmill Psalter* done at Canterbury. See Rex Wailes, »Some Windmill Fallacies,« *Transactions of the Newcomen Society*, XXXII (1959–60), 94 (hereafter Wailes, *TNS*). He dates the psalter at c. 1260, but the Pierpont Morgan Library has revised this to c. 1290. Rex Wailes, »A Note on Windmills,« Singer II, 623, fig. 561 reproduces the figure. Daumas, 470, figs. 38–42 reproduces several early illustrations of windmills.

[16] The tower mill, in which only the roof or »cap« of the mill turns to face the wind, is first represented in a psalter from c. 1420. See Wailes, *TNS*, 94. Both tower and post mills were used until this century, but the tower mill gradually predominated.

[17] A mill whose lower stone rotated (»underdrift«) would approximate the milling characteristics of a water-mill, which is always »underdrift.« Some millers might have preferred this.

stopped only by being »quartered,« i.e., turned so that the sails were parallel to the direction of wind flow and could catch no breeze.[18] The first mechanical brakes for windmills appear in Leonardo's work. As we noted in Chapter III, they may have developed from brakes on construction machinery like those shown on fol. 3v.

The drawing of a windmill by author B (fol. 47r) shows no interior gearing, but it does illustrate an improved sack hoist to lift bags of grain up to the level of the millstones. Crude sack hoists appear in some of the early windmill drawings. They take the form of a beam extending from beneath the mill's roof on which a pulley was mounted. A rope passed over the pulley was used to haul up the sacks.[19] Author B shows a shaft projecting from the mill with a rope wound around it. Turning the shaft inside the mill raises a sack of grain outside. Author B does not indicate the mechanism that turns the shaft, but it was surely only a rope looped around the shaft that the miller could pull by hand. The sack hoist was not powered by the windmill's gearing.[20] There seems to be no detailed discussion in the learned literature of the sack hoist system shown by B, but it appears in several marginal illustrations of windmills in manuscripts from the fourteenth century, and it may even date back to the thirteenth.

In all other respects our windmills conform to the normal pattern shown in the early drawings, but they show details of construction in greater detail. For example, we can see that the joint between the sails and the windshaft (see fol. 19r) is reinforced with iron for greater strength, and the sub-structure of the mills is made with diagonal bracing (double bracing on fol. 19r). The only exceptional feature is the platform that appears to support the mill on fol. 19r. There is no other extant example in which a platform rather than a post bears the mill's weight. Author A probably intended this to represent the bottom floor of the mill since it would be impractical to try to turn a mill resting on a broad platform.

Watermills

Water-driven mills appear only twice in our manuscript, both times in section A (fols. 18v and 23r). Author A seems to have sketched only watermills that were unusual to him in

[18] Stanley Freese, *Windmills and Millwrighting* (Cambridge: 1957), 3 mentions a tradition among old millers that the earliest windmills had no brakes. Rex Wailes, article »Windmills,« Singer III, 104 notes that primitive mills in Brittany still have no brakes and must be »quartered« to be halted.

[19] The Walsoken Brass, done c. 1346–49, now at St. Margaret's Church, King's Lynn, Norfolk and often known as the »King's Lynn Brass,« shows a windmill with this feature. See Rex Wailes, *The English Windmill* (London: 1954), 138 and Storck and Teague, 120, fig. 161 for reproductions. The so-called *Veil rentier de Audenarde* (Brussels, Bibliotheque Royale, Ms. 1175 done c. 1275) contains an illumination depicting a windmill with a hoist. See Daumas, 470, fig. 38. Wailes, Singer III, 105 incorrectly claims that sack hoists date from the sixteenth century.

[20] Ramelli, fig. 153 shows a sack hoist turned by a rope loop pulled by the miller. Hoists powered by the windmill's gearing appear in the eighteenth century in the *Encyclopédie* of Diderot. See Arnold, 17, fig. 30 for a reproduction of the engraving.

some way. The prime mover for a watermill, the waterwheel, appears in both China and Europe between c. 200 B.C. and 100 A.D.[21] Horizontal and vertical waterwheels both appear at this time.[22] The Romans used the vertical waterwheel mainly for grinding grain, and in this form it carried over into the middle ages. In medieval Europe the vertical waterwheel began to be applied to a number of tasks other than milling after the late ninth century.[23] The elaboration of such mills seems to have inspired A in fol. 23r. Since they could be made to do many different jobs, they could also be arranged to do multiples of the same job. As author A shows, one wheel could be geared to two grain mills and thus grind double amounts simultaneously. The history of the horizontal waterwheel is a more difficult matter. Indeed, the classical and medieval documents usually fail to distinguish between vertical and horizontal wheels, and thus the amount of unambiguous evidence is extremely small. It appears that in Asia the horizontal type was reponderant, if not universal; vertical waterwheels have been postulated, but never proven in any device from the Orient.[24] In Europe, on the other hand, no simple conclusions can be reached. Early modern and contemporary survivals give evidence that horizontal wheels were frequently used in parts of northern Europe (Scandanavia, Ireland, Isle of Man, the North Atlantic Islands) and in regions of southern and central Europe (the Alps, parts of France, Bulgaria and the entire Mediterranean).[25] Vertical mills appear to have been more frequent elsewhere, to judge from the same sort of evidence. From the middle ages themselves, the evidence is very scant; an Irish law code of the late ninth or early tenth centuries and a Celtic epic of the eleventh provide the only clear early proof of the horizontal wheel. Thereafter, unambiguous evidence is lacking until

[21] For Europe the evidence is straightforward. The earliest known reference to a waterwheel of any kind occurs in the second century B.C. and speaks of a *noria*. Watermills for grinding grain are mentioned by the geographer Strabo and his contemporary Antipater, about 30 B.C. It is uncertain exactly what type of mill is meant, but Vitruvius, writing at about the same time gives the first thorough description of a watermill with a vertical wheel and a horizontal axle. Excavations in Jutland have revealed a site dating from the time of Christ that contained a different type of waterwheel which was mounted horizontally and had a vertical axle. See White, 80–81. For China the evidence is more confusing. Paradoxically, the earliest reference (31 A.D.) mentions waterwheels in connection with a machine of such complexity that one assumes they must have had a long history of prior development. The form of the wheel is not specified, but it was probably horizontal. This was the normal waterwheel in later Chinese engineering. There is no explicit mention in Chinese sources of vertical waterwheels until the period of western influence. However, heavy trip hammers operated by waterpower appear in the first century and may have used vertical waterwheels as a power source. See Needham, 369 ff. and 390 ff.

[22] Watermills are described according to the orientation of their wheels. A vertical (or Vitruvian) mill has a vertical wheel, and a horizontal (or Norse) mill has a horizontal wheel. Windmills, on the other hand, are described according to the orientation of their axles, not their sails.

[23] Cf. White, 83–84; Bradford B. Blaine, »The Application of Water-Power to Industry During the Middle Ages« (Unpublished Ph. D. Dissertation, University of California, Los Angeles, Department of History, 1966).

[24] Needham, 368 ff., 390 ff.

[25] *Ibid.*, 368. Cf. Blaine, »Water-Mill,« and Louis C. Hunter, »The Living Past in the Appalachias of Europe: Water Mills in Southern Europe,« *Technology and Culture*, VIII (1967), 446–66.

the fifteenth century.[26] Regional studies indicate a large number of what appear to be horizontal mills in mountainous areas in the later middle ages, especially from the late thirteenth century onwards in Italy.[27]

The geographical distribution suggested by modern survivals has perhaps been taken too seriously by scholars; it may not indicate with certainty anything about medieval practice at all. Since we know that water power was applied to milling on a wide scale in medieval Europe (the famous Domesday Survey lists at least 5,624 watermills in post-Conquest England alone!), we should probably assume that both horizontal and vertical waterwheels were in use but that one type or the other predominated in certain areas. Generally, we seem to find the horizontal wheel in the more mountainous, less populous or less prosperous parts of Europe. This suggests (though it does not prove) that the vertical type was more extensively employed in the lowlands, where broad, slow-flowing rivers provided the right sort of power, and concentrations of population and economic activity made feasible the more powerful, but more expensive, vertical wheel.[28] If we are willing to make the further assumption that most of the extant pictorial and written evidence stems from the very regions where vertical waterwheels predominated, then we can understand the relative paucity of firm evidence for horizontal waterwheels in our sources.

Given the manifold uncertainties of this question so far, hesitation about further speculation is necessary. In particular, it is very difficult to assess the importance of the horizontal wheel's appearance in the fifteenth-century technological literature. Only two aspects deserve some note. First, more attention was paid to the horizontal wheel by the Italians than by the Germans, and this attention led directly to progressively more elaborate and imaginative applications.[29] Second, within the German tradition, the horizontal wheel remained a somewhat unusual device until the seventeenth century. The earliest drawings of horizontal waterwheels in both traditions (from the 1430's) illustrate this point. Taccola's Italian

---

[26] Blaine, »Water-Mill.«

[27] Felix F. Strauss, »Mills without Wheels in the 16th-Century Alps,« *Technology and Culture*, XII (1972), 23–42 discusses the mills that are attested in a census of the Gastein Valley in the Austrian Tyrol from 1550. Horizontal types are overwhelmingly in the majority, and many of these must date at least from the mid-fifteenth century. John Muendel, »The Horizontal Mills of Medieval Pistoia,« *Technology and Culture*, XV (1974), 194–225 finds a similar preponderance in Tuscany based on notarial records from 1276–1421. Significantly, both these monographs deal with mountainous regions.

[28] The model is suggested by Blaine, »Water-Mill,« who wisely adds that it »is useful only if used to launch further inquiry, not to close it.«

[29] Some of these may be seen in Paul N. Wilson, *Watermills with Horizontal Wheels* (Kendall, Eng.: 1960), *passim*; Ladislao Reti, »A Postscript to the Filarete Discussion: On Horizontal Waterwheels and Smelter Blowers in the Writings of Leonardo da Vinci and Juanelo Turriano,« *Technology and Culture*, VI (1965), 428–41. Devices using horizontal wheels appear too often to mention in the sixteenth- and seventeenth-century printed technological literature, but note should be made of Besson's fig. 29, which shows a spiral-vane conical wheel turning within an enclosure. This is the first known waterwheel to use both impulse and reaction to generate rotary power and it illustrates the course of evolution that horizontal wheels took. Such mill-wheels were in operation near Toulouse in Besson's day (c. 1570).

drawings show the waterwheel in a highly developed form at work grinding grain.[30] The contemporary German drawing, by contrast, shows a primitive paddle wheel driving what appears to be a vertical wooden pipe borer–an unusual and highly specialized application.[31] Author A's fol. 18v, the second known Germanic representation, shows an ordinary grain mill, but this text speaks of it in a strange way. He accentuates the obvious, i.e. that it needs no gearing, and he attributes its invention to a distant and powerful figure (»a pope from Rome«). In short, A treats the horizontal waterwheel as if he expected it to be unfamiliar to his readers. In sharp contrast, the nearly contemporary notebooks of Francesco di Giorgio Martini are filled with countless examples of sophisticated horizontal wheels equipped with elaborate mechanical transmissions.[32] This difference is explicable in terms of our earlier assumptions. It would seem that the German writers (presumably mainly lowland dwellers from beyond the Alps) were not as acquainted with the horizontal waterwheel as their Italian counterparts, for whom it was probably a familiar, traditional item in rural culture. Hence, author A's drawing provides further evidence of his deep interest in milling and the range of his technological experience.

Stone-Polishing Mill

Author A shows one other mill, a special one for polishing precious stones. His drawing shows a very simple machine, but it is interesting because it is one of the earliest illustrations of a lapidary polisher and because it shows an early application of a belt-drive. Author A is not the first person to depict belts or ropes used to transmit rotary power. This principle is first found in Europe in the late thirteenth century in the spinning wheel,[33] but A's drawing demonstrates clearly how this new mode of transmitting power was being used at a fairly early date in history.

Precious stones have been polished since antiquity to improve their appearance. This requires that the stone be moved rapidly across an abrasive surface to remove superficial irregularities. Without some sort of mechanical assistance, stone polishing is a very laborious process. Nevertheless, we have no evidence that rotating stones were applied to grinding or polishing operations before 816–34.[34] Medieval descriptions of stone-polishing, e.g., that of Theophilus Presbyter in 1122–23, are silent as to whether rotary stones or stationary

[30] For example, Taccola (Clm 197, II), fols. 56v [59v], 69r [72r], and 91r [92r].

[31] Vienna, Waffensammlung, Ms. P5014, fol. 11r, done c. 1437.

[32] See for example, Maltese, figs. 63–66, 70 and 75.

[33] White, 119. The earliest known spinning wheels appear in China in 1035. See Needham, 758 and fig. 405. The earliest western drawing showing a device other than a spinning wheel using a belt drive is Giovanni da Fontana's (fol. 7r). Cf. Chapter VIII for further discussion.

[34] Cf. White, 110. It appears in the *Utrecht Psalter*. D. B. Harden, article »Glasses and Glazes,« Singer II, 341 asserts that Roman glass cutters used rotary cutting wheels, but gives no evidence for this view. Feldhaus, *Technik*, 954 notes that no ancient sources show or describe rotary grinding wheels.

disks were employed.[35] We know that several German cities began to develop a stone-polishing craft industry in the fourteenth century,[36] but the earliest polishing mill seems to have been set up in Augsburg in 1389.[37]

Illustrations of machines to polish stones are rare. Aside from author A's, I know of only one other fifteenth-century German example,[38] and this depicts a wheel (or disk) turned by a simple crank with no gearing or other mechanism to increase its speed. When compared with this drawing, author A's appears remarkably well-developed. There is a contemporary illustration of a stone polisher that works on the same principle as A's in a French manuscript, and Leonardo depicted some belt-driven lens polishers before the end of the century.[39]

It would seem, then, that A's design originated outside Germany (which probably accounts for his reference to »the grand masters of Venice«) and that he included it in his work because it was obviously superior to the ordinary German machines. A stone-polisher operated by a belt-drive could develop high speed at the polishing disk. All that is necessary is that the pulley attached to the disk be much smaller than the one driving it through the belt. One revolution of the large pulley produces many revolutions of the small pulley. By contrast, a machine with only a simple crank would operate very slowly.

Author A's understanding of the stone-polishing craft does not seem to have been great. He neglects, for example, to specify that an abrasive must be used with the metal polishing disks. Also, his drawing does not seem to show any of the ancillary devices one would expect on a lapidary mill, e.g., brakes or clamps to hold the stone. Nevertheless, A did understand an important mechanical principle, the belt-drive through differential pulleys, which would be of benefit to the stone polisher; and he can claim credit for having introduced it to Germany.

---

35 Theophilus refers to »a smooth plate« *(tabula aequale)* of lead, wood or copper strewn with abrasive. From this one assumes that he meant stationary disks. Cf. Theophilus Presbyter, *De diversis artibus*, ed. and trans. C. R. Dodwell (London: 1961), 168 ff. and *idem*, *On Divers Arts*, trans. John G. Hawthorne and Cyril Stanley Smith (Chicago: 1963), 189 ff. Regarding the date, cf. Lynn White, jr., »Theophilus Redivivus,« *Technology and Culture*, V (1964), 230.

36 Christof Gottlieb von Murr, *Beschreibung der vornehmsten Merkwürdigkeiten in der H.R.Reichs freyen Stadt Nürnberg* (Nürnberg: 1778), p. 695 notes that »Demantpolierer« were active in that city in 1373.

37 Stettin, I, 141.

38 It appears in a woodcut in a *Book of Wisdom* (Urach: 1481); Feldhaus, *DM*, 270, fig. 180 reproduces the picture.

39 Paris, B. N. Ms. latin 7295, fols. 137 r and 138 r. The manuscript as a whole dates c. 1466, but the segment (by Jean de Meur ?) containing these folios may be slightly later. Fol. 137 r shows a stone polisher with a very large driving pulley that transmits motion through a belt to a small vertical axle with a polishing disk. This is probably what A had in mind. Gille, *Engineers*, 196 reproduces a similar design from an unidentified manuscript (it may be this folio). For da Vinci's belt-driven lens polishers see Uccelli, 66–70.

CHAPTER VIII

MOTION TRANSMISSION AND CONVERSION

We have now reached a point where we must shift emphasis. Previous chapters have discussed various types of machinery in the manuscript, but this will consider some of the means used by our authors to transmit and convert motion. Motion conversion is the heart of any machine. The availability or absence of certain motion conversion devices determines how machines may be built and prime movers applied to certain tasks. All machines convert or transmit motion, even the simple ones such as levers, screws or wedges. We shall not attempt to analyze all the machines our authors show, but instead we shall concentrate on gearing, cranks and crankshafts with connecting-rods, and driving belts. The final section of the chapter will try to evaluate the abilities of our authors to understand and apply these devices.

Gearing

One of the most interesting contrasts between authors A and B concerns how much attention each gives to gearing. Author A illustrates it on eleven folios (3v, 10r, 19r, 19v, 20r, 20v, 21r, 21v, 23r and 24r) while B shows it only once (fol. 47v). Most of A's examples use gearing in mills, a usage that dates back to Vitruvius.[1] The Hellenistic and Roman engineers obviously understood gearing quite well. All the basic forms—spur, crown and worm—as well as rack and pinion appear in the ancient treatises.[2] Yet the ancients may not have used gearing as extensively as the middle ages did.[3] For most ordinary purposes, such as lifting bur-

---

[1] In the Vitruvian mill gearing was necessary to convert the water-wheel's motion from a horizontal to a vertical plane and to increase the millstone's speed of rotation. *De architectura* describes the first use of right-angle gearing. See Drachmann, *MT*, 203.

[2] Cf. Drachmann, *MT*, 22–26, 32–34, 82–85, 200–03; Conrad Matschoss, *Geschichte des Zahnrades* (Berlin: 1940), 6–9.

[3] Drachmann, *MT*, 32 is of the opinion that the ancient use of gearing was confined to heavy water-powered mills and to certain automata. Archaelogical findings substantiate this; gears appear infrequently. Two major finds include the lantern pinion discovered at a Roman *limes* fort at Saalburg in the Taunus and the »Antikythera machine« found on a sunken vessel off Greece. The former was part of a large animal or slave-powered mill in the fort. It consisted of two oak disks affixed to an axle joined by six iron spindles each sixteen centimeters long. This is an interesting reversal of the medieval practice shown in our manuscript of using iron disks and wooden spindles. See Moritz, 23ff. and plate 14c. The »Antikythera machine« was a »computer«

dens, block-and-tackle would give satisfactory mechanical advantage without expensive gearing. It was with the profusion of machines using rotary motion, chiefly mills, in the medieval period that gearing became more common.

Medieval gears were always made of wood;[4] thus one could replace individual teeth as they became worn or broken. Lantern pinions had wooden spindles that could also be replaced. How long such gears lasted is open to question since medieval sources are silent on the point. Yet modern wooden gearing in windmills has often seen forty or more years of service.[5] Their longevity would have depended on the type of wood used and the skill with which the gears were made, but in general gears made of wood served the needs of the medieval miller well.

Our manuscript always shows the same type of gear train: a crown or spur gear meshes with a lantern pinion. This holds true even where the axes of the gears are parallel and where the modern eye expects to see two spur gears employed. The nature of wooden gearing mitigated against any such arrangement, however, for the peg teeth of a spur gear could mesh well only with the interstitial spaces created by the rods (author A's *tribeln* of fol. 20r) used in a lantern pinion. These are at right angles to the teeth of a spur gear. Only cast metal gearing, whose tooth profile combines in one element the projections of a spur gear and the depressions of a lantern gear, can accommodate the meshing of two spur gears in the same plane.[6] Author A, like all his contemporaries, makes a virtue of necessity by employing this »extra« gear to adjust the overall gear ratio of the drive system so that input and output velocities are in the proper relationship (see fols. 19v, 22r, and 22v).

The sequence of drawings on fols. 22r–23v is worthy of some comment. Each of the latter two drawings depicts two grain mills driven from two power sources. As far as we can now

---

showing the motion of the celestial spheres and used very complex forms of gearing. See Derek J. de Solla Price, »An Ancient Greek Computer,« *Scientific American*, CC (June, 1959), 60–67. It contains the earliest European example of all-metal gear-wheels. Price's more extended treatment appeared as *Gears from the Greeks: The Antikythera Mechanism–A Calendar Computer from ca. 80 B.C.* (»Transactions of the American Philosophical Society,« N.S. Vol. LXIV, Part 7; Philadelphia: 1974).

[4] O. Kammerer, »Die Entwicklung der Zahnräder,« *Beiträge zur Geschichte der Technik und Industrie* (now *Technikgeschichte*), IV (1912), 246 notes two fifteenth-century iron gears in a winch in Ragusa. His dating assumes that the winch is as old as the building housing it. Normally, iron gearing was not used in the west until the late eighteenth century. Chinese examples of bronze gears, however, date back to the second or first century B.C.; see Needham, 86.

[5] Wailes, *English Windmill*, 123ff. describes wooden gearing in some detail. He claims that a well-made set of such gears would run quietly and smoothly and would last a lifetime, and he also notes instances of iron wheels fitted with wooden teeth. The great advantage of wooden over metal gears was their lower cost and the fact that breakage of one tooth did not ruin an entire gear.

[6] The solitary exception to this rule is an American mill of ca. 1870 now in the Smithsonian Institution, Washington, D.C. To achieve spur to spur mesh, the gear tooth profile had to be specially shaped in crude imitation of all-metal gears. See Gleisberg, *Triebwerke*, 101 and fig. 119.

determine, this was a new concept in fifteenth-century technology,[7] since earlier machines are generally shown as discrete, complete entities in which one power source, let us say a waterwheel, powers one mechanical function, e.g. a grain mill. Groups of power sources, such as clusters of waterwheels set along the same stream, are known to have existed; but each wheel of such groups apparently powered one and only one machine. In many instances this habit must have meant an obviously superfluity of power input relative to the load, and fifteenth-century engineers first sought to use this surplus power by multiplying the number of loads on a given power source. Exactly how this principle came to be grasped is not known; perhaps clockwork gearing, with its multiple stage gear drives for the escapement and various dials, provided the model. Apart from clockwork, the earliest examples of multiple-stage gearing that I have noticed occur in Taccola's works. The earliest type, depicted in *De ingeneis*, involves placing two large crown gears on one main driveshaft; each crown gear drives its own subsidiary train. The second type appears somewhat later in *De machinis* and probably represents a rationalization of the first.[8] Here, in lieu of the duplicated crown gears, the main drive shaft contains only one spur gear, which meshes with two lantern pinions set 180° apart and with their axes parallel to that of the main shaft. Each lantern pinion in turn drives its own secondary gear train leading to the loads. This is the method shown by author A on fol. 23r, and it was the more popular of the two methods in later centuries, especially in the so-called »Panstermühle« much beloved by eighteenth-century writers on milling. A minor variation, depicted by author A on fol. 22v, employed a crown gear on the main drive meshing with two lantern pinions whose axes are set at right angles to that of the main shaft. Taccola's earlier system, with two crown gears on the main shaft, was never entirely abandoned; it appears in both Agricola and Veranzio, but apart from a few eighteenth-century windmills, it never was used widely in practical milling.[9]

Another minor novelty is seen in fols. 22r and 22v. There author A shows a horse (or human-powered) gin as the main drive for one and two mills respectively. The path that the animal would take is kept free of impediments by burying the gearwork in a pit (indicated in these drawings by a circular brick wall); the pit is to be understood as having a cover over which the gin is mounted. Normally, the gin and its gearing would have been mounted overhead and power transmitted to a mill outside the animal's path. Taccola is once again the earliest writer to show such a system; in *De ingeneis* he depicts a single mill with underground drive, while in *De machinis* he includes a double-mill version.[10] I suspect that both the underground drive and the use of multiple stage gearing may have been known to author A (at

[7] Both Feldhaus, *DM*, 241 and Gleisberg, *Triebwerke*, 63ff. claim that author A's drawings are the first to depict multiple stage gearing. Although this specific claim must now be rejected, the general time period is still apparently correct.

[8] *De ingeneis*, Clm 197, II, fol. 63r [66r]. *De Machinis*, Scaglia (ed.), 115–16, plts. 73 and 197.

[9] Regarding »Panstermühle,« see Gleisberg, *Triebwerke*, 65ff. and figs. 65–68. On Agricola, Veranzio and the windmills see *ibid.*, 60–61 and figs. 55–60.

[10] *De ingeneis*, Clm 197, II, fol. 63v [66v]. *De machinis*, Scaglia (ed.), 116 and plt. 73.

least in principle) on the basis of Italian practice and that he included these aspects in modified form as a unitary part of his work in order to demonstrate his familiarity with unconventional, but practical, gearing methods.

Crankshafts and Connecting-Rods

The most interesting motion-conversion devices in our manuscript are crankshafts and connecting-rods. These were new to the fifteenth century, and our authors' drawings of them are significant records of the ways in which such machine elements were understood and applied. Cranked motion is of great importance in machine design since it converts rectilinear to rotary motion. Simple cranks first appear in China in the late second century,[11] but the earliest European representation of a crank occurs in the *Utrecht Psalter* (816–34).[12] As White points out, the crank remained a dormant element in machine design for many centuries. From the ninth until the fourteenth century the use of cranks remained on a very simple level—they were applied to relatively few machines, and there was no attempt to make compound cranks (crankshafts) or cranks with ancillary linkages.[13] The first step from the simple crank to the crankshaft took place in the early fourteenth century, when cranks were first attached to either end of a common axle.[14] Linking such cranks in the middle of an axle produces a true crankshaft, the first example of which appears in Guido's treatise (c. 1335).[15] The earliest practical machine to use the crankshaft was the brace-and-bit, which appears in Flanders in the early 1420's.[16] The next step, the addition of connecting-rods to

---

[11] One is shown in a pottery model of a farmyard scene as part of a machine for winnowing rice with a fan; the model is now in the Nelson Art Gallery, Atkins Museum, Kansas City. Cf. Needham, 118, fig. 415; White, 104, fig. 4. The water-powered bellows of Tu Shih (31 A.D.) are often cited as the earliest example of cranked motion, but Needham suspects that these were probably worked by lugs and springs, not cranks. Needham, 377–78. While he does not exclude the possibility that Tu Shih's machine used cranks, he dates crank-operated bellows in the later Thang or Sung period, i.e., ninth to twelfth century. The earliest depiction of water-powered cranked bellows is found in Wang Chen's *Nung Shu* of 1313; *ibid.*, 373 ff.

[12] White, 110. He also argues against ascribing cranks or continuous rotary motion to any ancient device in Europe, including hand querns; *ibid.*, 103–109.

[13] *Ibid.*, 110–11 surveys the use of cranks in medieval Europe. Most of the early machines to use them were grindstones, hurdy-gurdies, wheels of fortune and winding reels for yarn. Theophilus specifies cranks for several lathes. Cf. Dodwell, 150 ff. and 162 ff.; Hawthorne and Smith, 167 ff. and 183 ff. None of these machines used compound cranks or linkages.

[14] White, 111 notes the *Luttrell Psalter* (c. 1340) as having the earliest representation of this. It shows a grindstone mounted on an axle with a crank on each end. An illustration in a *Book of Hours* done for the Duke of Bedford (c. 1423–30) shows a windlass worked in the same way; cf. Brandt, 246, fig. 331.

[15] Guido da Vigevano, fols. 10v and 14v. Cf. Singer II, figs. 594 and 659.

[16] Several Flemish paintings show the brace-and-bit, the earliest of which is Meister Franke's »St. Thomas Altar« (1424). Cf. White, 112; Gille, Singer II, 653, fig. 595 and pls. 12 and 30a. Fontana's treatise shows a brace-and-bit, but in a mechanically unworkable form. See White, 112–13 and fig. 1.

the crankshaft, was formerly ascribed to our manuscript on the basis of its supposed early date. Since this can no longer be maintained, we must re-establish the connecting rod's history on other bases.[17]

The first texts to show crankshafts equipped with ancillary linkages are from Taccola's pen.[18] He includes two distinct types; for use with pumps, he shows multiple-throw crankshafts from which ropes are slung to the pistons in the pump cylinders. The rotation of the crankshaft exerts an upward pull on the pistons, which descend of their own weight once the crank throw is past top dead center and the rope tension slackens. A second type appears in the form of a small grain mill (or in one instance a bucket hoist) powered by a single-throw crankshaft moved with a rigid connecting-rod, which Taccola twice calls the *manicho de* (or *di*) *paviolo*. The use of an obviously vernacular name for the device indicates that it was in common use in Taccola's day. This impression is further strengthened by his artistic treatment of the human figure shown operating the mill. As Scaglia notes,[19]

> the miller who has brought the crank of the millstone to its dead point near him ... is beginning to push it sideways and back; his hands, firmly gripping the crankrod, are still in dead-point position, but his body is swinging toward the machine.

The ability to capture such detail indicates strongly that Taccola had had the opportunity to observe men at work with machines like those he drew. The contrast between these obviously well-developed grain mills and the rather crude, rope-tied pump crankshafts, may be understood if we recognize that the grain mill stands firmly in the tradition of crank applications in the middle ages, i.e. power is supplied *to* the crank from a human operator in the form of reciprocal motion. The pump crankshafts, on the other hand, represent the first known attempt to visualize the opposite function, i.e. to translate the crank's rotary motion to a reciprocal form. Taccola's grain mills must represent a fairly recent, but still well-established aspect of Italian technology in the early fifteenth century. His crankshaft-driven pumps are part of a newer, and perhaps still speculative, extension of the mechanical possibilities inherent in the crankshaft.

The German evidence indicates a somewhat slower development. The earliest trans-Alpine text to show anything approaching a crank and connecting-rod is Vienna, Waffensammlung,

---

[17] We should stress at this point that the Chinese had the earliest connecting rods, linked to simple cranks. The earliest certain example is on a model of a quern taken from a tomb near Naking dating from 420–589. See Needham, 759 and fig. 413 for a later illustration of the same system. It is possible that the bellows of Tu Shih (see note 9) used cranks and connecting-rods also. Surprisingly, the Chinese never developed true crankshafts. They were introduced by the Jesuits in the seventeenth century; *ibid.*, 118.

[18] *De ingeneis*, Book II, fol. 82v and Book III, fol. 29r, both showing suction pumps, and Book IV, fol. 65r, showing a grain mill. See Prager and Scaglia, *Taccola*, plts. 15, 25, 89 respectively. *De machinis*, fol. 36v (reiterated in miniature, fol. 17v) showing a bucket lift, and fol. 46v, showing a grain mill. See Scaglia (ed.), 108, 86, 115 and plts. 45, 18 and 198 respectively.

[19] Prager and Scaglia, *Taccola*, 183.

Ms. P 5014 (c. 1437),[20] where we find a drawing of two men turning an Archimedes screw by means of a heavy beam with a hole in its middle. The handle of a crank attached to the Archimedes screw passes through this hole. The apparent intention of this unusual design was to allow two men to turn the crank without interfering with one another's movement. Two individual connecting-rods would achieve the same end in far less awkward fashion, but the anonymous author of the Viennese text seems to have been unaware of this alternative. Apart from this odd example, no other known German text shows crankshafts before our manuscript (see fols. 18r, 20r, 21r, 21v, 42r).[21] All the crankshafts shown in Clm 197,I transform reciprocal to rotary motion; there is as yet no hint of the converse principle as it appears in Taccola. In addition, all our crankshafts are shown in conjunction with grain mills (and this is true of most subsequent Germanic drawings as well). Despite the apparent conservatism of the Germans in this respect, there is evidence that techniques of using cranked motion had evolved somewhat since Taccola's day. Author A prefers, for example, multiple throw crankshafts to the single-throw designs used in the Italian grain mills; he also knows to employ flywheels as adjunct elements to crankshafts; and he can conceive of complex, »rocker arm« connecting-rods as shown on fol. 21v. In all this, author A's knowledge compares fairly favorably with that of his Italian contemporary, Francesco di Giorgio Martini. Comparing the German and the Italian evidence, we can conclude that the transition from the simple crankshaft to the crankshaft with connecting-rod took place in the early decades of the fifteenth century, most likely in Italy. These early linkages were employed mainly in grain mills, although more highly evolved applications had been conceived if not made part of daily practice by the 1430's. Different forms of crankshafts and connecting rods made their appearance as the century wore on, and this development took place at roughly the same pace in Germany and in Italy.

The extent to which the crankshaft was actually applied has been misunderstood somewhat. Bertrand Gille claimed that the system was not widely used until the later eighteenth century, and he considers Leonardo's invention of mechanical alternatives to crankshafts as symptomatic of a general, persistent aversion to their employment.[22] He also suggests that friction problems may have made crankshafts with connecting-rods difficult to use.[23] There is a grain of truth to Gille's observations. Leonardo did spend much ingenuity in devising gearing and linkages to perform the same tasks as a crankshaft and connecting-rod, and his

---

[20] The drawing appears on fol. 32v and may be related to a drawing of a cranked Archimedes screw that appears in Kyeser.

[21] Three machines using crankshafts and connecting-rods appear in manuscripts that date before ours, but in each case the style of the drawing differs greatly from the rest of the manuscript, suggesting that it was probably a later addition. For the present these cannot be considered to antedate our text: Cpg 787 (probably c. 1430), fol. 104r shows two grain mills each with double-throw crankshafts, flywheels and gearing; ÖNB 5518 (1420–50), fol. 1r shows a mill with a single-throw crankshaft.

[22] Bertrand Gille, »La naissance du système bielle-manivelle,« *Techniques et civilisations*, II (1952), 45–46. For some of Leonardo's alternative devices see Needham, 383–86.

[23] Gille, Singer II, 653–54.

followers often used some of them.[24] Nevertheless Gille has seriously overstated his case. Crankshafts were used continously from the time of our manuscript on. Even Ramelli, one of Leonardo's heirs in mechanics, used them in combination with connecting-rods. From the sixteenth through the eighteenth centuries their use grew steadily.[25] Gille also seems to have over-estimated the role of friction. Close examination of the crankshafts and connecting-rods in our manuscript reveals that all the cranks but one are made of iron (author A specifies this material on fol. 18 r), and the connecting-rods are wooden. The wood-to-metal bearing surface at the crankpin would cause less friction than wooden shaft with a wooden connecting-rod. Also, it would be quite easy to insert a metal bushing in the rod's eye to make a metal-to-metal bearing surface.[26] The frequency with which metal bearings appear in our manuscript indicates that this would have been within the capacities of A or his contemporaries.

Also, we should note that author A's crankshafts and connecting-rods have the look of real machines, not engineering fantasies. Even though cranked motion is very difficult to visualize, A made only two errors in depicting crankshafts.[27] Both could be rectified with only minor changes. He knew that iron shafts would have better characteristics than wooden ones (author B did not realize this; see fol. 42 r).[28] Author A could conceive of very complex connecting-rod motions (see fol. 21 v) and he was familiar with the chief problem of crankshafts with connecting-rods, the »dead spot« where the rods can exert no effective thrust on the

---

[24] Ramelli, for example, is particularly fond of one of Leonardo's devices, a wheel geared through half its circumference meshing with two lantern pinions on one shaft. Thus the shaft is given an alternating rotary motion that can be converted to linear motion through linkages.

[25] White, 171–72 gives an extensive list of examples to disprove Gille. He notes that Ramelli used connecting-rods eighteen times. For examples of their use in the decades immediately following our manuscript, cf. Maltese, figs. 62, 65, 70, 71 and 73; Bossert and Storck, fig. 32; Cpg 126, fols. 5 r and 9 r.

[26] It is difficult to say whether author A used metal bushings or not. All but one of his connecting-rods are shown mounted on the crank-pin, which obscures the bearing surface. The one connecting-rod shown dismounted (fol. 20 r) is ambiguous. The interior surfaces of the eye are darker than the rod itself, but they are not colored as metal. Author A nowhere specifies bushings in his text, but fol. 18 r states that »the supports where the iron [crankshaft] goes in should also be steeled ...« If one takes »supports« *(»sullen«)* in the sense of bearing, then this would indicate metal bushings. The text is too ambiguous to permit definite conclusions, however. Author B's connecting-rod is made of wood, but it has what may be a metal ring where it joins the crank-pin. The ring is colored to represent wood, however, as is the crankshaft. Only one of Taccola's crankshafts shows unequivocally an iron crank throw and a connecting-rod with an iron head. *De machinis*, fol. 46 v, which survives only in a slightly later copy datable to 1449–53. See Scaglia (ed.), 115 and 37 ff., plt. 198.

[27] In the drawing on fol. 21 r author A misaligned the ends of the crankshaft; it cannot work as drawn. This is a common error. Cf. White, figs. 6 (Fontana) and 7 (Taccola). Fol. 20 r shows a machine whose crank-pin is set so that a connecting-rod attached to it could not turn the shaft through a complete revolution. Author A seems to have recognized his error, for the text breaks off as it begins to describe that part.

[28] B's crankshaft (fol. 42 r) is made of wood. If the eye of the connecting-rod is also wooden (see note 26), then friction between the two parts would be great. Also, a wooden crankshaft would tend to deform when subjected to the torque caused by turning it with a connecting-rod.

crank-pin. He alludes to this indirectly once in his comments (fol. 21r): »Once the wheel is in motion, one should not let it come to a halt; otherwise it will be hard to start up again.« He takes pains to describe or illustrate means of overcoming »dead spots.« He recommends using two men and two connecting-rods (fol. 18r); he counsels that the gears should be more closely-toothed (fol. 21r); and he includes flywheels on some of his mills (fols. 20r and 21r).[29] All of this bespeaks a close acquaintance with the properties of crankshafts and connecting-rods. It is impossible to believe that anyone could have been this well versed in their use without having had practical experience with them. Author A's drawings of machines with crankshafts, connecting-rods and flywheels are the product of a technological milieu that was becoming more interested in using cranked motion for a wide variety of tasks, and they show us how sophisticated the later fifteenth century had grown in understanding such devices.[30]

Belt-Drives

Our re-dating of the manuscript affects not only the history of crankshafts and connecting-rods but also that of belt-driven devices. The use of belts to convert linear to rotary motion is extremely old, possibly Palaeolithic, and is represented in Egypt as early as 2500 B.C.[31] The most primitive tool for this was a bow whose string was loose enough to be looped about

[29] The flywheel is an ancient device. It first appears as part of the potter's wheel. Aside from this it was not applied to any machine in antiquity. Theophilus describes flywheels as part of more complex machines in 1123–24; cf. White, 115. By the fifteenth century flywheels were being used in such devices as windlasses (cf. Brandt, 246, fig. 331) and lathes. See Wilhelm Treue *et al.* (eds.), *Das Hausbuch der Zwölfbrüderstiftung zu Nürnberg* (2 vols.; München: 1965), Vol. I, 116; Vol. II, 55. The first theoretical discussion of the properties of flywheels was Jean Buridan's (d. 1358?) theory of impetus. He used the continuing motion of a rotary grindstone as an example. See White, 116. Gille's observation (»Manuscrit,« 83) that the flywheels of our manuscript show »le principe d'inertie avant la lettre« is inaccurate.
Author A did not apply a flywheel to the mill on fol. 21v because he probably felt that the complex motion of the connecting-rods made one unnecessary. This is incorrect, but it is an easy mistake to make (Gille, »Bielle-manivelle,« 45 made the same error).

[30] This sophistication is nowhere more evident than in the use of flywheels and flyweights to smooth out the motion of a crankshaft. Solid flywheels like those shown by A quickly gave rise to a new type, probably adapted from Tibetan prayer wheels. These have weights attached by short chains to arms mounted on the crankshaft. Fol. 42r shows a crude version of this. Cf. White, 116; Needham, 91–92 for a discussion of such devices. Weights mounted in this way appear in well-developed forms in di Giorgio's work. Cf. Maltese, fig. 327 and White, fig. 9. They acted not only as flywheels but also as governors, since increasing the crankshaft's speed forced the weights outward, lengthening the moment arm and exerting a braking force. Thus the weights acted (to a very limited extent) as a »negative feedback« regulator system. This action led Gille, *Engineers*, 111 to proclaim such flyweights as an anticipation of James Watt's ball governor for a steam engine. The most recent student of the history of regulating techniques denies that there is any connection between the two, however; cf. Otto Mayr, *Zur Frühgeschichte der technischen Regelungen* (München and Wien: 1969), 97.

[31] V. Gordon Childe, article »Rotary Motion,« Singer I, 188–90. He notes that claims for the bow drill in the upper Palaeolithic are conjectural, since we are not certain that the bow itself is that old.

a spindle. Moving the bow quickly back and forth gave the spindle an alternating rotary motion. A somewhat later system dispensed with the bow and used simply a strap pulled by two men.[32] These methods shared one fault: the rotary motion they produced was discontinuous and in alternate directions.[33] But for many purposes rotary motion needs to be continuous and unidirectional, and to produce this with a belt requires that it be a closed loop running over two different pulleys.

The earliest example of a closed loop system is the ancient *sāqīya* or »chain of pots« described by Philo of Alexandria and Vitruvius.[34] In this machine a loop of rope carrying buckets at regular intervals was passed over a pulley and used to raise water. Normally only one pulley was used at the upper end and the loop's lower end was left to hang freely in the water.[35] Another of Philo's devices uses a closed loop, this time with pulleys at either end. He described a repeating ballista which was spanned by means of a hook on an endless chain. However, he admitted that the device was impractical, and it was probably never built.[36] Neither of these ancient devices used belts to transmit rotary power from one pulley to another. The *sāqīya* transported water, while the ballista's endless chain transmitted linear motion.

The earliest machine to employ the belt-drive to transmit rotary motion was the spinning wheel, which appears in Europe in the late thirteenth century.[37] The spinning wheel marks a considerable advance in the history of machine design for it uses not only the belt-drive, but also the flywheel principle and the speed differential made possible by having pulleys of different sizes. In spite of this, however, no other machine attempted to use the belt-drive to transmit rotary power for over a century. Probably the slippage between the pulleys and the belt, which was made of rope or leather, was so severe that belt-drives could not be applied to heavy machines. It was once thought that our manuscript's drawing of a stone

---

[32] Feldhaus, *DM*, 77, fig. 45 reproduces an Egyptian figure from c. 300 B.C. showing a lathe turned by this system. He is uncertain how much older it may be.

[33] The pole-lathe, a thirteenth-century invention, shared this same limitation but it remained popular until the late eighteenth century nevertheless. See White, 118.

[34] See Needham, 352.

[35] Philo specifies a toothed wheel at each end of the loop, but Vitruvius mentions only one. The chain of buckets in the Lake Nemi ships have such sprocket wheels, Needham, 109. Later Arabic representations, such as al-Jazarī's (c. 1210) leave the lower end free; *ibid.*, 353. Probably the inconvenience of placing a toothed wheel underwater was so great that it was omitted in practice.

[36] *Ibid.*, 109.

[37] The earliest reference to spinning wheels in the west is c. 1280. See White, 119 and 173. In China, however, the earliest belt-driven quilling wheels (to wind thread onto a bobbin) appear in the Han (Needham, 266–69) and the earliest true spinning wheels appear by c. 1035 at the latest (*ibid.*, 100ff. and 758). Connections between Chinese and western spinning wheels are uncertain. It is not surprising that the Chinese were also the first to use the chain drive to transmit motion from one sprocket to another. It appears in an astronomical water clock of the late eleventh century and may date back to 978; *ibid.*, 457 and 471.

polisher (fol. 23v) was the next example of a belt-driven machine.[38] This matter must be re-examined in the light of our redating.

The successful model of the spinning wheel seems to have revived interest in closed loop systems. For example, Giovanni de' Dondi's *Astrarium* (completed in the early 1360's) used loops of linked plates moving over two sprocket wheels to display the dates of the movable feasts.[39] By the early fifteenth century loops of rope passing over two pulleys were employed to carry material; these used cranks to power them.[40] None of these were true belt-drive machines, for they used loops to transport material or information, not to transmit rotary power. Nevertheless, they were devices that demonstrated the practicability of such transmissions, and they included such devices as sprockets and cranks as part of their equipment. The earliest devices to take the next step and use belt-drive simply to transmit rotary power appear in Giovanni da Fontana's treatise (c. 1420) and an anonymous German manuscript (c. 1437).[41] It is not clear exactly what sort of belt-driven machine these manuscripts show, but there can be no doubt that they depict belt-drives transmitting rotary power. Even if, as seems likely, the machines in question were not practical and were never built, nevertheless they serve as clear demonstrations that the idea of using belt-drives had entered Europe's engineering consciousness. There is other evidence of increasing interest in this principle. Taccola, for example, shows a chain loop passing over a large, ribbed pulley and hanging slack beneath it, used as a pull-chain to drive a bucket-lifting mechanism.[42]

This type of interest ultimately found expression in a practical device that could utilize the unique properties of the belt-drive to full advantage, the stone-polishing mill. We have already noted (Chapter VII) that such devices appear first in France and shortly thereafter in Germany and Italy. The polishing-mill seems more closely related to the spinning wheel than the other devices noted above. It uses the flywheel principle and the speed differential obtainable through different-sized pulleys. Also, like the spinning wheel, it was a machine designed for light duty and thus slippage was less of a problem. It is not, as once thought, the

---

38 For example, Needham, 102.

39 Bedini and Maddison, pp. 15–16 and fig. 43.

40 ÖNB 3069 (c. 1411), fol. 89v shows a rope loop mounted on two pulleys spanning a gulley; a basket fixed to the rope carries material across it. Feldhaus, *Technik*, 1023 and Uccelli, 50, fig. 142 reproduce this folio. Vienna, Waffensammlung, Ms. P 5135 (c. 1435) shows a similar rope loop with baskets. The text describes it as a machine to aid in excavating ditches. These are the first machines I have seen to use cranks in combination with rope loops.

41 Fontana, fol. 7r; Vienna, Waffensammlung, Ms. P5014, fol. 36r. Both drawings show similar machines. A tall, rectangular tower has spools or pulleys mounted vertically on its top. In Fontana's drawing there are three such pulleys and they are connected by a single rope loop. In P 5014 there are also three, but two separate belts transmit power from the central pulley to the peripheral ones. The purpose of both machines is unclear. P 5014 lacks an explanatory text, and Fontana's is entirely in code.

42 Clm 197, II, fol. 20v; cf. Berthelot, 478, fig. 30 *bis*.

first machine after the spinning wheel to use belt drive, but it was the earliest practical one to appear. Why it evolved only in the later fifteenth century is an unsolved mystery.

The drawing of a polishing-mill by author A does not show an intimate acquaintance with belt-drives. He has not shown the rope under the necessary tension, nor has he made the driving pulley large enough in relation to the driven pulley to gain the desired increase in speed. Indeed, A's crude example seems to apply the sort of rope-loop found in material-transport devices to the polishing-mill. This helps confirm our hypothesis that he learned of such mills from abroad and attempted to draw one without ever having seen it. It is not surprising that A was unsure of what such a mill should look like, since belt-driven devices other than spinning wheels were a rarity in his day and remained somewhat unusual until the eighteenth century.[43].

General Evaluation

How well do our authors understand and apply devices to transmit and convert motion? Author A has a fairly good grasp of many such devices, while author B's abilities are generally poor. To take the simple case of block-and-tackle arrangements, we see that A can comprehend and reproduce complex pulley systems (cf., e.g., fols. 1v and 2r). He made several mistakes, (e.g., fol. 6r), but as a rule A was able to show them without growing confused. Author B, by contrast, includes far fewer hoists than A, and they are all much simpler. Two of his examples contain serious errors (fols. 38r and 42v), and they suggest that B was unable to visualize complicated pulley arrangements or to reproduce them without making mistakes.

In the area of geared devices, there is very little comparison between the two men. Author A shows a large number of them, while B shows only one (fol. 47v). B's drawing is unclear in its perspective and puzzling in its intentions. It shows ordinary right-angle gearing with flyweights on the cranked shaft that turns it. There is no machine which it powers, and the drawing seems to be a study of gearing in the abstract. Author A's gear trains are always part of a complete machine, and they are always carefully designed to play a role in the machine's functioning.

Author A did not use the more sophisticated types of gearing employed in Italy in the early fifteenth century, e.g., worm-and-spur gearing, but he does make effective use of the sort of

[43] Belt drives became somewhat more common in the sixteenth century in polishing mills and lathes. For example, Wolf, 537 describes a pewterer's lathe with a belt-drive illustrated in 1568. Franz Maria Feldhaus, *Die Geschichte des Schleifmittels* (Hannover-Hainholz: 1919), figs. 23–24 (hereafter Feldhaus, *Schleifmittel*) shows some seventeenth-century polishers driven by belts. Where heavy machinery was involved, belts were unsuitable because of slippage until modern rubber belts were invented. Chain drives were attempted in the sixteenth century. Ramelli, for example, shows them six times; cf. figs. 39, 79, 86, 126 and 168. However, industrially practical ones were not developed until c. 1770; the common bicycle chain appeared only in 1869.

gearing common in Germany in his day. Moreover, he could envision some new ways in which gearing might be utilized, viz., the use of two gear trains driven by one power source (fols. 22v and 23r). Similarly, A's knowledge of how to combine braking and escapement mechanisms with gear trains (fol. 3v) reveals a wider range of curiosity and ability than most of the other German treatises display.

Cranked motion, however, is the area in which A's sophistication is shown most clearly. Here again he is patently B's superior. Author B's only crankshaft (fol. 42r) is not even connected with the mill it purports to drive and it is made of wood, not iron. Author B obviously understood little of the problems of cranked motion. His only novel device, chain-linked flyweights, is rendered almost ineffective by having the weights linked too closely to the arms. Author A, on the other hand, is very well-informed of the properties of crankshafts of all kinds, and he shows or describes a number of ways to overcome the problems of using crankshafts and connecting-rods. In his understanding of such devices he approaches the Italian engineers. Indeed, author A makes errors in depicting crankshafts and connecting-rods, but they are minor ones and arise from the inherent difficulties of visualizing cranked motion properly (see fols. 20r and 21r). Few German engineers of his day could have equaled A's grasp of this subject.

Finally, in respect to belt-driven machines, A's drawing of a stone-polisher (fol. 23v) shows how he was able to grasp the principle of a new machine without understanding its details of operation. He drew a crude machine, lacking proper belt tension and size differential between pulleys. This is to be expected since A was illustrating a machine he had never seen. The crudeness of his illustration notwithstanding, it is important as a record of how the belt-drive principle was being applied to new machines, and it serves to show us that A was in touch with developments beyond the area in which he worked. His drawing demonstrates that technical information passed quite rapidly from one part of Europe to another in the fifteenth century, and that there were engineers in many areas who could grasp and reproduce new designs as they learned of them. The remarkable aspect of his drawing is not its crudeness, but the fact that its author was able to comprehend the principles on which it was designed.

CHAPTER IX

COMPARATIVE CONCLUSIONS

Now that we have looked at the manuscript from a number of specialized points of view, what general conclusions can we draw? In evaluating Clm 197,I *in toto* and in assessing its significance for the history of technology, we must be careful not only to keep in mind that our redating of the text alters its position significantly, but also that the meaning of one manuscript can be ascertained only against the background presented by other manuscripts from the same era. It cannot be emphasized enough that we are inadequately equipped in our present understanding of this background; we can only indicate in sketchy fashion the manuscript's position in the grand scheme of things.

Throughout this study, we have treated the manuscript as if it were a literal representation of the realities present in its author's environment. On the microscopic level, such an approach is valid enough to be usable. On the broader scale, we must take care not to carry this »literal« viewpoint too far. Our manuscript is a piece of technological literature, and, like all »literary« works, it mirrors reality in selective or at times distorted fashion. Even though Clm 197,I conveys (at least to this reader) a vivid sense of the technological realities of the fifteenth century, when we look at the work as a whole, some of its nonrealistic aspects demand attention. It may seem perverse to insist on this point in view of the undeniable fact that the number of direct links between our manuscript and the rest of the Germanic tradition is very small. Indeed, one of the major reasons why Clm 197,I is remarkable is precisely the distance that separates it from the rest of the tradition. It copies very little directly from other works, and other works in turn copy little directly from it. Nevertheless, we should not infer from this independence that the manuscript is therefore some sort of primitive act of technological journalism, reporting without bias what the authors saw. The naivete of such a viewpoint is apparent from Bertrand Gille's comment: »Perhaps the author of Clm 197,I jotted down drawings and notes taken during a course of lectures or journeys in a notebook, which, later on, was to become a ›Collection of Machines.‹«[1] Gille would have us see the manuscript through the lenses of modern technological practice, where journeys and lectures are commonplace. For fifteenth-century technicians the sense of how technology ought to be described and depicted was shaped by very different considerations, by conventions that made certain demands on any author. While we can see our manuscript as a set of true statements about the technological world, we cannot understand those statements unless we comprehend in a general way the conventions that underlie them.

[1] *Engineers*, 68.

Part of the grammar of conventions concerns the sense of order that our manuscript, like most of its contemporaries, displays. Section A in particular shows a fairly methodical division of its drawings into various categories, e.g. cranes, siege engines, aquatic devices, mills, etc. This order is not entirely consistent, but obviously an attempt was initially made to provide certain groups of drawings as the text was done. Section B shows much less grouping of its drawings, but even so, some of its individual folios contain related pairs of drawings on recto and verso, e.g. fols. 39 (siege engines) and 40 (crossbows). The habit of arranging the contents of technological texts in such a manner was part of the Germanic tradition from the time of Kyeser onwards, and few, if any, of the later texts abandon it completely. Kyeser's own categories (seven in all) were sometimes followed and sometimes ignored, but some sort of categorization was always prominent in the minds of technical authors.

Another conventional element is represented by the fact that all the drawings in our manuscript are done in what we might call a »formal« or »scriptorial« style (or at least a reasonable approximation thereof in section A); all are done according to an elaborate, and apparently rigid, color scheme; nearly all have a »finished« quality in their use of line, perspective (in the medieval, not the Renaissance sense) and presentation. A notebook, a series of unfinished sketches, or even a rough draft of a formal treatise (such as Taccola's *De ingeneis*) would lack this quality of formality. Indeed, much of the material available to us from Italy is informal in character, e.g. some of Leonardo's notebooks. In the north, however, the impetus provided by Kyeser—and doubtless reinforced by much older artistic custom—apparently forbade such an unfinished working style. A »complete« drawing (symbolized above all by the use of color) was the only acceptable fashion until very late in the century, when Italian practice modified German artistic concepts. In a similar vein, the absence of text in section B is not a personal idiosyncracy of its author, but is also rooted in the tradition of textless »picture manuscripts« on technological subjects that dates back at least to the 1430's. Likewise, the stylized manner in which author A uses words (»Item: das ist ein ...«) hints at a formal, almost stereotypic, concept of how technical descriptions ought to be written. Both sections of Clm 197,I conform to the demands of tradition in matters of style, although they do this in different ways.

Regarding content, the two sections differ strongly in their relationship to convention. Author B was much more strongly influenced by traditional ideas of what constituted the proper subject matter of technology than was author A. B shows in a very highly developed form both the focus on matters military and the somewhat »fantastic« elements that originally appear in *Bellifortis*. The military devices that appear in section B outnumber the non-military by slightly more than two to one (26 vs. 12 drawings); moreover, many, though not all these devices, are mere repetitions of well-worn types that go back seventy years or more. The elements of fantasy that show themselves in section B are not magical as in Kyeser's work (although Kyeser certainly would not have considered magic as fantasy), but rather take the form of overly-elaborate, ill-conceived devices that ignore many practical considerations, e.g. heavily-amored, spike-bearing siege engines with six or more wheels, often arranged in odd patterns. While these look impressive to the untrained eye, many of them are riddled with elementary design errors.[2] From a purely modern point of view, section B looks like the work of an »armchair engineer,« a man who had little or no practical experience with the devices he depicted. Seen against the background of other fifteenth-century manuscripts, however, B stands out only as an extreme case of the conventional concerns, showing well-established tendencies within its tradition.

Section A, on the other hand, has a sufficiently different relationship to tradition to justify some of the favorable comments made by Gille about the entire manuscript. To some extent, its »intellectual craftsmanship is better than that of Kyeser.«[3] While we ought to demur slightly in judging author A »a practical man, a craftsman who had begun to collect evidence about his craft and whose curiosity is always on the alert,«[4] still, A's work is remarkable in comparison with other Germanic works for its keen sense of the practical. Since A's sources are unknown to us, we can only speculate on the degree of his indebtedness to others. On a qualitative basis, we can claim that, no matter how much he may have copied, he copied very well. His borrowings were intelligent, and he related what he absorbed to the realities of engineering as he understood them. His drawing of a belt-driven stone polisher (fol. 23v) is a case in point. Author A must have learned of it from abroad; certainly he did not invent it himself. But he recognized the mechanical value of such a device, and he was able to draw a crude, but workable, version of it. Author A's human failings are numerous: he could not write well, nor was he a talented draftsman. He was not above a bit of exaggerated credulity, e.g. in his attribution of an »automobile« siege engine to Erkinger von Seinsheim (fol. 16v). He made some mistakes in his drawings, e.g. the mill on fol. 20r, where A misunderstood the mechanical relationship between axle, crank-pin and connecting-rod. But even in error author A shows a breadth of vision not always found in the other Germanic manuscripts. His strong interests in mills, for example, represent a new departure within the trans-Alpine writers. We can fully agree with Gille on one point: author A's mind was »drawn to solving technical problems, ... able to appreciate another person's machine or imaginative faculty.«[5]

It was not the apparent intention of either author to create an entirely new or original work on technology; rather, both seem to aim at compiling a sort of miniature *summa* of technical devices along the general lines laid down by Kyeser. The differences between the two sections come about mainly because of their respective authors's approach to tradition. Section B is the work of a man steeped in the »classical« German texts on military technology, espe-

[2] For example, fol. 46v, where B drew an excellent picture of a wagon with fold-down sides (a Hussite innovation, by the way), but showed the wagon's six wheels in such a position that the sides could not fold down. Without exception, B's siege engines are heavy, cumbersome vehicles, burdened with too much iron reinforcement and studded with unnecessary spikes. Most of these are shown with so many wheels, or with wheels in such strange patterns, that they would have been impossible to turn. They could only have been dragged about by brute force. In short, B's siege engines are usually highly impractical.

[3] *Engineers*, 73.

[4] *Ibid.*

[5] *Ibid.*, 69.

cially the *Bellifortis*. We have already noted that some of B's drawings are almost direct copies of those in other works.[6] Author B's work is completely traditional, always within the influence of the *Bellifortis*. Section B bears witness to the strength of Conrad Kyeser in the development of technological thought in fifteenth-century Germany. Almost a century separates section B from the *Bellifortis*, yet one feels that Kyeser would have been quite comfortable were he to use almost any of the devices in B's drawings. There are some exceptions to this, of course. Kyeser might have been puzzled by crankshafts with flyweights and breech-loading cannon, but he would have been very much at home with the other machines. Neither in his range of interests nor his treatment of his subjects does B wander far from the canon established by Kyeser. Section A shows a strong respect for tradition, but modulates this with an unusual concern for novelty and variety. Some of its contents, such as cannon, siege engines and siege devices, are responses to the demands of tradition, while others, such as mills and cranes, are unusual features for a German manuscript in the 1470's. Even the traditional sometimes shows an admixture of the novel; A's cannons, for example, are downright old-fashioned, but they are shown mounted on fairly novel stands and carriages. It is this concern for novelty within tradition that sets section A apart from the majority of the Germanic manuscripts of its day.

Both sections of Clm 197,I are of major importance for our understanding of late medieval technological developments. Section A especially gives us several important pieces of information about the evolution of applied mechanics in this period. We have already discussed crankshafts and connecting-rods as well as belt-drives in some detail in Chapter VIII. We only need to reiterate here that, since section A dates c. 1472–85 rather than c. 1430, the development of both these systems was more gradual than has been commonly believed. In addition, close inspection of author A's drawings and comments about connecting-rods indicates that Gille's hypothesis concerning the role of friction in retarding their widespread adoption is incorrect. There are many other techniques whose history is illustrated by section A. Cranes of all types appear frequently, and our discussion of them has pointed out the ways in which such machines were made and used in the fifteenth century. Brakes make their initial appearance in the west in our manuscript (see fol. 3v), not, as was formerly thought, in Leonardo's works. Stamps are illustrated twice (fols. 10v and 17v), and the latter drawing is one of the earliest to depict recumbent trip-hammers in Europe. Our discussion of this device shows the ways in which it developed and its origins in Chinese engineering practice. Author A's drawing of a water-pipe borer (fol. 15r) has been identified with a similar machine described in Endres Tuchers's *Baumeisterbuch*, and our analysis of A's drawing reveals that the application of water-power to such machines may have come about earlier than was previously thought. Another sort of water-powered device, the horizontal water-

[6] The trebuchet on fols. 32v–33r is certainly heavily influenced by one in Kyeser and is probably copied from a *Bellifortis* manuscript. The cart running on rails and used for fortification construction (fols. 35v–36r) is a direct copy of a drawing by Johann Formschneider. Quite possibly many of B's other drawings are copies or adaptations from works now lost.

mill also appears in section A (fol. 18v) and we have pointed out the renewed interest in horizontal waterwheels on the part of many fifteenth-century engineers. Finally, some of author A's mills show that new methods of gearing were beginning to emerge in the fifteenth century. In particular fols. 22v–23r illustrate the most popular means whereby two or more machines could be driven from a single prime mover.

Section B contains less material of significance, but it is not devoid of interesting illustrations. Author B's two drawings of vehicles moving on rails (fols. 35v–36r and 48v) show that this new method was gaining rapid acceptance in the fifteenth century. Our identification of Johann Formschneider's work as the source for the former drawing allows us to date such carts in the final third of the fifteenth century rather than in the sixteenth century as was previously thought. Also, author B's drawing of a crossbow-spanning device (fol. 29v; cf. A's version on fol. 18r) calls attention to an alternative machine for this purpose that may antedate the crossbow windlass and *cranequin*. Finally, the crankshaft with chain-linked flyweights (fol. 42r) shows clearly how widespread the understanding and application of the flywheel principle had become by the late fifteenth century.

Section A and section B contain drawings of windmills (fols. 19r, 19v and 47r). Those from section A show us the earliest known representations of the interiors of such mills and it is particularly interesting to note that the illustration on fol. 19v shows gearing designed to drive the lower millstone rather than the upper one, in contrast to common practice in windmills. Author B's illustration shows only the exterior of a windmill, but depicts clearly an advanced type of sack hoist consisting of a rotating shaft around which a rope winds. Such hoists were previously thought to appear first in Ramelli's work. Finally, both sections contain many drawings that illuminate one other aspect of technical development, the use of iron parts in wooden machines. This has been a curiously neglected subject. Only A. R. Hall seems to be aware of its importance. He first pointed out that Guido da Vigevano's drawings show a prolific use of iron parts and that Agricola's illustrations show the same tendency.[7] Our identification of the ways in which our authors use iron (cf. also commentaries to individual folios in Part II) reveals that this was not an isolated trait, but instead seems to have been a common practice among fifteenth-century engineers. For these and many other reasons Clm 197,I is an important document for our understanding of late medieval technical developments.

This work is only the first step toward a general study of the northern tradition of technology in the later middle ages. Its subject, Clm 197,I, is only one of many technical manuscripts from the period illustrating trans-Alpine technical developments. We have tried constantly to see our manuscript in relation to as many of its contemporary texts as possible. The vast majority of them, however, are still in need of thorough investigation to clarify their contents and their role in the northern tradition. Monographic literature is poor or non-existent in this field, and the chapter in Gille's *Engineers of the Renaissance* is the only attempt at a comprehensive study of the German school in relation to the Italian line of

[7] Hall, »Vigevano,« 968.

development. Much work remains to be done even to classify and date the German manuscripts, to say nothing of elucidating their contents. Chapter I and the following appendix indicate some of the difficulties that await the researcher in this field. Some of the opportunities need to be stressed as well. Microfilm techniques make it fairly easy to compare drawings by different men to determine lines of transmission. Several new works, chiefly Quarg's edition of Kyeser's *Bellifortis,* have laid a foundation for the study of other, similar works, and the recent appearance of scholarly editions to the treatises of men like Taccola and Francesco di Giorgio offer unparalleled opportunities to compare the German with the Italian engineers. The further study of the northern tradition will shed new light on the early history of such subjects as machine design, military technology, applied chemistry and hydraulic engineering, to name but a few. It will clarify the relationship between the German and Italian schools. It will provide new information about the sources used by later technicians such as Agricola, Besson, Ramelli and Veranzio. Finally, it will make possible more penetrating analyses on the role of technology in the social changes of the late middle ages, e.g., in the growth of cities and in the formation of new types of armies. In short, late medieval northern technology is a virgin field with many possibilities. It lacks only the attention of interested scholars.

APPENDIX

SELECT LIST OF MANUSCRIPTS

SELECT LIST OF MANUSCRIPTS

The further study of fifteenth-century technology depends to a considerable extent on the resolution of a number of outstanding textual problems. The following list of manuscripts is intended to help isolate and identify some of these problems and to provide a partial foundation for further work. It is not merely an appendix to the main work, nor is it designed to buttress a case that has already been argued there. It is broader in scope than the study of Clm 197,I would seem to require (though it is far from comprehensive). It is not directed at the casual reader but rather at those who are interested in manuscript problems, in the frank hope that it might help to stimulate interest leading to some solutions of these problems. The list is limited mainly to works which appeared originally in the region of Europe where German was the everyday language. It covers primarily works that have not received adequate treatment in the existing scholarly literature. (Exceptions to this criterion include several Italian texts that have not been edited yet and Kyeser's *Bellifortis*.) Texts from the sixteenth and later centuries are included only if they show strong connections with a fifteenth century work, e.g. a treatise on military technology from the sixteenth century which includes portions of the *Feuerwerkbuch* (see below). In general I have studied the listed texts myself, either at first hand or microfilm. Excluded are works which have come to my attention through older bibliographies, but which I have been unable to trace to their present resting places. The same is true of all texts apparently destroyed during World War II. (A list of spurious, doubtful and lost works would be useful, but very difficult to compile.)

A bibliography of this sort treats a very special class of texts with highly idiosyncratic problems, some of which should be made known to the reader. As noted above, the men who created and copied technical manuscripts in the fifteenth century did not share our modern notions of authorship, proprietary rights to ideas, or even the literary man's notion of a »correct« text. In addition, of course, they lived in a world which did not afford them copyright or patent rights to protect ideas or words. As a result, they often treated works in a fashion that is very strange to us. They borrowed from one another freely and without credit; they modified existing texts to suit their own needs without apparent hesitation; they mixed older and newer texts and drawings without distinguishing between the two; they often neglected to date their works; and they frequently did not indicate either titles or authorship. Many of these tendencies were shared by those who collected and preserved these works. In some cases what is today one manuscript will contain works by several different authors, all bound together under one set of covers with few or no distinctions between the various parts. Faced with such an untidy situation, generations of bibliographers and librarians–men who shared a common *horror vaccui* in these matters and who could scarcely have been trained to recognize technical works in any case–have added to the problem by assigning the works vague and misleading titles, by attributing them to anyone whose name appears in the text, and by ascribing dates based on the slightest scraps of evidence. The situation, in short, is highly confused and badly in need of further work. I have tried to construct this list in such a manner as to reduce the confusion to its absolute minimum.

The following list is arranged alphabetically by city; all spellings of German place-names are given in the original, i.e. Wien, not Vienna, and Köln, not Cologne. Multiple holdings of a single library are listed in ascending numerical order without regard to subclasses of the collection, e.g. *fol.*, *qu.*, *Ms. lat.*, or *Ms. germ.* Manuscripts containing more than one work are so indicated, and wherever possible the folio numbers of the individual segments are given. Works that appear in these manuscripts but that are far beyond the scope of this work are omitted, e.g. sets of alchemical recipes, treatises on veterinary medicine, and the like.

Titles represent a special problem. As a rule I have eliminated artificial or misleading titles assigned by librarians and have replaced them with brief descriptive titles in English. My titles are intended only for the reader's convenience and have no special status or validity. Exceptions here include Giovanni da Fontana's *Bellicorum instrumentorum liber* and the *Feuerwerkbuch*; in both cases the title has become conventional. The designation *Feuerwerkbuch* is used here in a restricted sense to mean that treatise on gunnery and pyrotechnics which first appeared c. 1420 and was first printed in 1529. There are numerous other *Feuerwerkbücher* (usually later) whose relationship to the original is yet to be determined. When such works are included, they are designated with descriptive titles in English.

I have sought to be extremely cautious with respect to attributions of authorship, dates and interrelationships between manuscripts. Authorship has been indicated specifically only when the author's identity is unquestionably clear and when I have confirmed this matter through my own observations. False or dubious attributions of authorship are indicated in the notes. The dates of these texts are in many cases very difficult to determine. When a work in the list is given a specific date (or a precise and narrow range of dates) this means either that the work is explicitly dated in the manuscript or that it can be dated beyond reasonable doubt by internal or watermark evidence. In cases where such evidence is lacking or where the necessary labor of pursuing to its conclusion the available evidence has not been performed, I have given only the most probable range of dates with very broad latitude (40–50 years in some cases). The relationships between various texts remain highly problematical at this time. In the indication of contents or in the notes I have sought to give only the most obvious connections or resemblances; no attempt to establish manuscript traditions has been made and none is implied. For the reader's convenience, an index of the major text types, e.g. *Bellifortis*, *Feuerwerkbuch*, is given on pages 210–212.

Manuscripts from the sixteenth and later centuries present their own special difficulties. Even a superficial inspection of some of the major texts shows resemblances to the works of the fifteenth century. Were the later texts well-studied, it would be possible to indicate the extent of their dependence on earlier works. Unfortunately, they have not been, and thus this problem must remain for the future. To avoid propagating misinformation, I have restricted notice of later works to those which I have studied in some degree of depth and which reveal explicit instances of borrowing from fifteenth-century works. I suspect that the full extent of this borrowing, once thoroughly investigated, will turn out to be substantially larger than indicated here.

Finally, by way of notational conventions, I have tried faithfully to follow the custom of the holding library with regard to manuscript signatures. *Ms., Hs., Cod., fol., qu.* etc. are all used here in conformity with the system of the library holding the manuscript in question. Only in instances in which the library's custom is unclear to me have I reverted to the standardized Anglo-Latin *Ms.* The personal names appearing in the notes refer the reader to additional works which touch on the manuscript listed. In general, I have omitted references to esoteric, obsolete or unavailable secondary works, as well as those that only touch tangentially on the manuscript. The following bibliography refers to the manuscript list only.

BARACK — Barack, Karl August. *Die Handschriften der fürstlich-fürstenbergischen Hofbibliothek zu Donaueschingen.* Tübingen: 1865.

CAROLSFELD — Schnorr von Carolsfeld, Franz. *Katalog der Handschriften der königlichen offentlichen Bibliothek zu Dresden.* 4 Vols. Leipzig: 1882–1906.

CSAPODI — Csapodi, Csaba. »Ein ›Bellifortis-Fragment‹ von Budapest,« *Gutenberg-Jahrbuch 1974.* Mainz: 1974, pp. 18–28.

DEGERING — Degering, Hermann. *Kurzes Verzeichnis der Germanischen Handschriften der Preussischen Staatsbibliothek.* 3 Vols. Leipzig: 1925.

EIS — Eis, Gerhard. *Mittelalterliche Fachliteratur.* Stuttgart: 1962.

GILLE — Gille, Bertrand. *Engineers of the Renaissance.* Cambridge, Mass.: 1966.

GINSBURGER — Ginsburger, M. »Les Juifs et l'art militaire au moyen âge,« *Revue des Etudes Juives,* LXXXVIII (1929), pp. 156–66.

GROSZ and THOMAS — Grosz, August and Thomas, Bruno. *Katalog der Waffensammlung in der neuen Burg.* Wien: 1936.

HAAGE — Haage, Bernard. »Zum Kunstbuch des Hans Hassenwein in Hermannstadt,« *Sudhoffs Archiv,* LI (1967), pp. 268–69.

HAGENMEYER — Hagenmeyer, Christa. »Kriegswissenschaftliche Texte des ausgehenden 15. Jahrhunderts: Schermers Basteibau, Wagenburgordnung, Feuerwerksrezepte,« *Leuvense Bijdragen: Tijdschrift voor Moderne Filologie,* LVI (1967), pp. 169–97.

HALL — Hall, A. Rupert. »The Military Inventions of Guido da Vigevano,« *Actes du VIIe congres internationale d'histoire des sciences: 1956.* Florence: 1958, pp. 966–69.

HASSENSTEIN — Hassenstein, Wilhelm. *Das Feuerwerkbuch von 1420.* München: n. d. (1941–43).

JÄHNS — Jähns, Max. *Geschichte der Kriegswissenschaften, vornehmlich in Deutschland.* Vol. I. München and Leipzig: 1889.

MARX — Marx, E. »Bericht über ein Dokument mittelalterlicher Technik,« *Beiträge zur Geschichte der Technik und Industrie,* XVI (1926), pp. 317–21.

MENHARDT — Menhardt, Hermann. *Verzeichnis der altdeutschen literarischen Handschriften der Österreichischen Nationalbibliothek.* 3 Vols. Berlin: 1960–61.

NEWALD — Newald, R. Article: »Hartlieb, Johannes,« in Stammler, Wolfgang (ed.). *Die Deutsche Literatur des Mittelalters: Verfasserlexikon.* Vol. II. Berlin: 1933, col. 195.

QUARG	Quarg, Götz (ed.). *Conrad Kyeser aus Eichstätt, Bellifortis.* 2 Vols. Düsseldorf: 1967.
QUARG, *TG*	Quarg, Götz. »Der *Bellifortis* von Conrad Kyeser aus Eichstätt, 1405,« *Technikgeschichte,* XXXII (1965), pp. 293–324.
RETI	Reti, Ladislao. »Review of Quarg, Götz (ed.), *Conrad Kyeser aus Eichstätt, Bellifortis,« Technikgeschichte,* XL (1973), pp. 77–82.
SARTON, III	Sarton, George. *Introduction to the History of Science:* Vol. III, *Science and Learning in the Fourteenth Century.* Baltimore: 1948.
SCHILLING-SWARZENSKI	Schilling, Rosy (Swarzenski, Georg, ed.). *Die illuminierten Handschriften und Einzelminiaturen des Mittelalters und der Renaissance im Frankfurter Besitz.* Frankfurt: 1929.
STEINSCHNEIDER	Steinschneider, Moritz. »Jüdische Literatur und Jüdisch-Deutsch,« *Serapeum,* XXV (No. 3; 15 February, 1864), pp. 33–46 and (No. 6; 31 March, 1864), pp. 81–103.
STREIDL	Streidl, Hans. »Geschichte der Hebraica-Sammlung der Bayerischen Staatsbibliothek,« in Franke, Herbert (ed.). *Orientalisches aus Münchner Bibliotheken und Sammlungen.* Wiesbaden: 1957, pp. 1–35.
TODERICU	Todericu, Doru. *Preistoria rachetei moderne: Manuscrisul de la Sibiu (1400–1569).* Bucharest: 1969.
–	»Raketentechnik im 16. Jahrhundert: Bemerkungen zu einer in Sibiu (Hermannstadt) vorhandenen Handschrift des Conrad Haas,« *Technikgeschichte,* XXXIV (1967), pp. 97–114.
UNTERKIRCHNER	Unterkirchner, Franz. *Inventar der illuminierten Handschriften, Inkunabeln und Frühdrucke der Österreichischen Nationalbibliothek.* Part 1. Wien: 1957.
WEGNER	Wegner, Hans. *Beschreibendes Verzeichnis der Deutschen Bilder-Handschriften des späten Mittelalters in der Heidelberger Universitäts-Bibliothek.* Leipzig: 1927.

Signature	Contents	Notes
Augsburg, Stadtarchiv Ms. Schätze 137f.	Cannon and ordnance inventory (unillustrated). 1463.	Eis, 21 Jähns, 414
Besançon, Bibliothèque Municipale Ms. 1360	Kyeser, *Bellifortis* (illustrations only; lacks text) Probably mid 15th century.	Quarg, II, xxx Gille, item 14
Berlin, Staatsbibliothek Preussischer Kulturbesitz Ms. germ. fol. 94	1. Vegetius in German translation; printed edition of 1529 fols. 1r–90v. 2. *Feuerwerkbuch*; printed edition of 1529 fols. 91r–101v. 3. Johann the Elder, Count of Nassau, Collection of texts and illustrations from 15th century German works on technology fols. 103r–251r. Last item probably second half 16th century.	1. Illustrations drawn from Kyeser and Valturio. 3. Illustrations drawn from »Primitive« types, Kyeser, Valturio. Hassenstein, 85 Degering, I, 11 Gille, item 26 (errs in citing Ms. germ. qu. 94)
Ms. germ. qu. 128	1. Anon., *Von der Büchsenmeisterei und vom Kriegsfeuerwerch* (fragment) fols. 2r–17v. 2. Anon., *Rüst- und Feuerwerchbuch* (illustrated) fols. 18rff. Both probably first half 17th century.	2. Contains *Feuerwerkbuch* fols. 22rff. Hassenstein, 85 Degering, II, 21
Ms. germ. fol. 132	Anon., Illustrations and descriptions of tools, weapons, torture devices, animal traps, and various household devices and recipes. 1505. Nürnberger provenance.	Drawings highly detailed showing sophisticated grasp of techniques. Degering, II, 22
Ms. germ. qu. 621	1. *Feuerwerkbuch* fols. 1r–48r. 2. Drawings from *Bellifortis* fols. 49r–60v. Entire codex 15th century.	Degering, II, 112 (attributes to Hans Schulte; cf. Ms. germ. qu. 867) Hassenstein, 85

Signature	Contents	Notes
Ms. germ. qu. 867	1. Hans Schulte, »Kunst pussenpulver to maken« (i.e. *Feuerwerkbuch* in Low German) fols. 1r–41r.	Hassenstein, 85 Jähns, 394
	2. *Feuerwerkbuch* (several hands; Upper German) fols. 48r–96v.	Degering, II, 150 (cf. Ms. germ. qu. 621)
	Entire codex 15th century.	
Ms. germ. qu. 1018	1. Anon., Dialogue between »Feuerwerker« and »Büchsenmeister« fols. 12r–26v.	Hassenstein, 85 (errs in attributing to Jacob Pinchwangler, whose name appears in bill-of-sale fol. 76r)
	2. *Feuerwerkbuch* fols. 32rff.	
	3. Drawings of machinery, mainly hoists fols. 88vff.	Degering, II, 169
	Entire codex probably second half 15th century, Provenance Memmingen.	
Ms. germ. fol. 1117	*Feuerwerkbuch* fols. 294r–322r only. 1463. Provenance Swabian.	Hassenstein, 85 Degering, I, 155
Ms. germ. qu. 1187	1. Johann Hartlieb, *Onomantia* (name magic) fols. 1r–13r.	Degering, II, 206 Hassenstein, 85 Jähns, 393
	2. *Feuerwerkbuch* fols. 13v–45v.	
	3. »Achilles Thabor« (Marcus Greacus), *Liber ignium* fols. 45vff.	
	4. Directions and recipes for military devices, fols. 69r–71r.	
	Entire codex 15th century. Bavarian provenance.	

Signature	Contents	Notes
Ms. germ. qu. 1188	Anon., Work on military techniques and gunpowder; incipit: »Ein Buch durch einen gelehrten Kriegsverstendigen zusammen gezogen ...« (Contains portions of *Feuerwerkbuch* fols. 209 ff.). Probably first half 17th century.	Many works with this or similar incipit from 16th and 17th century contain portions of *Feuerwerkbuch*.
Budapest, Bibliotheca Academiae Scientiarum Hungaricae. Latin cod. fol. 14	Kyeser, *Bellifortis* (fragment; 8 fols.). Probably 15th century.	Csapodi Reti
Chantilly, Musée Condé Ms. 348 (633)	Kyeser, *Bellifortis* Probably mid-15th century.	Quarg, II, xxx Gille, item 11
Colmar, Bibliothèque de la Ville Ms. 491	Kyeser, *Bellifortis* Date: c. 1473.	Quarg, II, xxxi Gille, item 7
Donaueschingen, Fürstliche Fürstenbergische Hofbibliothek Ms. 860	Kyeser, *Bellifortis* Probably mid-15th century.	Quarg, II, xxx Gille, item 8 Barach
Dresden, Sächsische Landesbibliothek Hs. App. 463	*Feuerwerkbuch* (unillustrated), Date: 1458.	Hassenstein, 85 (errs in dating 1468) Carolsfeld
Hs. C 113	Anon., Untitled illustrated treatise on military techniques. (Contains portions of *Feuerwerkbuch* fols. 3 ff.). Probably first half 16th century.	Hassenstein, 88 Carolsfeld

Signature	Contents	Notes
Erlangen, Universitätsbibliothek Ms. 1390	Ludwig von Eybe zum Hartenstein (the Younger), *Kriegsbuch*. Large (305 fols.) richly illustrated encyclopedic collection drawn from German 15th century works on technology and warfare (contains drawings from Kyeser, Valturio *et al.*). Done for Pfalzgraf Philipp bei Rhein. Date: c. 1500.	Quarg, II, xxxi Gille, item 24 Jähns, 272 ff. Sarton, III, 1564
Frankfurt, Stadt- und Universitätsbibliothek Ms. germ. qu. 14 (formerly Ms. II. 40 and/or Ms. Ausstellung 48)	1. *Feuerwerkbuch* (illustrated) fols. 20v–42r. 2. Work on cannons, (illustrated) fols. 44r–68v. 3. »Rüst boch« (illustrated) fols. 69r–128r. 4. Work on instruments (illustrated) fols. 133r–143v. 5. Work on decorative fireworks and surveying (illustrated). fols. 145r–161r. Entire codex probably 1490–1510.	1. Some drawings descended from »Primitives«. 2. Some drawings from »Formschneider« tradition. 3. Drawings from Kyeser, Valturio and unidentified sources. 4. and 5. Illustrations of undetermined source. Hassenstein, 88 Gille, item 25 Schilling-Swarzenski, 229
Ms. ger. qu. 15	Kyeser, *Bellifortis* (illustrations only, no text). Date: 1460–80. Provenance: Alsatian.	Schilling-Swarzenski, 210.
Freiburg i. B., Universitätsbibliothek Hs. 362	*Feuerwerkbuch* fols. 73r–89r only. Dates c. 1445.	Hassenstein, 88 Newald (errs in attributing to Hartlieb, Johannes).
Göttingen, Universitätsbibliothek Cod. phil. 63	Kyeser, *Bellifortis* (1405). Earliest complete deluxe parchment copy.	Quarg (edition based mainly on this Ms. Reproduces in full in facsimile).

Signature	Contents	Notes
Cod. phil. 64	1. Kyeser, *Bellifortis*. Crude, *Urtext* version. 1402. 2. *Feuerwerkbuch* fols. 94r–146v.	1. Quarg, II, xxvff. Gille, item 19 (errs in calling German text) 2. Appears somewhat later than Kyeser.
Heidelberg, Universitätsbibliothek Cod. palat. germ. 126	Philibs Monch, Untitled work on cannon and military devices. Illustrated. 1496.	Sarton III, 1554 Jähns, 271 Wegner Gille, item 34
Cod. palat. germ. 130	Ulrich Bessnitzer zu Landshut, Work on cannon, incipit: »Zumerken der gezewg mit seiner zugehöringe.« Illustrated. Probably c. 1485.	Sarton III, 1553 Gille, item 33 Jähns, 412–13 Wegner
Cod. palat. germ. 502	*Feuerwerkbuch* fols. 1–45 only. Probably end 15th century.	Wegner
Cod. palat. germ. 562	1. Hans Schermer, *Ueber Basteibau* fols. 1–6r. 2. *Feuerwerkbuch* fols. 6v–12v. 3. Magical recipes and formulae fols. 12vff.	Jähns, 431 Hagenmeyer
Cod. palat. germ. 787	1. *Feuerwerkbuch* fols. 2r–27r. 2. Fragment on use of quadrant in artillery fols. 27v–33v. 3. *Feuerwerkbuch* (dialect differs from 1). fols. 34r–45r. 4. Kyeser, *Bellifortis* fols. 46r–106r. Crude, *Urtext* version (cf. Göttingen Cod. phil. 64) 1402.	1. Probably c. 1430 Hassenstein, 88 4. Quarg, II, xxvff. Gille, item 27
Innsbruck, Bibliothek des Tiroler Landesmuseums – Ferdinandeum Co. 16.0.7.	Kyeser, *Bellifortis* (but highly garbled and mixed with additional, unidentified material). Provenance Ingolstadt, c. 1435.	Quarg, TG, (Q. omits from Ms. list in edition) Gille, item 6

Signature	Contents	Notes
Karlsruhe, Badische Landesbibliothek Ms. Durlach 11 (formerly Ms. Durlach 241)	Kyeser, *Bellifortis* (small amount of extraneous material). c. 1410.	Quarg, II, xxx (dates near end 15th century) Gille, item 9 Jähns, 257
Kassel, Murhardsche Bibliothek der Stadt Kassel und Landesbibliothek Ms. math. qu. 14	*Feuerwerkbuch* (date and provenance unknown).	Hassenstein, 88 (dates this c. 1500)
Köln, Stadtarchiv Ms. 232 (formerly Ms. W fol. 232)	Kyeser, *Bellifortis* (Incipit: »Dieses ist ein buxen buch und hat gemachet augustinus dachssberger von münchen ...«). 1443.	Quarg, II, xxx Gille, item 10 Jähns, 259–60
Leipzig, Universitätsbibliothek Ms. 1597	*Feuerwerkbuch*. Probably first half 15th century.	Hassenstein, 88 Jähns, 393–94
London, British Museum Ms. Additional 34113	Anon., Collection of drawings, mainly technological, done by anonymous Sienese after Francesco di Giorgio *et al.* Probably c. 1480.	
Department of Prints and Drawings Ms. 197, 621 (formerly Harleian Ms. 3281)	Francesco di Giorgio, *Opuscula de architectura.* c. 1472.	
München, Bayerische Staatsbibliothek Cod. icon. 222	Bartholomeus Freysleben, *Zeugbuch* (i.e. pictorial inventory of cannon and firearms). Done for Maximilian I. c. 1495–1505.	
Cod. hebr. 235	Anon., *Collectiana hebraica ad machinis aliaque curiosa pertinentia.* Contains *interalia*: 1. Drawings from Kyeser and Valturio, fols. 4–67, 89–90, 112, 114. 2. Portions of *Feuerwerkbuch* fols. 92 v ff. Probably late 15th century.	NB. Language Yiddish, not Hebrew. Steinschneider Streidl Ginsburger

Signature	Contents	Notes
Cod. icon. 242	Giovanni da Fontana, *Bellicorum instrumentorum liber.* Codex ca. 1420–30.	Text largely written in code. Sarton, III, 1551 Gille, item 40
Cod. germ. 356	1. *Feuerwerkbuch*, pp. 1–24. 2. Treatise on artillery and military technology. Source unidentified. Illustrated with drawings from Kyeser, Valturio and other unidentified sources. 1492–1512.	1. Hassenstein, 86 Gille, item 28
Cod. germ. 399	*Feuerwerkbuch* fols. 1r–48v only. Probably early 1480's.	
Cod. germ. 599	1. Drawings of cannon and war machinery from late 15th century fols. 1–47. Source unidentified. 2. Recipes for pyrotechnic compounds fols. 48–62. 3. Miscellaneous technical drawings fols. 63–65. 4. Martin Mertz, *Kunst aus Büchsen zu schiessen.* Entire codex probably 1470's.	2. Hassenstein, 86 (lists as as *Feuerwerkbuch*) 4. Gille, item 32 4. Sarton, III, 1553.
Cod. germ. 600	»Primitive« text. I.e. short picture manuscript of saltpeter and gunpowder preparation with descriptions. Probably 1410–25.	Gille, item 2 (G. coined term »Primitives« on assumption this type antedates *Bellifortis*) Jähns, 229–36 (dates c. 1350) Hassenstein, 86.
Cod. germ. 734	1. *Feuerwerkbuch* fols. 1r–59v. 2. Drawings and descriptions of cannon and war engines fols. 60v–71r (fol. 60v signed »Johann Formschneider, büchsen meister und gutter aben teurer«).	Hassenstein, 86 Gille, item 29 Jähns, 411 2. Formschneider active in Nürnberg 1460–70

Signature	Contents	Notes
Cod. germ. 734 (cont.)	3. Notes and drawings of pyrotechnic devices in Latin and German with some cipher fols. 73r–81r (fol. 78r »Nota ego Martinus Mertz ...«). 4. Drawings of military devices from Italian sources, mainly Valturio fols. 82r–151v. Entire codex probably 1460's–1480's.	3. Cf. Cod. germ. mon. 599
Cod. lat. 960	Anon., Drawings of lifting apparatus and siege engines fols. 1–9 only. Source unidentified. Probably c. 1481.	
Cod. germ. 973	Christof Seselschreiber, Untitled collection of technical notes and drawings, many after 15th century originals. 1524.	Hassenstein, 86
Cod. germ. 4902	*Feuerwerkbuch*, copied by Conrad Kauder von Schongau. Earliest dated Ms. Incomplete. 1429.	Hassenstein, 86
München, Bibliothek des Deutschen Museums Ms. 1949/258	Anon., Untitled collection of drawings and descriptions of cannon and siege engines. Fragment. Probably 1470's or early 1480's.	NB. Very similar to Formschneider segment of Cod. germ. mon. 734.
New York, New York Public Library Ms. Spencer 58	Kyeser, *Bellifortis*. Probably mid-15th century.	NB. Both Latin and German texts.
Ms. Spencer 104	1. *Feuerwerkbuch* fols. 1r–55r (56r). 2. *Bellifortis* fols. 1r–91r (note separate foliation). 3. Drawings in different (later?) style of cranes, pile drivers, cannon and siege engines fols. 92v–110r. Entire codex probably mid-15th century.	

Signature	Contents	Notes
Nürnberg, Germanisches Nationalmuseum Hs. 719 (formerly Kr. 300)	Anon., Untitled collection of drawings and descriptions of cannon and siege engines. Fragment (?). Probably second half 15th century.	NB. Very similar to Formschneider segment of Cod. germ. mon. 734 and Deutsches Museum Ms. 1949/258. Hassenstein, 87 Jähns, 392
Hs. 1480	*Feuerwerkbuch.* 1452.	
Hs. 1481 a	1. »Marcus Graecus,« *Liber ignium* fols. 1–12. 2. *Feuerwerkbuch* fols. 12–25. 3. *Feuerwerkbuch* (variant version in different dialect) fols. 25 ff. Probably early 15th century.	Hassenstein, 87 Jähns, 390
Hs. 25, 801	Anon., Drawings and descriptions of cannon and siege engines. Fragment (21 fols.). 1400–1410.	Gille, item 1 (G. considers as one of »Primitives«) Jähns, 259
Paris, Bibliothèque Nationale Ms. latin 11015	Guido de Vigevano, *Texaurus regis Francie acquisicionis terre sancte* fols. 32–55. c. 1335.	Hall
Ms franc. 20006 (formerly Ms. St. Germain franc. 1914)	Anon., Incipit: »Ce livre est un ramas de vielles machines dont de servis les anciens qui ne sont plus en usage aujour d'huy ...« Drawings derived from 15th and 16th century sources, including Kyeser (?) Valturio, and Ramelli. Probably second half 17th century.	NB. Some drawings apparently derived from »Formschneider« type, i.e. Cod. germ. mon. 734 or Deutsches Museum, Ms. 1949/258.
Sibiu (Rumania) Municipal Archives Ms. Varia II 374	1. *Feuerwerkbuch* fols. 1r–36v Hans Hassenwein von Landshut, copyist. 1459, but purporting to be from original dated 1400.	Todericu Haage

Signature	Contents	Notes
Ms. Varia II 374 (cont.) Sibiu (Rumania) Municipal Archives	2. A mixture of drawings and descriptions from various authors on military devices and methods. Includes some chapters from Vegetius. Drawings derived from Taccola, Valturio and German sources. Fols. 37r–183v. Probably from 1460's or early 1470's. 3. Conrad Haas, Treatise on pyrotechnics and rocketry fols. 184r–391v. Begun 1529. Completed 1569.	2. Some drawings very similar to »Formschneider« type, i.e. Cod. germ. mon. 734 and Deutsches Museum, Ms. 1949/258.
Strasbourg, Bibliothèque Nationale et Universitaire Ms. 2259	1. *Feuerwerkbuch* fols. 1r–18v. 2. Kyeser, *Bellifortis* fols. 18v–132v. Dates *post* 1420; probably first half 15th century.	Quarg, II, xxx Gille, item 17
Weimar, Nationale Forschungs- und Gedenkstätten der klassischen Deutschen Literatur Ms. fol. 328	Anon., »Das Ingenieur Kunst- und Wunderbuch« a large (326 fols.) parchment manuscript containing drawings taken from all known technical authors active in Germany in 15th century. A deluxe encyclopedic collection, virtually without text. Was once thought to have belonged to Skanderbeg, the Albanian general. Includes material drawn from Clm 197, I. Probably dates from very early 16th century.	Quarg, II, xxxi Jähns, 274–75 Sarton, III, 1552 (errs in claiming that folios after fol. 98 depict only »scenes from court life«) Marx
Ms. qu. 342	1. Anon., Drawings and descriptions of cannon and siege engines. fols. 1–47. 2. *Feuerwerkbuch* fols. 55–86. Probably mid-15th century.	1. Similar to »Formschneider« type i.e. Cod. germ. mon. 734 and Deutsches Museum, Ms. 1949/258. Hassenstein, 88 (attributes to Hans Henntz of Nürnberg, whose name appears on flyleaf, but written in 16th century hand) Gille, item 30 Jähns, 263–64

Signature	Contents	Notes
Wien, Oesterreichische Nationalbibliothek Cod. 2952	*Feuerwerkbuch* (illustrations drawn from *Bellifortis*) fols. 3r–95v only. Dates: c. 1450–70. Provenance Swabian.	Menhardt Quarg, II, xxx Gille, item 18 Hassenstein, 87 Jähns, 321–23
Cod. 2987	*Feuerwerkbuch*. 1468–69. Provenance Bavarian.	Menhardt Hassenstein, 87 Jähns, 394
Cod. 3062	1. *Feuerwerkbuch* fols. 1–25v. 1437. 2. Johann Hartlieb, *Onomantia* (name magic) fols. 29r–39r. 3. Kyeser, *Bellifortis* fols. 41v–240r. Entire codex 1435–41.	1. Explicit: »Anno domine tragesimo septimo per Johannem Wienensis scripsit.« Menhardt Quarg, II, xxx Gille, item 15 Hassenstein, 87 Jähns, 261
Cod. 3064	Anon., *Büchsenbuch* (contains segments of *Feuerwerkbuch* together with other material of unidentified source). Dates c. 1430–45. Provenance Bavarian-Austrian.	Menhardt Jähns, 394
Cod. 3068 (formerly Ambras 230)	Kyeser, *Bellifortis*. Probably 1415–41.	Menhardt Quarg, II, xxx (Q. considers this closely related to *Urtext* and to date 1402–05) Gille, item 20 Hassenstein, 87 (errs in attributing to J. Hartlieb)

Signature	Contents	Notes
Cod. 3069 (formerly Ambras 231)	Anon., Drawings without descriptions in »Primitive« style of gunpowder, cannon and siege engines. Some drawings based on *Bellifortis*. (Fols. 92r–94v contain drawings in later style). Date: 1411.	NB. Attributed to Johann Hartlieb, but dates prior to H.'s period of known activity (1431–c. 1474) Menhardt Gille, item 4 Jähns, 261, 371
Cod. 5278 (formerly Ambras 228)	Kyeser, *Bellifortis* (with some additional material of minor importance). Date: c. 1400–1428.	Menhardt Gille, item 16 Quarg, II, xxv ff. Jähns, 256
Cod. 5518	Kyeser, *Bellifortis*. Probably c. 1420–50.	NB. Excluding fols. 1r–6v, which include drawings in later style. Similar drawings found in Weimar Ms. fol. 328 (»Wunderbuch«). Quarg, II, xxx Gille, item 12
Cod. 10,855	Anon., Incipit: »Ein furtrefflich künst buech aus vielen haimlichen alten geschribnen Buechern ...« Contains elements of the *Feuerwerkbuch*; other material of unidentified source. Date: 1514–c. 1540.	Unterkirchner
Cod. 10,895	Franz Helm, Treatise on military techniques. Contains elements of *Feuerwerkbuch* (fols. 5r ff.). Date: c. 1550. South German provenance.	Hassenstein, 87 Jähns, 394, 608 ff. Unterkirchner
Cod. 10,917	Franz Helm, Treatise on military techniques. Contains elements of *Feuerwerkbuch* (fols. 2r ff.), Probably second half 16th century. South German provenance.	Unterkirchner

Signature	Contents	Notes
Cod. 10,953	Franz Helm, Treatise on military techniques. Contains elements of *Feuerwerkbuch* (fols. 281 ff.). Date: 1585. South German provenance.	Unterkirchner
Cod. Ser. nova 12,769	Anon., Treatise on gunnery and pyrotechnics. Contains elements of *Feuerwerkbuch* (fols. 115 ff.). Date: 1600. South German provenance.	Unterkirchner
Wien, Waffensammlung des Kunsthistorischen Museums Ms. P 5014 (formerly Ambras 52).	Anon., Drawings without descriptions in »Primitive« style of military and other technology. A few drawings based on *Bellifortis*. Contains pictorial dedication to Emperor Sigismund. Date: c. 1437.	Gille, item 3 Grosz and Thomas
Ms. P 5135 (formerly Ambras 51)	Anon., Treatise on cannon gunpowder, and siege engines. Illustrated. Portions based on *Feuerwerkbuch*, also *Bellifortis*. Dates: c. 1410–c. 1430.	Jähns, 382–87 Sarton, III, 1551 Hassenstein, 87 Grosz and Thomas
Ms. 5342 A (formerly Ambras 50).	Kyeser, *Bellifortis*. Probably dates from 1430's.	Quarg, II, xxx (Q. thinks derived from ÖNB, Cod. 5278). Gille, item 13 Grosz and Thomas
Ms. P 6562 (formerly Ambras 49)	Kyeser, *Bellifortis* (fragment) Dates: c. 1430–40.	NB. Includes a few non-*Bellifortis* drawings. Jähns, 258 Grosz and Thomas
Zürich, Zentralbibliothek Ms. Rh. hist. 33 b	Anon., Drawings with brief description in »Primitive« style of military and other technology. A few drawings based on *Bellifortis*. Probably 1420's.	NB. Very closely related to Wien, Waffensammlung, Ms. P 5014.

PART II

THE MANUSCRIPT

TRANSLATORS NOTE

In transcribing and translating author A's words I have tried to follow as closely as possible the principles set forth by A. Hübner in *Deutsche Texte des Mittelalters*, XXVIII (1934). Since the reader has a facsimile of the original available to him, no attempt has been made to reproduce the text in diplomatic transcription. Rather, I have made every attempt to make the German text as legible as possible for the reader who is familiar only with modern German, without however altering too seriously the words of the original and never changing their sense. This presents some peculiar difficulties, since only one additional text is available for collation or comparison, since the commentary is short and fragmented and since the writer shows little regard for consistency of spelling or even grammar. I have not tried to standardize A's spellings where two or more clear and distinct forms of the same word appear. Spelling has been standardized in instances in which A's hand is ambiguous and there seem to be no differences in meaning between two equally possible forms, e.g. *sal* and *sol*.

By way of convention, the following rules are observed throughout. Vowel *v* and *j* are converted to their modern forms, *u* and *i*, e.g. *und* not *vnd* and *Item* not *Jtem*. (*Item* is omitted in translation.) Abbreviations are expanded without indication, except in such doubtful cases as dealt with in the notes. Punctuation and capitalization of initial letters of sentences and proper names are added in accordance with modern usage. Hard *z* is indicated by A's *cz* and will be so transcribed throughout, e.g., *czwai, czwg* and *uncz*; soft *z* is usually indicated by a simple *z* in A's hand and is so transcribed throughout, e.g. *daz*. Author A uses *i* and *y* interchangeably throughout, and his spelling has been followed in transcription, e.g. *schiff* and *schyff*. Double consonants are transcribed as in the text, even when inconsistencies exist, e.g. *puchsen* and *puchssenn*. *W* is used to indicate both the consonant and the vowel sounds, e.g. *czwg (Zug)* and *czwo (zwei)*. It should be noted in this connection that author A shows some tendency to write vowel *w* as *uu*, whereas the consonant *w* is usually *w*; cf. e.g., fol. 2v *czw (zu)* and *czwen (zwei)*. However, his vowel *w* is apparently interchangeable with *u*; cf. fol. 1v *czwg* and fol. 2v *czug*. For this reason, it seems preferable not to distinguish between vowel and consonant *w* in transcription. The original text generally follows modern patterns of word separation. In the interests of legibility, I have corrected the few instances in which this is not the case, e.g. *cze machen* not *czemachen* and *inwending* not *in wen ding*. Such emendation will always be noted. In the apparatus, readings from Weimar, Hs. fol. 328 are indicated thus: W. Finally, for ease of reference to the facsimile, the lines of text in transcription will always correspond precisely to the lineation of the original.

With regard to translation, I have tried throughout to render A's prose into a form of English that should be comprehensible to the average reader of modern English who has not had a technical education. I have modified A's words as little as possible in translation, generally only by rearranging them adding or deleting parts of speech and punctuating them in accordance with the demands of English grammar. The translation thus tries to strike a balance between the awkwardness of a literal rendering and the granting of a fluency to A's prose in translation that it lacks in the original. The guiding principle throughout has been clarity, not literary elegance. The reader should note that there are two blank folios in the manuscript, fols. 14v and 26r. Finally, I would like to express my appreciation once again to Professors Arthur B. Groos, Jr., Stephen P. Schwartz and Wilma Iggers for their helpful criticism and generous assistance in preparing this work.

SECTION A

Folio 1r[1]

COMMENTARY

A construction crane: it is typical of all such cranes in the manuscript. A large vertical mast has a short cross-bar braced near its top. Two square frames are joined to the mast near its bottom. They are offset slightly to allow the windlass axle to pass behind the mast. Two wheels with what appear to be holes for handles are fitted to the axles by means of spokes.[2] The pulley system is very simple, indicating that the crane is to be used for light duty only. Iron is used in this machine for the anchor hook, the suspended block with its hook and reinforcement collars where sheaves are built into the mast. The sheaves themselves are wooden. There are no bearings on which this particular crane can pivot, but the overall design—the grouping of all elements on one mast—indicates that author A intended for it to do so. Other cranes in the manuscript are shown with bearings in which to pivot, e.g., fol. 8v.

[1] Beck, 271, fig. 307, Gille, Singer II, 656, fig. 600 reproduce this folio.

[2] Spoked wheels used in this way are unusual. Normally spokes are inserted directly into the windlass axle to form X-handles. Having short handles on wheels, such as A shows here, would increase the moment of force slightly, but the stress on the wheels would be considerable. Nevertheless, author A depicts this arrangement in most of his construction cranes.

Folio 1v

TRANSCRIPTION AND TRANSLATION

Das ist der czwg der von Munchen,
Der gehort czw der puchssenn

This is the windlass like those in Munich.
It belongs with the cannons.

Folio 1v[1]

COMMENTARY

A cannon hoist: it is designed for extremely heavy duty. The two ancillary masts and the main, bifurcated mast are joined at the top. Power is supplied by a windlass mounted permanently between the short legs of the main mast. Note that X-handles and ratchets are used on the windlass. The block-and-tackle consists of two discrete sets of ropes and pulleys linked in series. This will give considerable lifting power to the hoist, but it limits how high a cannon can be raised. Iron is used extensively in this hoist: reinforcement collars, hooks, sheave block, windlass bearings and ratchets are all made of this material. Note that two of the pulleys are built into the main mast. A's perspective is such that one cannot be certain whether the ratchet pawls are properly drawn or not.

Author A's identification of this hoist with Munich may indicate that he had seen one like it at work there. The same block-and-tackle system appears in several manuscripts of south German or Austrian provenance.[2]

[1] Berthelot, 439, fig. 1; Beck, 272, fig. 309 reproduce this folio.

[2] Cgm 734, fol. 61 v (by Johann Formschneider); Sibiu, Municipal Archive, Ms. II Varia, 374 (cf. Todericu, *PRM*, 174, fig. 76); Vienna, Waffensammlung, Ms. P 5014, fol. 42 v; Erlangen, Ms. 1390, fol. 105 v. Seselschreiber, (Cgm 973), fol. 68 r includes it, but he might have copied it from A.

Folio 2r[1]

COMMENTARY

A cannon hoist: it is lighter and more portable than the one on fol. 1 v. Two masts are used, one of which is shaped like an inverted Y. The windlass is mounted between the short legs of this mast. The block-and-tackle system in this hoist is conventional, with all pulleys mounted in iron blocks. Iron collars are used for reinforcement, but less extensively than in the former example. The overall impression is that this hoist could be dismantled and transported quite easily.

[1] Berthelot, 441, fig. 2; Beck, 272, fig. 308; Canestrini, 252, fig. XC' reproduce this folio.

Folio 2v

TRANSCRIPTION AND TRANSLATION

Item: Daz ist ein czug mit czwain kasten, und der ainer auff der ander ab. Czw Nurenberg[1] *auf*[2] *dem graben den treyben czwen pfard.*[3]

This[a] is a windlass with two containers; one goes up, the other down. Two horses drive it in Nuremberg at the moat.

[1] Cf. fol. 15 r *supra Nuernberg*. W: *Nuernperg*. [2] Text: *auch*. [3] W: *ain ross*.

[a] Author A uses the word *Item* frequently. It serves the same grammatical function in German as in Latin, i. e., as an introductory device, particularly in technical or legal works. One could translate it into English as »next« or »also,« but it will be omitted in these translations for the sake of simplicity.

Folio 2v[1]

COMMENTARY

The »double-acting crane« of Nuremberg: the drawing shows two slewing cranes powered by a horse-gin. The crane consists of two vertical masts which pivot in a common frame. Between them is a vertical winding drum around which a continuous rope passes. The drum is turned by a capstan beneath it (both handles of which project toward the viewer in this drawing). The illustration does not correspond with the description in all details. There is insufficient room between the masts for the capstan's handles, to say nothing of the »two horses« that should turn it. Moreover, the simple pulleys and masts do not appear strong enough to bear loads heavy enough to require two horses. Also, horses would have to be unhitched and turned around to reverse the crane's motion.[2] In redrawing this folio for Weimar, Hs. fol. 328, fol. 138v, the copyist corrected some of these details. There only one horse is specified, and the space between the masts is made larger. It is possible that such a crane was in use in Nuremberg in the 1470's and 80's, but I can find no evidence to confirm this supposition.[3] The drawing indicates that iron is used only for caps on the winding drum where the handles enter on the bottom, for hooks and pivots. The rest is made entirely of wood, including the pulleys.

[1] Berthelot, 444, fig. 3; Beck, 272–73, fig. 310; Gille, »Manuscrit,« 80, fig. 1 reproduce this folio.

[2] Taccola shows a crane with a reversing gear driven by a donkey to eliminate this problem (Prager and Scaglia, 71–73 and fig. 28). A similar crane may have been used in the cupola's construction in Florence. Possibly the Nuremberg crane is a crude imitation of the Florentine machines, cf. Erlangen, Ms. 1390, fol. 112v for a similar double acting crane with reversing gear.

[3] Double acting construction machinery appears in later texts, often in elaborate forms. See e.g. Weimar, Hs. fol. 328, fol. 2r, 4v–5r, and 6r.

Folio 3r

COMMENTARY

An earth-moving machine: like the crane on fol. 2v, it uses the double-acting principle. The windlass in the frame at the bottom carries a continuous rope which passes over two pulleys mounted in the separate frame at the top. Two containers are fastened to the rope's ends. This is the simplest earth-mover seen in the manuscript and it was probably a common type in A's day. The coloration reveals that no iron parts are used in this machine; it is made entirely of wood.

Folio 3v

TRANSCRIPTION AND TRANSLATION

Daz ist ein czug doman
turne mit abtregt oder

ander paw abtregt. Get ein truch auff dy ander ab.

That is a windlass with which one razes towers or other buildings; one trough goes up, the other down.

Folio 3v[1]

COMMENTARY

A rubble-removing machine: it uses a fan-escapement and a rope brake. As author A states in the text, this machine is designed to remove material from a building being demolished. Thus it does not lift material but instead must lower it safely to the ground. It uses two containers. As the loaded one descends it raises the empty one to be filled. The entire structure resembles a tree rising from a single trunk. At the base of the trunk we see an iron pivot bearing on which the machine rotates so that the empty container always faces the precise spot being demolished. (The board seen above the pivot is not the machine's base, but a stabilizing bar; the base itself is not drawn.)

The mechanical heart of the machine is the brake and escapement system visible at the top. Note that the rope connecting the two containers passes over (or is looped around) a ribbed axle or a lantern pinion at the apex of its path. Moving the containers rotates this axle. A large spur gear on this axle engages a lantern pinion on another axle directly above the first. On one end of this axle is a fan whose blades are set perpendicular to the direction of rotation so that they will act to slow the movement of the axle. On the other end is an hourglass-shaped ribbed drum. A second rope passes over this drum and both its ends are fixed to a set of weights hanging below. Finally, a lever which runs behind the machine (and is mounted on a pivot there) presses against the ribbings of that same drum. Ropes hang from each end of the lever.

When a filled container begins its descent, it rotates the axles at the top. The fan then acts as an escapement to slow its fall. The weighted rope looped over the ribbed drum acts as a second brake to further reduce the speed of the descent. The feather-duster shaped object hanging from the cubical weight may be a frame with removable small weights to adjust braking pressure. If a complete halt is desired, the operator pulls on one of the ropes hanging from the diagonal pole, forcing the opposite end against the ribbing of the drum and halting the machinery. To release this he pulls on the other rope.[2] Finally, when the loaded container nears ground level, the machine is halted, and the structure is rotated to present the elevated empty container to the demolition crew. The filled container is emptied at ground level while the empty one is being filled above. Then the whole process is repeated.

[1] Beck, 273, fig. 312; Feldhaus, *Technik*, 520–21 and *TdAdm*, 348, fig. 370; Uccelli, 24, fig. 61 reproduce this folio.

[2] Beck, 274 maintains the opposite, i.e., that the pole served to lift the weighted rope and to speed up the process. The drawing is unclear on this point and Beck's version has parallels in the action of windmill brakes

at a later date. However, we should realize that, given the small braking surface offered by the ribbed drum, the weight must have been quite heavy to have been effectual, and extremely heavy to have stopped the descent of the container altogether. I question whether it would have been desirable to have released the brake or whether a man could have lifted the brake-weight with the leverage the drawing shows. Moreover the ribbing on the drum would have enabled the pole to act as a pawl engaging the ribs to stop the descent, whereas Beck's explanation ignores the ribbing completely. Feldhaus ignores the brake in his accounts.

Folio 4r

COMMENTARY

A crane: note that the windlass and the mast are not structurally connected. This example is quite simple and would function well as drawn. Author A intended that the reader should visualize the crane's mast mounted on a building under construction and the windlass on the ground beneath it. Such a crane would be easier to use than one with a built-in windlass under certain conditions. Cranes of this sort appear in medieval illuminations depicting construction scenes.[1]

[1] Gille, Singer II, fig. 580 shows a thirteenth-century illustration of such a crane at work. For a fifteenth-century example similar to this drawing, cf. Toussaint, 37.

Folio 4v

COMMENTARY

A construction crane: it is virtually identical to the one on fol. 1r. Its pulley system is slightly more complex and offers greater mechanical advantage. Author A has designed the windlass with the same wheels and the same sort of frame as shown on fol. 1r. In this case, however, the windlass axle penetrates through the mast rather than being offset to pass behind it. Iron is used for the suspended block with its hooks and for reinforcement collars. Here again no bearings are provided on which this crane might pivot, but the general structure indicates that it was intended to slew.

Folio 5r *supra*

TRANSCRIPTION AND TRANSLATION

Item: Daz hayssent dy dumler dy man in trucken graben wurfft,
und dy halczer dan anschlecht,[1] *und ein prugk*
daraus[2] *macht, daz man uber den graben get.*

These are called »dumlers«[a] which one throws into dry ditches, then attaches the wooden pieces[b] and makes a bridge out of them, so that he can cross the ditches.

[1] Text *an schlecht*; W: *holtz dar an schlecht.*

[2] Text *dar aus*; W: *dar auff.*

[a] From the drawing it is clear that what author A calls *dy dumler* function as what is known in English as »trestles.« In both Middle High and Middle Low German, however, a *dummeler* (or *tummeler*) normally means a siege machine that hurls stones, i.e., a trebuchet or catapult. Cf. Grimm, II,1515–16 and Schiller and Lübben, Vol. IV,629. Since trebuchets were used to hurl material to fill in a moat (cf. Chapter IV), Author A may have generalized the meaning of *dummeler* to include any object used to traverse a moat or ditch. The word later took on other connotations and was used to describe any person, animal or object that reeled about or went head-over-heels. Cf. English »tumbler.« Perhaps author A had this sort of meaning in mind.

[b] I.e., the planks shown above the trestles in the drawing.

Folio 5r *supra*[1]

COMMENTARY

A portable bridge: the drawing shows two trestles and two beams with holes in each end. The device is to serve as a means of crossing small, dry gullies. As the text explains, the trestles are put in the gully and the beams fitted atop them. The doughnut-shaped iron rings at each apex of each trestle seem to be for reinforcement as well as protection against bumps or jolts. Similar rings are seen in triangular frames in the *Bellifortis*, but neither author A nor Kyeser explains their function.[2] In spite of A's remark about »dried-out« ditches, these rings may have been made large deliberately in order to keep the trestles from floating in a small stream.

[1] Berthelot, 444, fig. 4 reproduces this folio.

[2] Quarg, *Kyeser*, I: 110v–111r; II: 84.

Folio 5r *infra*

TRANSCRIPTION AND TRANSLATION

Daz ist ein czug daz czwen man ein puchsen auff ainen wagen heben dy sex centen[1] *swar ist*[2]

This is a windlass with which two men lift a cannon weighing six hundred pounds onto a wagon.

[1] *centner* Berthelot, 444; Canestrini, 252. This would be the most frequently-used form, but *centen* is a normal variant; cf. Schmeller, II,1139.

[2] Followed by slash mark in text; 5r *infra* written with a much broader pen-point, and is considerably darker than 5r *supra*. The handwriting is the same, however.

FOLIO 5r *infra*[1]

COMMENTARY

A cannon hoist: it is supposed to lift a heavy cannon. The main mast is arched, and pulleys are set into it in tandem. The only conceivable advantage to this design is the greater structural strength of the large main mast. Its arched shape will contribute nothing, since it is incomplete and is supported on one end by a bipod. The rope could scarcely occupy the position shown in the drawing; under loaded conditions it would slip off the mast. The drawing is hastily done and incomplete. The windlass lacks a bearing on its left side and the lower pulley in the suspended block is not round. Iron is used for the hooks and the block as well as for reinforcement collars where the mast bifurcates. Note that the weight specified in the text, about 300 kilograms, indicates that this is a hoist for siege pieces, not for field weapons.

[1] Berthelot, 444, fig. 4 *bis*; Canestrini, 252, fig. XC reproduce this folio.

FOLIO 5v

TRANSCRIPTION AND TRANSLATION

Item: Mit der kunst czwcht[1] *man ein vall prugk auff. Dy tor*[2]
mit dreyen[3] *schlossen verschlossenn ist, da schlecht man*
dye krachen[4] *oben in*[5] *dy pruck, und unten an dy*[6]
ring da pint[7] *czway starckew*[8] *pfar neben*
einander, und ruck dy pruck mit einander
auff. Daz ist pewart. Ein solichs kost ein [halben]
guldenn[9].

With this device one raises a siege-ladder.[a] The gate is secured with three locks,[b] so one forces the hooks into the top of the ladder.[c] Then underneath, at the rings, harness two strong horses next to one another and shove the ladder up together. This has been found efficient; such a one costs a half gulden.

[1] Reduced form of MHD *ziuhe* (= *ziehen*); q.v. Benecke, Müller, Zarncke. Cf. FNHD *zucken* (= »gewaltsam ziehen«) und *aufzucken* (= »in die Höhe reißen«); q.v. Götze.
[2] W: omits.
[3] W: omits.
[4] W: *hacken.*
[5] Not »m̄« *(mit?)*; cf. W: *in*; cf. fol. 15r *infra dexter* »stellt man in.« Author A's pen slipped as he dotted his »i« here.
[6] Text: *andy*; W: *an der.*
[7] Text: *dapint*: W: *sind.*

[8] The *ew* is a reduced form of MHD *iu* endings. Author A employs this form here and on fols. 15r *inf. dex.*, 20v, 21r and 26v *infra*. W: *starke*.
[9] W: *kum ein guldenn*.

[a] The word can also mean »draw-bridge.«
[b] I.e., the gate is securely shut and cannot be breached. »Secured with three locks« sounds proverbial. Grimm, II, 1371 notes the sacred character of the number three and that it carries the connotation of completeness; cf. also the Latin proverb »Omne trinum perfectum est.« The copyist of W omitted the entire phrase.
[c] I.e. one should catch the upper rungs of the ladder with the hooks so firmly that the ladder cannot fall back while being raised.

Folio 5v

COMMENTARY

Three ladders: (from the left) 1. a conventional ladder with a hooked pole to lift it; 2. a single-pole ladder (the »Copenhagen ladder«); 3. a rope ladder with wooden rungs and hooks at its head. The text describes the first ladder. Note the eyelets near the bottom of the pole for harness for the horses. Although the use of hooked poles to lift ladders is common, employing horses in this context is unusual. The idea is feasible, but the horses would have to be well trained to perform their tasks under enemy fire. The second and third ladders are types that appear frequently in contemporary manuscripts. The hooks at the head of each ladder or pole are made of iron.[1]

[1] Weimar, Hs. fol. 328, fol. 136v reproduces this drawing, omitting the Copenhagen and rope ladders.

Folio 6r[1]

COMMENTARY

A cannon hoist: the block-and-tackle arrangement is mechanically misunderstood. The framework is generally similar to that shown on fol. 2r, but it is carelessly drawn and lacks a bearing on the left side of the windlass and fastenings at the top. The block-and-tackle is large and complex. Author A may have wanted to illustrate a six-fold system or two parallel three-fold ones, but in either case he failed. The rope is continuous and winds onto the windlass axle, where both ends are anchored, from opposite directions. There is an extra sheave in the upper block over which the rope passes. Thus, as the windlass is turned the rope will wind out over all the sheaves and back in on the windlass. There will be no contraction of the rope's overall length and the load will not be raised. The coloration indicates that the sheaves, blocks and reinforcement collars are all made of iron.

[1] Berthelot, 445, fig. 5 reproduces this folio.

Folio 6v

COMMENTARY

A construction crane: it is similar to those on fols. 1r and 4v, but uses a different pulley system. Only one wheel is shown on the windlass axle, and the axle penetrates through the single vertical mast. Two ropes pass through a compound pulley system contained in two iron blocks. These ropes wind onto the windlass axle in opposite directions, so that as one winds in the other winds out. Only one rope will exert any lifting force. This misunderstanding parallels that on fol. 6r.

Folio 7r

TRANSCRIPTION AND TRANSLATION

Item: Daz ist ein heber damit man ein hebt
vier gadenn hoch, der da siecht waz man
in einer stat tüe[1]*, Dye da leyt auff einer oben,*
da sicht man waz man darinnen tüt. Daz ist
pewart und gerecht.

This is a lifter with which one lifts someone four stories high, who then sees what is being done in a city. One person supports it,[a] while another one aloft sees what is going on inside. This has been tested in practice and is useful.

[1] Blurred *e* in text.

[a] I.e., the machine is designed so that the observer can be held aloft by someone standing on the ground and holding the rope at the end of the lever. *Auff leyen* is New High German *auflegen,* »to support.« Cf. Grimm, I,683–85.

Folio 7r

COMMENTARY

An observation tower: it is used to look over the walls of a besieged city. A tall mast is fitted with a lever on a pivot near its top. The lever has a hook from which a basket carrying an observer is suspended. The splice joint in the lever indicates that it is longer than the average tree trunk and is constructed of two poles joined together. Author A claims that it can lift the observer »four stories,« but we should understand this to mean at the most twenty to thirty meters, not four modern stories. The device is simple and easily portable.[1]

[1] Thorndike, 16 notes this as »a woman being lifted in a box by a sort of crane.«

Folio 7v

COMMENTARY

Two separate drawings:[1] the upper drawing shows a pulley system and the lower one a pole with two wheels carrying a bucket on a slip ring. The upper drawing is hastily done (the left pulley is entirely missing).
It has an iron suspended block and hook and iron reinforcement collars at the sheave axle bearings. The lower drawing is better. The hooks and slip ring linking the pole and the bucket are made of iron. I suspect that this device was to be used to haul water from a stream to the camps of an army unit.

[1] Thorndike, 16, claims these are related, i.e., »tackle to carry a bucket of water on a pole and two wheels.« There is no reason to assume this.

Folio 8r[1]

COMMENTARY

A pile driver: the drawing is hastily done and poor in its perspective. It shows a pile driver of the type most common in the fifteenth-century texts. The driver-head, the square block on the right of the drawing, is raised by a windlass having two spoked wheels.[2] The connection between the rope and the driver-head is a C-shaped hook. This should be interpreted as a slip hook to release the head to fall freely on the pile below (not shown in this drawing). Author A depicts no release mechanism, but a simple pull-rope would suffice. There are no guides for the driver head; it will fall freely. Iron is used for the pulley at the top, for the slip-hook and for reinforcing bands around the driver-head.

[1] Beck, 281, fig. 329 and van Tyghem, 288, plt. 102 reproduce this folio.
[2] Beck, 280 asserts that the rear wheel on the axle is a flywheel. Both wheels are hastily drawn and lack the proper spokes, but the rear wheel is intended to serve the same function as the front one, to lift the head. It is not a flywheel. Van Tyghem suggests that the drawing might have intended to show treadwheels.

Folio 8v

COMMENTARY

A construction crane: it is clearly a pivoting or slewing crane. The structure is similar to all the other cranes of this type (e.g., fols. 1r, 4v and 6v) with each of the crane's elements grouped on a single mast. In this example, however, the mast protrudes below the windlass frame and is capped by an iron ring. This is incomprehensible except in terms of a slewing crane. The pulley arrangement is simple and correctly drawn. Iron is used for reinforcement collars, the fastening on the cross bar and the block with its hooks.

Folio 9r[1]

COMMENTARY

A screw-lift cannon hoist: the screw remains stationary while the nut is rotated about it. A raised platform on four legs supports the nut, which is shown in the shape of a long rectangle. It is slightly longer than the longest dimension of the platform, and two handles descend from its ends. To operate the hoist two or more men would move with the handles in a circle around the platform, turning the nut and lifting the screw. The color of the platform is darker than the normal wood tone. Since this same color appears on an axle and bracing on fol. 11r, it probably indicates a hard wood or wood subjected to heavy wear. The only iron parts of this machine are the hooks and collar at the lower end of the screw. Interestingly enough, the handles on the nut are wooden.[2] Cf. fol. 37r.

[1] Berthelot, 446, fig. 6; Beck, 274–75, fig. 313; Canestrini, 258, fig. XCII; Kellermann and Treue, 84; Treue, 90 reproduce this folio.
[2] Beck, 274 errs in describing them as iron.

Folio 9v

COMMENTARY

A well-hoist: it is quite simple, but employs the double-acting principle. Motive power is provided by a spoked wheel of the same sort found on author A's cranes. It is attached directly to a ribbed axle. A rope winds around the axle and carries a bucket at each of its ends. The colorless rectangle shown at the right of the drawing is probably meant to be the pillar supporting the axle's forward end, shown here in improper perspective.

Folio 10r

COMMENTARY

A well-hoist: it is more complex than the one on the preceding folio. It also employs the double-acting principle. Power is supplied by a spoked wheel, the spokes of which are shown parallel to the axis of rotation. On the same axle there is a lantern pinion engaging a spur gear on a short axle directly above. The rope carrying the two buckets passes over a drum mounted on the second axle. Author A should have shown the rope looped around the drum several times. Note that the gearing is designed to reduce the effort in lifting a bucket, not to increase the speed. A small handle projects from the masonry support next to the spoked wheel. This may be a latch to hold a filled bucket suspended until it could be emptied, but exactly how it works is not clear. The coloration shows that the end disks of the lantern pinion are made of iron and the spindles are wooden.

Folio 10v

TRANSCRIPTION AND TRANSLATION

Item: Daz ist ain stampff damit man pulver stost,[1] *und dye stempff gent all drey in ain loch,*[2] *ainer auff der ander ab etc.*

This is a stamp with which one crushes powder. The pestles each fit in a hole. One goes up, the other down, etc.

[1] Text: *o* written over an *a*.
[2] *Sic,* not *lach*.

Folio 10v[1]

COMMENTARY

A stamping mill: it is designed for use on gunpowder. The mill consists of three pestles mounted in collar bearings at the top of the mill frame. They are lifted by cams set into an arbor turned by a single crank. The arbor has two flywheels. The mill is badly drawn in several ways. Author A has neglected to include a second set of collar bearings near the bottom of the pestles to hold them straight. He has drawn the two pestles on the right as if they passed through the arbor instead of behind it. Also the text speaks of alternating motion on the pestles, but the cams are set to lift and release all three prestles simultaneously. The ends of the pestles are made of iron or perhaps they are sheathed with it.

[1] Beck, 279, fig. 325; Feldhaus, *Technik*, 915 and 1077, fig. 600; Needham, 394, fig. 619 reproduce this folio. Berthelot, 448, fig. 7 also reproduces it, but with the legend for fol. 17v.

Folio 11r

COMMENTARY

Two separate drawings: the upper one shows an axle with two wheels and the lower one a movable trestle for heavy material. The upper device has wheel spokes attached to the axle in clasp-arm fashion, indicating that it is intended for heavy burdens. The lower device is obviously intended to assist in the movement of large cumbersome pieces of material or equipment. Note that it has wheels at the top and the bottom to make loading and unloading it easier. The deep brown color we saw in fol. 9r appears here again in the axle of the upper drawing and in the cross arms in the lower. This probably indicates heavier, stronger wood. Both machines represent aids in moving heavy items.[1]

[1] Gille, »Manuscrit,« 80, »Il est possible que certaines images répondent à des élévateurs de mines (par example f° 11 et f° 12 r°).« There is no evidence for this assumption.

Folio 11v[1]

COMMENTARY

Two drawings: both show bridges with folding leaves. The upper drawing illustrates a set of panels in pictorial perspective. The iron spikes and the ropes at the end of each folding panel are to secure the bridge to the banks of the stream being spanned. The lower drawing shows a similar bridge in a side elevation. It is constructed of two thick planks linked together by iron bands. Small dowels (shown extending slightly from the right side of the central panel) help latch the leaf in an extended position. The small round protrusions at either end of the same panel are probably bumpers or handles.

[1] Cf. 11 r, note 1. A similar bridge appears in the *Bellifortis*; cf. Quarg, *Kyeser*, I : 61 r; II : 39.

Folio 12r[1]

COMMENTARY

A pontoon bridge: it is supported by inflated leather bladders, shown here as globular projections on either side of the bridge. The bellows indicates the method of inflating the bladders. The central panel between the parallel ropes, the floor of the bridge, is wooden, and the side panels which extend upwards from it are leather. This design is intended to keep the bridge from being flooded if a section of it should momentarily sink below the water level. The ropes steady the bridge and apparently keep its side panels upright.

[1] Cf. 11 r, note 1.

Folio 12v

COMMENTARY

Tools to break into a fortified building: they are hastily drawn and almost unrecognizable. They were copied, however, by Christof Seselschreiber, who named each one.[1]

The uppermost figure shows what Seselschreiber calls a »Preich Schrauf« (»breaking screw«). How it functions is not clear from the drawing, but it probably bored its way through wooden doors. The square protrusion at the right seems to be a nut on which a wrench could fit.

The next figure from the top is a form of crowbar with a spoon-shaped end. Selselschreiber's name for it, »Preich Neiber« (»breaking borer«)[2] indicates that it gouged into a wood or masonry surface.

The third figure shows another crowbar with a forked end. Seselschreiber refers to it as a »Ganzfues zum prechen« (»goose-foot for breaking«). »Goose-foot« is the fifteenth-century

German idiom for crowbar (see fol. 15r). The hole in the square shank can accomodate an extension lever.

The bottom figure is the most difficult to interpret. Seselschreiber calls it a »Prech Berfen« (»breaking handle«).[3] It was apparently intended to be inserted between bars to pry them apart. Seselschreiber's name indicates that it was twisted or turned to accomplish this.

All the drawings on this folio are poorly done; note the stray lines connecting the top three figures. They are also poorly colored. Although one would expect iron as the material for such tools, they are colored here as wood. I suspect that author A's intention was to show them as iron, but they were mistakenly colored wooden. This folio (cf. also fol. 15r) shows only the crudest versions of breaking-and-entering tools. Seselschreiber included more advanced devices on the same folio as his copies of A's designs, and he indicates that he considered A's tools old-fashioned.

[1] Seselschreiber (Cgm 973), fol. 126r.

[2] *Neiber (Neber)* = a borer; cf. Schmeller, I, 1713.

[3] *Berfen* (*Werfen*; also *Werfel* or *Warfel*) = a handle for turning an object; *ibid.*, II, 994.

Folio 13r[1]

COMMENTARY

Four related drawings: two ladders, a flotation collar and weighted boots. The collar occupies the central portion of the folio; the others are ancillary devices. The collar is designed to be buckled about a man's waist. The strap in the center would pass under his crotch, while the inflating tube would extend upward toward his mouth.[2] The boots are probably weighted to keep the swimmer upright.[3] The spikes on the insteps represent buckles and the ring at the heel of the right boot is to assist the swimmer in climbing the ladders seen on the right, which are obviously to help him into or out of the water. The collection of apparatus is intended for someone burdened with heavy weights, perhaps body armor and weapons.[4]

[1] Berthelot, 448, fig. 8; Beck, fig. 330 reproduce this folio.

[2] Seselschreiber (Cgm 973), fol. 139v copies this drawing and admonished his reader to »blow it up hard and buckle it on tightly.«

[3] Berthelot, 448 claims that they are metal, but they are colored yellow brown, indicating tanned leather.

[4] *Ibid.* seems to think this shows a diver's helmet and equipment. Thorndike, 16, refers to this and the following folio as illustrations of types of »costume, such as shoes with spikes on the instep and an armored coat.«

Folio 13v[1]

COMMENTARY

A flotation ring:[2] the design is unclear from this drawing. Weimar, Hs. fol. 328, fol. 134r reproduces this drawing with the note »Das ystrament gehört zu dem pfertt.« It must therefore be a flotation device for horses, probably to be used in conjunction with the flotation collar for humans depicted on fol. 13r. The upper, U-shaped portion of the collar would seem intended to fit around the horse's breast, thus placing the webbed segment under his belly. The two upper straps and the buckles would then reach over the horse's neck and back, while the two lower straps would have buckled around its flanks, fastening beneath the tail. (The Weimar codex elongates the flotation tube considerably in this portion.) Although a device of this sort would doubtless have been somewhat cumbersome, it could have added significantly to a horse's natural buoyancy, allowing him to keep his head above water while bearing a heavily-armored rider. The position of the inflation tube indicates that the rider was to blow up the ring while mounted. Note that author A has omitted one buckle from the left side and one strap from the lower right. Like the collar on fol. 13r, this is made of leather.

[1] Berthelot, 448, fig. 9; Gille, »Manuscrit,« 86, fig. 10 and *Engineers*, 70 reproduce this folio.
[2] Gille, »Manuscrit,« 86 and *Engineers*, 70 believes this is a helmet for the diver illustrated on fol. 14r.

Folio 14r[1]

COMMENTARY

An underwater diver: he is shown completely outfitted and engaged in recovering sunken items. The diver's suit is entirely of leather, including the helmet.[2] The eye pieces are glass. The leather breathing tube is held above the water's surface by means of two small leather bladders.[3] The diver wears sandals which are probably weighted to keep him upright in the water. His work is not warlike, but the recovery of goods that would otherwise be irretrievably lost.
The suit is simple in its construction, but nevertheless practical for shallow depths. Author A should have provided it with some sort of air pump—a simple bellows would have sufficed. It is interesting to note that Seselschreiber's copy of this drawing refers to it as »the old means of raising something from the water.«[4] As we noted in Chapter VI the methods in vogue in his day (1524) were scarcely more advanced.

[1] Berthelot, 448, fig. 10; Beck, 291, fig. 330; Gille, »Manuscrit,« 86, fig. 10 and *Engineers*, 70; Uccelli, 553, fig. 112; Feldhaus, *Technik*, 1119 reproduce this folio.
[2] Not metal as claimed by Feldhaus, *Technik*, 1119.
[3] Not cork floats as claimed by Feldhaus, *TdAdM*, 349 nor a double orifice as Berthelot, 448 claims. Note the similarity between these bladders and those supporting the bridge on fol. 12r.
[4] Seselschreiber, (Cgm 973), fol. 135v.

Folio 15r *supra*

TRANSCRIPTION AND TRANSLATION

Daz ist ein gestell damit man mit roren port,[1] *das haben*
dy von Nuernberg[2] *eins gemacht. Damit port man alltag*

xv roren daz yeglichen[3] *achczehenn*[4] *schuch lanck ist. Von den roren macht man prunen mit, etc.*[5]

This is a device with which one bores through pipes. The people of Nuremberg have made one, with which one bores fifteen pipes daily, each of which is eighteen feet long. From the pipes one constructs wells etc.

[1] Text: *prot.*
[2] W: *regensburg*; cf. Clm 197, I, fol. 2v *Nurenberg.*
[3] Text: *ic* ligatured. Possibly originally *yeglaichen* but altered to reduced form, *yeglichen.*
[4] W: *xx oder xiiiiiiii schuch.*
[5] W: omits *macht man prunen mit.*

Folio 15r *supra*[1]

COMMENTARY

A drill press: it is used to bore out logs to make water pipes. A log rests on the roller, which is turned by the lily-shaped handles. A drill bit rotated by the wheel at the right bores into one end of the log. The log is advanced against the bit by moving it across the roller. The four projections jutting from both sides of the log are handles clasped to it for the workmen to grasp to guide it. The wooden rod extending along the longitudinal axis from the left end of the log serves the same purpose.[2]
As noted above in Chapter VI, there is evidence to link this drawing to the boring machine for muncipal water pipes that was at work in Nuremberg after 1462. The drawing does not clearly indicate how the borer was powered, but it seems likely that it was water-driven. Even though the machine depicted on this folio would have required considerable skill to keep the log steady and advance it against the bit, such a device would have represented a considerable advance over the conventional method of boring logs, i.e. the T-shaped auger illustrated by Agricola.[3] The drill is made entirely of wood except for the bit, the four handles on the log and the end disks of the roller.

[1] Berthelot, 453, fig. 11; Beck, 280, fig. 337; Gille, »Manuscrit,« 84, fig. 7, *Engineers*, 69 and »LdV,« 81, fig. 10; Uccelli, 58, fig. 168; Feldhaus, *Technik*, 123 and *TdAdM*, 349; Gille, Singer II, 655, fig. 599 reproduce this folio.
[2] Gille, *Engineers*, 72 maintains that it represents a second bit boring into the log from the opposite end; but no power source is shown for it and it is colored wooden, not iron. A copy of this folio in Weimar, Hs. fol. 328, fol. 272r shows the wooden rod with a T-handle, indicating that it served as a grip for the workmen.
[3] Hoover (ed.), 177 and accompanying figure.

FOLIO 15r *infra sinister*

TRANSCRIPTION AND TRANSLATION

Das sint gans[1] *fues damit*
man ain gatter aus hebt oder
quader stain oder ein taras.[2]

These are goose feet with which one pries out a heavy gate[a] or building-stones[b] or fixed cannon.[c]

[1] Text; ambiguous; W: *gens.*
[2] Text and W: *tar* (abbreviation).

[a] *Gatter* is used in a military sense to mean a heavy, lattice-work gate in a fortification. Cf. Grimm, IV, 1502–07. Tucher, 210–11 [140a] twice refers to *gatter* in a defensive work around Nuremberg's walls.
[b] *Quader stain* is a possible plural form.
[c] *Taras* was the soldier's nickname for a type of cannon mounted on a fixed frame for use in a defensive fortification or a wagon-fortress. The full form of this word appears in several languages, e.g., Czech »tarasnice;« German »Tarasbüchsen;« and Latin »tarraspixides.« Cf. Durdík, *HH*, 95–98. It is possible that author A meant this as *tara*, a »weight of packed goods.« Berthelot, 453 reads it in this way. But this does not correspond with the verb *aus hebt*. It is more likely that author A meant a *taras*, which would require a lever in order to pry it from its mount.

FOLIO *15r infra sinister*[1]

COMMENTARY

Crowbars: they are used primarily for breaking into fortifications. The bars are made of iron and have forked or slotted ends. The text is revealing regarding the ways in which author A views such tools. He lists three uses for them, all of which are military. They may be used to lift a barred gate from its hinges to force an entry, to dismantle a wall stone by stone or to move a fixed cannon. It is interesting to note that Author A does not think of crowbars in a civilian context. He included them in his work solely because of their military value. Many treatises contain devices for this purpose, and these often used screws for greater mechanical advantage. Such tools evolved ultimately into adjustable wrenches, e.g., in Ramelli's work.[2]

[1] Berthelot, 452, fig. 11 *bis* and Weimar, Hs. fol. 328, fol. 273r reproduce this folio.
[2] Cf. Feldhaus, *Technik*, 990; Ramelli, figs. 155–58; Keller, *Machines*, 86–93 reproduces Ramelli's plates and notes instances where such instruments were used as well as extant artifacts.

FOLIO 15r *infra dexter*

TRANSCRIPTION AND TRANSLATION

Item das gestell das
stellt man in ain[1] *stain*[2]
pruch damit man stain aus
wigt.[3] *So stosset*[4] *man ain stang*
an das lenger eysen, da wigt
ain man hinter mer wann vier
nyder mit.[5] *Nu ist es gar ein*
nwczew[5] *arbayt.*[6]

One puts this frame in a stone quarry in order to break stones loose. When one places a staff on the longer iron piece, then one man behind[a] breaks through[b] more than four others without this machine.[c] Now that is really a useful bit of work.

[1] Text: *a* written over *e*; W: *ainen*.
[2] Text: erased *p* follows.
[3] W: *man in stain auff wigt.*
[4] W: *ain man hinden mer dan iiii in dem mit.*
[5] W: *nitzues*; cf. Clm 197,I fol. 5v Transcription, note 2. Author A uses *ew* ending with *ein* and feminine nouns also on fols. 20v, 21r and 26v.
[6] Text: slash mark follows.

[a] I. e., at the crank(?).
[b] *Mit* is used with *wigt* to mean »breaks through.« Cf. fol. 15r *supra* for similar use of *mit* with *port* and fols. 15r and 16r for use of *mit* with *machen* to mean »fabricate« or »produce.«
[c] This follows the sense of both text and W. Author A's intention is clear: the machine illustrated allows one man to do the work of four. Berthelot, 453, gives a smoother translation, but distorts the meaning of the original. »On met cet appareil dans unce carrière pour soulever des pierres; on attache un levier au fer le plus long et un homme appuie sur la manivelle, à l'arrière. Avec quatre hommes, on fait un très bon travail.«

FOLIO 15r *infra dexter*[1]

COMMENTARY

An apparatus for breaking stones loose in a quarry: the machine and its accompanying text are both unclear. The machine itself is a cranked arbor pierced by two iron wedges. The text, which is difficult to translate, speaks of a machine for use in a quarry, but it is difficult to imagine how the machine illustrated could have been employed. The directions for using the machine state that one should place a longer staff on one iron piece, which would reduce the machine's mechanical advantage. Furthermore author A's claim that one man with this machine can do the work of four is incredible. If the text and drawing are related, then we must assume that A's enthusiasm for cranked motion led him to believe that a crank offered the same sort of mechanical advantage as a lever. In this case the application of a crank has produced an impractical machine.

[1] Berthelot, 453, fig. 11 *ter* reproduces this folio.

Folio 15v

COMMENTARY

A cannon shield: the drawing is defective. The shield is mounted on two upright posts (only one of which appears in the drawing) and it is pivoted to move aside when the cannon it protects is to be fired. The lower frame is made of two longitudinal members with wheels, indicating that the shield is portable. The drawing shows what appear to be wheels mounted into the back and front edges of the shield, as well as a curious right-angle brace underneath. These features seem to have no purpose, but the overall drawing is reasonably clear. Similar shields are commonplace items in the Germanic texts.

Folio 16r

TRANSCRIPTION AND TRANSLATION

Item[1]*: cze machen*[2] *ein bedeckten steyg czeug doman mit wol auff*[3]
und ab steygt an schadenn damit sich ainer bedeckt. Man sol machen
in ein glid mit siben lochern, und drey schuch ains vom andern[4] *sey. Daz ist*
xxi schuch hoch und dy selben palchen[5] *sollen*[6] *all hol sein. Mit dreyen glideren*
da steygt ein man lxiii schuch hoch etc.[7]

To make a covered ladder so that one can climb up and down easily and without harm because one is shielded, one should make a limb with seven holes, each three feet from the other—that is a total of twenty-one feet high. The same beams should all be hollow. With three such limbs one man may climb sixty-three feet high, etc.

[1] Text is inverted relative to the drawing and the rest of the codex.
[2] Text: *czemachen*.
[3] Text: *wolauff*. W: *da mit man wol auff*.
[4] W: *ain 3 schuh von dem andern*.
[5] W: *lechern*.
[6] Text: *sullen*.
[7] W: *mit*.

Folio 16r

COMMENTARY

Four ladders: (from left to right) 1. a Copenhagen ladder in two sections, one of which has a small shield; 2. an articulated conventional ladder; 3. an articulated pole with no rungs; 4. a Nuremberg shears. The text is apparently directed toward the first ladder, but it does not describe well any of the ladders here illustrated. The text uses the word »limb« in the singular, indicating a Copenhagen ladder. The first figure, however, has more rungs than A's text indicates. The reference to a »shielded« ladder must refer to the shield shown on one segment of the first ladder. Unfortunately author A has drawn the shield on the middle segment of this ladder, so it will provide very little protection from enemy fire. Also, A's instruction that all the limbs be made hollow (to reduce their weight) is not carried out in the drawing, which shows solid segments linked by mortise and tenon joints. Given these discrepancies and the fact that the text is written upside-down in a hasty hand, we must assume that it is a very ill-considered addition to the drawing and may describe an entirely different ladder than the one illustrated. The hooks on all these ladders are made of iron.[1]

[1] Weimar, Hs. fol. 328, fol. 135v reproduces this drawing, omitting the Nuremberg shears.

Folio 16v

TRANSCRIPTION AND TRANSLATION

Item: Den schirem hat her Arckinger[1] *vor Sacz gehabt, Da genn*
hundert man wol darunder sicher. Der haspel ist inwen-
ding,[2] *und wan man czw der stat kumt, so czeucht man den*[3]
schirem auff, und schiust und lat in den wider czw gien.
Wint dan den[4] *haspel wider hintter sich, so get der schirem*
wider von stat und dye lewt[5] *stien darhyntter*[6] *an schad.*

Herr Arkinger had this shield before Žatec. A hundred men can easily go safely behind it. The winch is inside[a] and when one arrives at the city, he raises the shield and shoots and then lets it fall shut again. The he turns the winch inside[b] again so that the shield departs from the place and the soldiers stand behind it without harm.

[1] W: *erkinger.*
[2] Text: *in wen ding.*
[3] Text: *den* (stricken) *den.*
[4] Text: *den den;* W: *dan den.*
[5] Text: *dye/lewt.*
[6] Text: *dar hyntter.*

[a] Not »the winch may be moved« as Gille, *Engineers*, 69 maintains, nor »le treuil peut être mis en mouvement« as in Berthelot, 455 and Gille, »Manuscrit,« 81.
[b] Or, »he turns the winch backwards again.«

Folio 16v[1]

COMMENTARY

A siege engine: author A ascribes it to Erkinger of Seinsheim and claims it was used at the siege of Žatec (September, 1421). The drawing itself shows a triangular siege engine, heavily armored and having two iron wheels on an iron axle near its front. In the middle is a cannon resting on an axle with X-handles on either end outside the chassis of the vehicle. The drawing does not resemble what the text describes, a siege engine with a hinged shield, powered by an internal capstan and large enough for a hundred men plus a cannon.[2] Author A's drawing seems to show an ordinary cannon-carrying siege engine of the sort frequently sketched in section B. It is unlikely that this drawing shows any siege engine ever built; instead it represents an attempt to envision a »perfect« siege engine, one that offers an attacking force complete security.[3] Cf. Chapter IV.

[1] Berthelot, 454–55, fig. 12 reproduces this folio.

[2] Weimar, Hs. fol. 328, fol. 138r reproduces this folio with the same discrepancies between text and drawing.

[3] Weimar, Hs. fol. 328, fol. 271r shows a remarkably similar siege engine with the laconic note: »This is a mobile standing bulwark which can be moved wherever one wishes.« *(Item das ist ain zu treibens bolwerck oder ain still stends das dort fornan auff wann man will.)* The appearance of two such similar devices argues against ascribing any uniqueness to the »Erkinger« engine, and thus casts further doubt on A's identification and location.

Folio 17r

COMMENTARY

Three ladders: (from left to right) 1. a Nuremberg shears; 2. a rope ladder with iron hooks and wooden rungs; 3. a double conventional ladder made of interlocking sections. In the last-mentioned example the sections are joined with iron mortise and tenon joints and the outer vertical members each have an extra joint for added stability. The lower two segments have iron handles to aid in lifting the ladder. All the ladders have iron hooks and the third one has iron feet as well.

Folio 17v *supra*

TRANSCRIPTION AND TRANSLATION

Item: Daz ist ein schiff[1] *daz get auff stillen wasseren, und hat*[2] *4 vettig. Da gehoren*
4 man czw, czwen hinten und czwen foren, und daz mag wol xx wappen tragen,
und dy vier man dy daz schiff czychen. Und dy vettigch gen in daz wasser und inwen-

dig[3] *hat ydlich vettich ein wendl den man umbtrybt*[4] *inwendig*[5] *in dem schiff. So mag man faren auff dem wasser ab und czw. Und daz schiff sol verdeckt sein, daz man dy leyt*[6] *nicht geschen müg, und sol fornen ein sthachln*[7] *spicz haben, und an yedlicher seytten ein nebenn spicz und*[8] *ain puchsen. Daz hayst ein streytt schyff doman dye von Katalon*[9] *all andern schiffen obligen.*

This is a ship that travels on quiet waters. It has four pinions[a] and requires four men, two aft and two fore. It could easily carry twenty armed men as well as the four who drive it. The pinions go in the water and inside the ship each one has a crank which is turned from within the ship. In this way one can travel back and forth on the water. The ship should be covered over so that the people inside might not be seen. It should have a steeled point on its bow and secondary points on each side and also a cannon. It is called a warship; the people of Catalonia use this type to prevail over all other ships.

[1] Text: *i* ligatured to *ff*.
[2] The *h* apparently corrected from *g*.
[3] Text: *in wen dig*; cf. fol. 16v *in wen ding*.
[4] Text: *umb treybt*.
[5] Text: *in wendig*.
[6] Cf. fol. 16v *lewt*.
[7] W: *stechlin*.
[8] Missing in text and W.
[9] W: *damit die kathelenn*.

[a] I.e., paddle wheels. The literal translation suggests the role that the names for animals' body-parts play in the development of technical vocabularies.

Folio 17v *supra*[1]

COMMENTARY

A paddle-wheel boat: the paddles are turned by single-throw crankshafts. The boat and its propulsion mechanism are crudely drawn and do not correspond well with the text. The text speaks of a much larger craft, capable of carrying twenty armed men under cover and a cannon and requiring a crew of four to operate it.[2] If the drawing had been done in conformity with the text, however, the crankshaft mechanism would have been hidden by shielding or dwarfed by the sheer size of a boat filled with that many men. Author A's illustration is meant to highlight the propulsion system, which would be unfamiliar to most of his readers. They could easily have envisioned the proper proportions and shielding.
Author A's remark attributing boats of this type to the Catalonians is suspicious; we have no evidence that such vessels were used there or elsewhere in the fifteenth century (see Chapter VI). In all probability author A was repeating tales he had heard about such boats in an attempt to justify the practicality of an unusual type of device.

[1] Berthelot, 457, fig. 13; Beck, 281, fig. 332; Canestrini, 342, fig. CXXVIII; Uccelli, 535, fig. 52; Needham, fig. 632 (opp. p. 412); Horwitz, »Schaufelradantrieb,« 312, fig. 3 reproduce this folio.
[2] The claim by Horwitz that the description fits Valturio's paddle-wheel boat better than A's is not accurate. He ignores A's specifications regarding the rams and cannon and the number of paddle wheels. Weimar, Hs. fol. 328, fol. 137 r, reproduces this folio, but with a modified drawing that corresponds more closely to the text.

FOLIO 17v *infra*

TRANSCRIPTION AND TRANSLATION

Item: Daz sein[1] *stampff damit man korn stampft wan man nit mul gehaben mag. Den sie*[2] *treybt ein man umb an den*[3] *rad. Ains get auff der ander nyder etc.*

These are stamps with which one stamps grain, if one does not have a mill.[a] One man drives it at the wheel. One pestle goes up, the other down, etc.

[1] W: *send.*
[2] Text: *den sin*; W: *die.*
[3] Text: *anden*; W: *an dem.*

[a] I.e., a rotary grain mill with millstones. Author A is always careful to distinguish between a »stamp« and a »mill«.

FOLIO 17v *infra*[1]

COMMENTARY

A grain mill: it crushes rather than grinds grain. Four recumbent pestles are mounted on a beam supported by bearings on either end. The pestles are tripped by cams mounted on an arbor which is turned by a spoked wheel. The mill resembles those used in making gunpowder, but author A specifies that it is to be used for grain in cases where a normal mill is not available, e.g., with an army on the march. The text speaks of alternating motion by the pestles, but here, as on fol. 10v, author A has arranged the cams so that all pestles move simultaneously. Indeed, since they are all mounted through the same beam, they must all move at once. The entire machine is made of wood; the mortars are shown black, perhaps to indicate wear or stone linings. An obvious copy of this folio[2] modifies the pestles to a conventional vertical orientation and provides them with individual suspensions.

[1] Berthelot, 457–58, fig. 13 (with the legend for fol. 10v); Beck, 279, fig. 326; Feldhaus, *Technik*, 1077 reproduce this folio.
[2] Weimar, Hs. fol. 328, fol. 271v.

Folio 18r

TRANSCRIPTION AND TRANSLATION

Item: Wer dy mul machen will,[1] *der sol nemen czwa*[2] *schin teuchel eysen und sol machen dy form*[3] *des eysen als sy da gemalt ist.*[4] *Und sol oben, sein haupt, and untten stechln sein. Dy sullen da das*[5] *eyssen inen geht sol auch stachln sein, und dy mulstain dreyer*[6] *schuch prayt. Und man sol*[7] *sy thychen mit den czwayen czugenn czwyen man. So meld*[8] *sy vertig.*

Whoever wants to make the mill should take two *shin* of iron rod[a] and he should make the iron in the form as it is depicted here. Above, at its head, and beneath it should be steeled, and the supports where the iron goes in[b] should also be steeled. The millstone should be three feet wide and one should run it slowly with two pull-rods and two men;[c] then it mills well.

[1] Text: *i* ligatured to *ll*.
[2] Text: *a* open at top, resembling *u*.
[3] Text: *farm* (?); again *a* open at top.
[4] Text: slash mark.
[5] Text: *da da*.
[6] Text: *r* corrected over *y*.
[7] Text: *sal*.
[8] Text: *e* indistinct, ligatures to *l*.

[a] There are several possibilities for the two words *shin* and *teuchel*. Beck, 275, translates this as »Eisenschienen« (»a small staff or tube of iron«). Author A's use of *czwa* before *shin* indicates that the latter word is a unit of measure. *Shin* can mean the length of a miner's tool (cf. Schmeller, II,425) and Tucher, 97–98 [52a–53a] uses *shin* to mean a length of iron weighing about a quarter of a kilogram. If *shin* is a unit of measurement, then *teuchel* must mean a tube or staff (cf. Schmeller, I,582). Tucher, 97–98 uses the word in this sense. We do not know exactly how long a *shin* was, but it would seem to have been about a half meter.

[b] I.e., bearings, supporting surfaces. Author A could mean either the bearings at the top and bottom of the crankshaft or an iron bushing inserted in the eye of the connecting-rod.

[c] Beck, 275, transcribes *thychen* as *tzychn* and translates it as »ziehen.« *Thychen*, however, is probably a form of *deichen, dicheln, dauken or dichseln*, all of which mean »to go softly or smoothly,« »to slink« (cf. Schmeller, I,482, 583 and 484 respectively). This would fit the context, particularly since fifteenth-century engineers might well have wondered how well a rotary machine would work with a push-pull power input. Author A's *czugenn* must mean »pull-rods« even though the drawing shows only one such rod. The text indicates that if one wants this mill to grind grain as well as a water-mill, one must run it slowly using two connecting-rods and two men to power it. Such an arrangement would make for much smoother operation of the mill than the single connecting-rod shown in the drawing. Beck's translation, »ziehen mit den zwei Zügen je zwei Mann,« does not make sense here.

Folio 18r

COMMENTARY

Four illustrations: 1. (top left) a series of blocks along which two ropes pass through loops; 2. (top right) a crossbow spanner; 3. (bottom left) a stamp (for making gunpowder?); 4. (bottom right) a grain mill operated by a crankshaft and connecting-rod. The text describes the last drawing.

The first figure is inexplicable. It seems to show some system of lifting building stones by means of iron rings fastened to them, but exactly what author A intended to depict is not clear.

The second figure is a crude sketch of a lever-operated crossbow spanner. A stirrup on the front of a crossbow would fit over a hook on the front of the spanner's frame (not shown in this drawing). The hook on the lever just in front of the pivot would engage the bowstring. By pushing down on the long arm of the lever, the bowstring would be drawn back to the trigger. See fol. 29v for a clearer illustration of the same device.

The third figure shows a simple stamp operated like a see-saw. The pestles at either end of the beam are iron. The machine is worked by hand, or by an operator who would stand on the horizontal beam and rock to one side or the other. Weimar, Hs. fol. 328, fol. 270v shows a similar device complete with such an operator. The text to the Weimar drawing indicates its use: »This is a tread stamp with which one stamps corn or grain.« [1]

The fourth figure shows a grain mill turned by a crankshaft and connecting-rod.[2] The upper part of the mill with the millstones and the hopper is conventional, while the rest of the design strongly resembles Taccola's drawings. The text consists mainly of directions for making the transmission. Author A specifies that the crankshaft is to be made of iron and that the upper and lower bearing surfaces (and possibly the eye of the connecting-rod) are to be steeled iron. Author A does not specifically describe the connecting-rod, but the drawing shows that it is made of wood. The text shows A's familarity with transmissions of this type, for it recommends operating the mill with two connecting rods and two workers. This would help to smooth out the mill's motion and to overcome the »dead spot« where a single rod can exert no effective thrust on the crankshaft.

[1] *Item: das ist ain dreit stampff damit man kerner stampt oder getraitt.* Cf. also Clm 197, I, fol. 17v.

[2] Beck, 275, fig. 314; Gille, »Bielle-manivelle,« 44, fig. 2 and Singer II, 653, fig. 596; Keller, »Linkages,« 323, fig. 2 reproduce this figure.

Folio 18v

TRANSCRIPTION AND TRANSLATION

Item: Das ist ain wasser mull da das rad nach der prayt
auff dem wasser leyt,[1] *und sein well pawm stet uber sich,*
und ist auch ein gerechte mul, und sy darff chain
chomp rad.[2] *Ain pabst von Rom der hat sy erdacht.*[3]

This is a water-mill in which the wheel lies on its side on the water and its axle stands over it. It is also a regular mill and it needs no gearing. A pope from Rome devised it.

[1] W: *gett oder leitt.*

[2] W: *kampf rad*; cf. Clm 197, I, fols. 20r *komp rad* and 20v *kompt rad.*

[3] Text: final *t* lost in binding.

Folio 18v[1]

COMMENTARY

A grain mill: it is turned by a horizontal waterwheel. The mill itself is the simplest type known, and needs no gearing of any kind. The shape of the water-wheel in this drawing is not, as some have claimed, an early version of the modern turbine,[2] but instead a »tub-« wheel. This sort was common in the Alpine area in the sixteenth century.[3] Author A's statement that a »pope from Rome devised it,« together with the drawing of a dwelling and a mountain in the background, led Reindl to speculate that this might be a drawing of the monastery of St. Georgenberg in the Austrian Tyrol.[4] This association is most unlikely, however.[5] Author A's »pope from Rome« may reflect the new interest shown by Italian engineers in horizontal water-wheels during the fifteenth century, (see Chapter VII) but it is probably a fiction or a folk-tale. The copy of this device in Weimar, Hs. fol. 328, fol. 272v does not place the mill in any landscape setting.

[1] Beck, 274, fig. 324; Gille, »Manuscrit,« 83, fig. 6; Feldhaus, *TdAdM*, 349; Gleisberg, 31, fig. 7; Arnold, 14, fig. 23; C. Reindl, »Die Entwicklung der Wasserkraftnutzung und der Wasserkraftmaschinen,« *Wasserkraft Jahrbuch* (1924), p. 6, fig. 2; Strauss, 37 reproduce this folio.

[2] Beck, 279 claims it is similar to a spiral-vaned »Besacle mill;« Forbes, Singer II, 595 calls it a kind of Pelton wheel; Feldhaus, *TdAdM*, 349 calls it a »Turbinenrad.«

[3] Cf. Strauss, *passim*, esp. 37 ff.; the other major type of wheel was the »Floder« or flat-blade wheel.

[4] Reindl, 5–6 argues that A's sketch resembles the monastery, where in 1097 a race and mill were granted by Henry IV. He claims that A's reference to the »pope« derives from an (unspecified) visit to the monastery by a pope. There is no evidence for his claim.

[5] Feldhaus, *TdAdM*, 349 claims that he has searched in vain for any mention of such a mill among papal documents. He further claims that the mill at St. Georgenberg was in reality a water-raising wheel.

Folio 19r[1]

COMMENTARY

A windmill: the drawing shows the interior and exterior construction in great detail. The gearing shown is the standard type for a windmill, i.e., a crown gear on the same shaft that carries the sails (the »wind-shaft«) engaging a lantern pinion on the spindle that turns the upper millstone (the »stone spindle«). The sails themselves are also standard. They are non-adjustable frameworks through which canvas panels (not shown in this drawing) would be interlaced. Iron bearings are shown at the ends of both shafts and an iron collar reinforces the windshaft where the sails join it.

This mill, like most windmills of the period, is a post or »bock« type in which the entire mill housing pivots on a post in order to face the wind. Note the long pole extending from the mill housing to the left; this would be used to turn the mill. Our drawing shows the mill resting on what appears to be a platform rather than a post, but this is probably meant to be the floor of the mill, not a support platform. Note that there is no brake of any kind shown in the mill. It could be brought to a halt only by turning it 90° so that the sails were parallel to the wind's direction.

[1] Beck, 278, fig. 321; Gille, »Manuscrit,« 83, fig. 6 and *Engineers*, 69; Feldhaus, *Technik*, 1328; Uccelli, 9, fig. 24; Usher, 174–75, fig. 57; Gleisberg, 50, fig. 16; Arnold, 17, fig. 30 reproduce this folio.

Folio 19v[1]

COMMENTARY

A windmill: only the interior and the sails are shown. This appears to be a post mill, however; note the bearing at the bottom of the structure for a post. The gearing is quite unusual. A lantern pinion on the windshaft engages a crown gear on a vertical shaft. At the bottom of this shaft is another lantern pinion, which engages another crown gear on a horizontal shaft. At the opposite end of this shaft a third crown gear engages a third lantern pinion on the stone spindle. This gearing seems overly complex, but it is designed to drive the lower millstone rather than the upper one, as was common in windmills. Author A seems to have chosen his gears carefully to make the over-all ratio between wind-shaft velocity and millstone velocity the same as it would be for a conventional over-drift windmill. Thus the mill would have the advantages of a windmill but the same milling characteristics as a water-mill (which is always under-drift). The sails are shown dismantled from the shaft. Their inclusion in the drawing is necessary only to identify this positively as a windmill. All the shafts have iron pivot bearings on their ends. Here, as on fol. 19r, no brake of any kind is shown.

[1] Beck, 278, fig. 322; Feldhaus, *Technik*, 1329; Matschoss, 17, fig. 20 (from Beck); Uccelli, 9, fig. 25 reproduce this folio.

Folio 20r

TRANSCRIPTION AND TRANSLATION

Item: cze machen[1] *ein mul. Man sol*
machen ein mul gestell x schuch

lanck und acht schuch weyt (und daz
gestell sal acht schuch wayt sein), und
dy slegel dy sol man auff seinen well
pawm gen[2]. *Mitten an dem well pawm*
da sol ein komp rad[3] *sten vier schuch*
hoch, eng kompt mit lii komen, und
sein fur getrib mit sex tribeln, und sein
steg[4] *und sein mul eysen gestalt als ein*
ander mul eysen. Und sein mul stein, der
obrist, dreyer schuch[5] *prayt und eine hant*
prayt und aines halben schuch[6] *dick. So*
wirt es ein recht vertigen mul dye
da melt als vil als ain halbe pach
mul.
Item: Sein well pawm ist stossen[7]
an ainem gevirten czappfenn
der sol[8]

To make a mill, one should make a mill frame ten feet long and eight feet wide (and the frame should be eight feet wide) and he should mount hammers on their axle. In the middle of the axle there should stand a cog-wheel four feet high closely set about with sixty-two cogs. Its fore-gear[a] should be set with six rods.[b] Its support beam and its spindle should be formed like any other spindle.[c] Its upper millstone should be three feet and one hand wide and a half foot thick. Thus it will be a very complete mill that mills half as much as a water-mill.
Its axle is mounted on a square, tapered block[d] which should ...

[1] Text: *czemachen.*
[2] W: *die schlegel süllenn aussen an sein welpaumb gestossenn an ainen gefuerten zapfen der sol auff seinen welpaum gemittelt sein.*
[3] Cf. fol. 18v *chomp rad* and fol. 20v *kompt rad.*
[4] Text: *e* indistinct resembling *o.*
[5] W: *mül stein ebenn dreyer.*
[6] W: *ains schuch.*
[7] Text: *ist* faint, ligatured to blurred *s* of *stossen.*
[8] Text breaks off here.

[a] I.e., the lantern pinion which transmits power from the crown gear to the stone spindle.
[b] *Tribeln* survives as a dialect word describing a child's game in which two sticks are struck together in such a way as to cause one to fly end-over-end through the air (Schmeller, I, 641–42). A later form, *Trieben* meant the lantern gear of a windmill, i.e., the entire gear of which the author here mentions only a part. The later words *Triebstrecken* or *Triebstocken* describe what is meant here, the individual rods which mesh with the teeth of the crown gear.

[c] I. e., the so-called »stone spindle« or shaft which directly drives the millstones; it was iron in this case.
[d] *Czapfenn* could refer to anything tapered or conical. A »squared *czapfenn*« is difficult to imagine in this context, but probably refers to the upright blocks or beams which support the axle bearings. Woodsmen call the thinner end of a woodblock the *Zapfen* in Bavarian dialect (cf. Schmeller, II, 1142) and this meaning would make sense here, cf. W's reading note 2 *supra*.

Folio 20r[1]

COMMENTARY

A grain-mill: the transmission is mechanically misunderstood. The millstones and hopper are conventional, as is the gearing (a crown gear engaging a lantern pinion). On the right we see the main shaft with four flyweights mounted on it with rigid arms. On one arm is a small iron hook. A connecting-rod (shown parallel to the driveshaft and above it) would fit over this hook and turn the shaft; mechanically this is equivalent to a crankshaft and connecting-rod. Unfortunately, author A has drawn the flyweights on the shaft in such a position that the connecting-rod cannot turn the shaft through a complete revolution. He seems to have recognized his error, for he broke off the text just as he was beginning to describe that part of the mill.[2] Had author A placed the flyweights on the end of the shaft and in front of the bearing at the right, his design would have functioned well. In spite of his failure it is interesting to note that he was attempting to make a mechanism that would combine the action of a crankshaft and connecting-rod with that of a flywheel and to make a single mechanical unit which could be constructed mainly out of wood rather than iron. The text gives the dimensions for the mill and the ratio of the gears. By making the ratio 1:0.096 author A has insured that all the teeth in each gear will wear evenly.

[1] Beck, 276, fig. 316 reproduces this folio.
[2] This is the only example of an incomplete text in the manuscript.

Folio 20v

TRANSCRIPTION AND TRANSLATION

Item: cze machen[1] *ein trett mul. So sol man machen ein mul gestell*
xiii schuch,[2] *und sein trett rad xii*[3] *schuch hoch,*[4] *und sein kompt*
rad[5] *vi schuch und sein iiii schuch prayt.*[6] *Wil man daz es*
als vil moll,[7] *so sal man czwen man daran stelln. Daz ist*
ein vertigew[8] *mul, daraus sich ain yedlicher muller woll verrichtt.*[9]

To make a tread mill, one should make a mill frame thirteen feet high and its tread-wheel twelve feet high and its cog-wheel six feet high; its [lantern gear?] is four feet wide. If one wants to mill a great deal, he should place two men on it. This is a practical mill which any miller could operate well.

[1] Text: *czemachen.*
[2] W: adds *hoch.*
[3] W: *vii.*
[4] W: *brett.*
[5] Cf. fol. 20 r *komp rad* and 18 v *chomp rad.*
[6] W: omits *und sein ... prayt.*
[7] Apparently no ellipsis intended; W: *will man das es vill mall.*
[8] Cf. fols. 5 v, 15 r, 21 r and 26 v; W: *wendig.*
[9] Text: slash mark follows; W: *dar auff waiss sich ain yeglicher müller woll ze verrichtenn.*

Folio 20v[1]

COMMENTARY

A grain mill: it is powered by a treadwheel. The upper part of the mill, with the millstones and hopper, is conventional, as is the gearing (a crown gear engaging a lantern pinion). The prime mover is a treadwheel, shown here being turned by one man who grasps the handles at the top of the frame for support. The mill is functional as drawn, although the treadwheel seems clumsy. The description given in the text is unclear on one critical point when it states, »sein iiii schuch prayt.« What is to be »four feet wide« can only be guessed at.[2] The only significant mechanical element to which this could refer would be the lantern gear on the stone spindle. If we assume that author A meant this gear, then the dimension given (ca. 1.3 meters) is far too large to be practical if taken literally. I suspect that A was trying to indicate that in a treadmill the stone spindle's gear should be larger than would be normal for a water-mill. This would increase the torque on the stone spindle, decrease the speed of the millstones and ease the burden on the treadwheel operator. Author A's remark that two men could be employed to drive the machine if much milling is to be done indicates that he was aware of the problem of operator fatigue in such machines.

[1] Gille, »Manuscrit,« 82, fig. 4; Matschoss, 16, fig. 17 reproduce this folio.
[2] Weimar, Hs. fol. 328, fol. 275 v copies this text and drawing, but omits the entire passage in question.

Folio 21r

TRANSCRIPTION AND TRANSLATION

Item: cze machen[1] *ein mul*[2] *dy da melt als vil als ein wasser mul.*
So sol man machen ein gestell xii schuch prayt[3] *und achter*
weyt, und ein schwanck[4] *rad xii schuch hoch und ein*[5] *schuch dick.*

Und den selben welpawm da sol an sein ein chumps[6] *eysen*
daz czech[7] *ist, daz sal vier schuch lanck sein, und das sal man an*
czyechen mit ainer stangen, uncz daz daz rad an sein lauff chumpt,[8]
so sol man es nicht lassen stil stan, anders ez wirt hart an cze
pringen. Und sein mul werck sol sein aines czwerches[9]
enger gechempt sein wan ein[10] *wasser mul. So hat man ein*[11]
guttew[12] *wasser mul, dy ein guttn*
natturfft melt damit man
bewart ist.[13]

To make a mill that mills as much as a water-mill, one should make a frame twelve feet tall and eight wide and a flywheel twelve feet high and one foot thick. On the same axle there should be a cam-iron which is hard.[a] It should be four feet long. One should pull on it with a rod. Once the wheel is in motion, one should not let it come to a halt; otherwise it will be hard to start up again. Its mill work should be a cross [gear] with more teeth than a water-mill's.[b] Then one has a good water-mill[c] that mills the good necessities [of life][d] with which one is provided.

[1] Text: *czemachen.*
[2] W: *die vor gemalt müll.*
[3] W: *hoch.*
[4] W: *kampf.*
[5] W: *iiii.*
[6] Text: *ch* written over *k.*
[7] W: *ein krumes ziecheisen.*
[8] Text: *m* with descending final stroke.
[9] Text: illegible and partially striken word follows, probably *enger;* W: omits.
[10] Text: *sein;* W: *wan anden.*
[11] Text: *n* final stroke blurred, connected with abbreviation over *gutten.*
[12] Text: cf. fols. 5v, 15r, 20v and 26v; W: *gütu.*
[13] Text: slash mark follows.

[a] I.e., a crankshaft made of hardened metal; cf. Götze, qv. *zech.*
[b] I.e., the mill's gearing should resemble that of a water-mill, but should have more teeth in the lantern pinion to lower the gear ratio. This would make the mill run more slowly and easily; cf. A's comment on fol. 18r regarding running such mills slowly.
[c] I.e., a mill as good as a water mill.
[d] *Natturfft* = *Notdurft* in metaphorical sense.

Folio 21r[1]

COMMENTARY

A grain mill: it is driven by a double-throw crankshaft with two connecting rods. The grindstones, the hopper and the gearing are all conventional. The crankshaft (which cannot function as drawn here) is made of iron; the connecting-rods, however, are wooden. A flywheel is mounted on the main shaft. The dimensions given for the flywheel in the text, about 3.8 meters, seem disproportionately large. Otherwise the mill is well designed. Author A shows a keen appreciation of the mechanical properties of a crankshaft and connecting-rod transmission. He knows that it is difficult to start such a machine from a stand-still and that the gear ratio should be lower than a conventional mill's in order to make it easier to run. Author A's remarks bespeak a practical familiarity with the operation of mills driven in this way. A copy of this folio in Weimar, Hs. fol. 328, fol. 274v repeats the text, but fails to show the crankshaft. I suspect that the copyist did not understand its function.

[1] Gille, »Manuscrit,« 82, fig. 4 and *Engineers*, 71 reproduce this folio.

Folio 21v

TRANSCRIPTION AND TRANSLATION

Daz[1] *ist ain mul dy man*
lewt mit den[2] *sayllen an den*
stangenn.

This is a mill which one operates[a] with the ropes on the shafts.

[1] Text correctly oriented with remainder of codex but figure inverted.
[2] Superscripted.
[a] *lewt* (modern *läut*). In this context, »tolls bells.« The usage is significant for the operation of the mill, which is powered by pulling on the ropes as one would when ringing bells. This can be translated only in the general sense of »run« or »operate.«

Folio 21v[1]

COMMENTARY

A grain mill: it is powered by means of a double-throw crankshaft and two connectings-rods. Mechanically this mill is quite similar to the one on fol. 21r (although here the crankshaft is drawn properly), but it lacks a flywheel. The wooden trestle lying on its side underneath the mill must be imagined in an upright position with the connecting rods extending across the trestle's top. The connecting rods would then be roughly parallel to the ground and the ropes at their ends would hang down like bell ropes. By pulling on the proper rope at the proper time a rocking motion would be imparted to the connecting rods, which the crankshaft would transform into rotary motion. Of course, the motion of the rods themselves would be complex, since they would not only rock up and down, but would also be driven back and forth across the trestle over a distance equal to twice that of the throw in the crankshaft.[2] Such a system would be inefficient (and noisy) by modern standards, but it would work. This is the only way to explain A's allusion to »tolling bells« in his description of how the mill is run. He might have assumed that this arrangement obviated the »dead spot« and the need for a flywheel.[3] It does not, however, for at the top and bottom dead center of the crank's motion, the rods can still exert no effective thrust. When necessary the operators of the mill could overcome this by pulling laterally on the ropes. The remainder of the mill is conventional. The crankshaft and the stone spindle are made of iron, but the connecting rods are wooden. Author A has neglected to show any means of preventing the rods from slipping off the sides of the trestle while moving. There is no direct evidence that mills powered in this way were built, but drawing number 39 in the Jesuit-produced Chinese technological encyclopedia *Chhi Chhi Thu Shuo*, done in 1627 by Fr. Johann Schreck and Wang Chêng, shows a mill driven in the same way.[4] This indicates that interest in mills of this type continued for at least a century and a half after our manuscript.

[1] Beck, 277, fig. 316; Gille, »Bielle-manivelle,« 44–45, fig. 3 and Singer II, 653, fig. 597; Usher, 171, fig. 50; Uccelli, 76, fig. 238; Feldhaus, *Technik*, 593 reproduce this folio.
[2] This explanation follows Beck, 277; Gille, Singer II, 653, caption to fig. 597 claims that the »two connecting rods are apparently supposed to be linked to a pair of treadles ...« This is incorrect.
[3] Gille, »Bielle-manivelle,« 45 states that there is no »dead spot« in this system. The complex motion of the rods obscures the fact that there is one.
[4] Needham, 217.

Folio 22r[1]

COMMENTARY

A grain mill: it is driven by a gin. A set of gears and a millstone is shown in a pit (indicated by the circular brick walls). Above them at ground level is the gin. The gin's main shaft has a large spur gear which engages a lantern pinion on a second vertical shaft. On the same shaft a second spur gear engages another lantern pinion on the stone spindle. Why author A included the intermediate shaft with two gears is unclear but he probably wanted them to regulate the ratio of velocity between the gin and the millstones. The drawing is done in two different perspectives: the gin and gearing are shown in side view, while the pit is seen from above. The stone spindle is made of iron and the gin's main shaft has pivot bearings at its top and bottom which should be iron, but have been mis-colored as wood. The design is simple and perfectly functional.

[1] Beck, 277–78, fig. 318 reproduces this folio.

Folio 22v[1]

COMMENTARY

Two grain mills: both are driven by a single gin. Like the example on fol. 22r, this machine is built in a pit and its gin extends above it at ground level. The general arrangement of the gears and the gin as well as the double perspective are similar to fol. 22r also. Here, however, a large crown gear on the main shaft engages two lantern pinions on intermediate horizontal shafts. These carry power to the separate stone spindles of the two sets of millstones. All shafts other than the main one are made of iron, and the main shaft has iron bearings at either end. The machine is functional as drawn, but since there is no means to disengage one set of millstones, both must run simultaneously. This means that if the mill is to be run at all, there must be sufficient grain for both sets of stones to grind. Otherwise one of them will run dry and overheat while the other is working.

[1] Beck, 278, fig. 319; Usher, 17, fig. 48; Matschoss, 16, fig. 19; Gleisberg, *Triebwerke*, fig. 100 reproduce this folio.

Folio 23r[1]

COMMENTARY

Two grain mills: both are driven by a single vertical water-wheel. The water-wheel shaft carries a large spur gear which engages two lantern pinions mounted on parallel shafts on either side of the spur gear. On the other ends of these shafts crown gears are mounted, and they engage lantern pinions on the stone spindles of the two sets of millstones. The drawing shows the gearing systems and the mills atop one another, although author A's obvious intention is that they should be located on either side of the main shaft. This distortion leads to some mechanical oddities; for example, the stone spindles appear to have bearings set into the main shaft. Author A's conceptual abilities outstripped his drawing skills here. Only the stone spindles are made of iron. All other shafts and gears are wooden, and no iron bearings are shown. Once one makes allowances for the distorted perspective, this machine appears functional. But since no means of disengaging either set of millstones is shown, this mill will encounter the same problem as the example on the preceding folio, i.e., both sets of stones must grind grain whenever the mill is run.

[1] Beck, 278, fig. 323; Matschoss, 16, fig. 18 reproduce this folio.

Folio 23v *supra*

TRANSCRIPTION AND TRANSLATION

Item: Daz ist der Hussen wagenpurgk[1] *daraus*[2] *dy Hussen*
vechten. Dy ist gut und gerechtt.

This is the Hussite wagon-fortress from which the Hussites fight. It is a good and proper device.[a]

[1] Text: *wagenpurgk.*

[2] Text: *dar aus* (= *woraus*), not *darauf* as in Berthelot, 458 and Beck, 283.

[a] Author A uses similar phrases on several fols.; cf. 7r for example. Here it seems to mean »tried and true« and is not the sort of statement he would make about a novel weapon.

FOLIO 23v *supra*[1]

COMMENTARY

A Hussite wagon-fortress *(Wagenburg)*: the drawing is quite crude. Indeed, it would scarcely be recognizable as a wagon-fortress if it were not for the accompanying text. Two wagons are shown linked together; what appears to be a shield on crossed braces protexts the junction.[2] The most important feature to be depicted are the palisades protecting the side of each wagon down to the hubs of the wheels.[3] Other elements known to have been a part of such wagons are lacking here, chiefly the additional shielding running underneath the wagon behind the wheels. I suspect this drawing is based on scanty knowledge or confused descriptions. (Cf. also Chapter V.)

[1] Berthelot, 458, fig. 14; Canestrini, 182, fig. LXI; Gille, »Manuscrit,« 84, fig. 9 reproduce this folio; Beck, 283, considers only the text.
[2] Such shields protected the gaps between every fifth or sixth wagon in formation. Beck and Canestrini claim that this drawing shows a capstan in this location, and that the wagon is internally driven with it. What they assumed was the X-shaped handle of a capstan drawn in incorrect perspective is in fact the bracing of a shield.
[3] The four circles on each plank indicate the fastenings holding the palisade together, not »meurtrières« as Berthelot claims.

FOLIO 23v *infra*

TRANSCRIPTION AND TRANSLATION

Item: Daz ist ein palier mull als dy hochn maister czw
Venedig haben,[1] *darauff man allerlay gestain paliert. Dy bedarff*
iii scheiben: dy erst ist pleyen,[2] *dy ander czynen,*
dy tritt kuppfernn.[3]

This is a polishing mill like that which the grand masters of Venice have, on which one polishes all sorts of stones; it requires three polishing surfaces,[a] the first is lead, the second tin, and the third copper.

[1] Text: *hab*; W: *haben*.
[2] Text: *pleyen* repeated.
[3] Text: *kupfrenn*.
[a] Literally »disks;« the word can also mean a »pulley.«

FOLIO 23v *infra*[1]

COMMENTARY

A stone-polishing mill: it is run by a belt-drive. The mill has an upright, hourglass-shaped drum, the top and bottom edges of which carry the polishing surfaces.[2] They should probably be imagined as covering more of the drum's surface than author A has shown here. The drum has a rope looped about its middle, which passes over two idler pulleys and a sheave at the bottom turned by a crank. The rope should also loop around the cranked sheave. The sketch is crude, but it shows that author A had a fair grasp of the new principle of belt-drives. The drum is smaller than the cranked sheave and will thus turn faster than the sheave; author A should have made the difference in size greater. The text speaks of three polishing surfaces, but A has alloted space for only two in his drawing. The attribution of this machine to Venice, a center of fine lapidary work, is natural, but precisely who the »grand masters« might have been is not clear.[3] Author A presumably meant the master stone polishers of that city.[4]

[1] Beck, 282–83, fig. 323; Berthelot, 458, fig. 14; Gille, »Manuscrit,« 84, fig. 9 and *Engineers*, 71; Feldhaus, *Technik*, 954–55, *DM*, 240 and *Schleifmittel*, 17–18, fig. 6; Uccelli, 65, fig. 202 reproduce this folio.

[2] The claim by Feldhaus (*DM*, 240) and Gille (*Engineers*, 71) that only the machine's transmission is shown is incorrect.

[3] Feldhaus, *Technik*, 955 and *Schleifmittel*, 18 notes that the word »Hohenmeister« often refers to the Teutonic Knights, who did have economic establishments in northern Italy in the fifteenth century. He thinks it unlikely that they were active in the production pf precious stones, however.

[4] Weimar, Hs. fol. 328, fol. 139v copies this folio, but with certain errors in the drawing that suggest the copyist did not understand the device.

Folio 24r[1]

COMMENTARY

A mill: the drawing is unclear, probably incomplete and impossible to interpret fully. Author A's intention was probably to depict a type of water-mill; it is inconceivable that this could represent a windmill. The millstones are driven by a stone spindle to which a lantern pinion is mounted.[2] A spur gear on a separate vertical shaft drives the spindle, and it is turned by a crown gear on horizontal shaft at the mill's base. The gearing will work, but no power source is shown to drive it. The main horizontal shaft extends beyond the mill housing and carries what appears to be another lantern pinion. It terminates in an iron pivot bearing inserted in a strange framework which has no other mechanical connection with the mill. This could not be a set of windmill sails, because it does not resemble any known type of sail and because all western windmills have their windshafts above the millstones, never beneath them.[3] It is possible that author A may have meant this as some sort of water-mill. Since the drawing is probably incomplete (note that only one pivot bearing is shown) and has no descriptive text, we can only guess at what A meant to illustrate.

[1] Gille, »Manuscrit,« 85, fig. 9 reproduces this folio and claims that it shows a windmill.

[2] The peculiar curve to the left which appears at the bottom of the stone spindle is in fact a separate brace on the corner of the mill housing. The spindle terminates at the horizontal beam in the middle of the housing.

[3] Wailes, Singer II, 624 notes that a rare Ukranian type of windmill does have the wind-shaft mounted below the stones in the lower corner of the mill housing. Mills of this type have never been seen outside the modern Soviet Union, however.

Folio 24v[1]

COMMENTARY

A trebuchet: it is illustrated at the moment after release of the missile. This is the simplest sort of trebuchet. Its counter-weight is an integral part of the main beam. Note that only one rope on the sling is fastened to the beam. The other should have a loop on it which would fit over a hook at the end of the arm (not shown in this drawing). Author A has omitted several other important details, including the spanning ropes and the latching and trigger mechanisms. Only the spanning windlass is shown. The same brown color seen on fols. 9r and 11r appears on the diagonal struts in this drawing, indicating that harder wood is to be used for these members.

[1] This drawing is inverted in the manuscript.

Folio 25r

COMMENTARY

An earth-moving hoist: it would seem to have been difficult to operate in practice. Hoists of this specific type are rare, but both this drawing and a slightly more elaborate version appear in Weimar, Hs. fol. 328, fols. 139r and 12v respectively. The former includes a descriptive text omitted in our manuscript:

> To make a hoist with which one lifts earth from an excavation [so well that] ten men can do the work of fifty. The horizontal arm should be fifty feet long so that it descends twenty-five feet [when depressed]. One should hang a container on the arm [holding] six hundred pounds. With this in place, one can lift a bucket which has meanwhile been filled. Two men do the lifting, each is capable of moving three hundred pounds, which is twelve hundred pounds [sic] total.[1]

Presumably the final figure refers to the total maximum weight involved on both sides of the hoist, i.e. 600 lbs. in the container and 600 lbs. in the counterweight. Even so, one must allow for substantial exaggeration here.

In principle this hoist is workable. It is obviously intended to be placed at an excavation site and to be moved as the work progresses; hence the use of a heavy counterweight rather than a fixed anchor. The design seeks to make use of the pivoted horizontal arm to assist in moving the load to a place convenient for unloading. Assume that the container is unloaded and the counterweight resting on the ground at the start. Then the operators would lower the empty container into the pit where excavation was under way. The bucket, when filled, would first be left standing while the arm was raised to a level position (or slightly elevated on the load side). This would be accomplished by winding in the counterweight windlass while winding out the load windlass. When the arm was in the desired position, the load would be raised by taking in the right windlass. All this presumes that the counterweight would always be heavier than the filled container, since if it were not, the entire machine would run in reverse, lifting only the counterweight. Author A seems also to have wanted his hoist to slew about, for he shows the main mast with a pivot foot. This theoretically possible, but it would require very skilled operators to maintain just the right amount of tension on the ropes during the pivoting. Perhaps realizing this difficulty, the Weimar copyist did not show any slewing capacity in his version. This hoist is an interesting example of the interest shown in the fifteenth century in mechanizing construction operations.[2]

[1] *Item: ze machen ain zug da mit man ertreich auff ainen graben zeuch das allmall x menschen als vill ziehenn als sunst L menschen tuen. So sol man machen die wag L schuch lang so tuet sy funff und zwanntzig schuch dieff. So sol man hinten ann die wag henchen vi zentner gewicht da mit man ain kasten auff ziehen die weil so man gefelt hat und zwen man die ziehen auff so gepirt ainen man iii zentner da send die xii zentner.*
[2] See also Erlangen, Ms. 1390, fols. 112r and 114v for somewhat similar hoists.

Folio 25v

COMMENTARY

A cannon hoist: it is simple, portable and well-designed. Here, as on fol. 2r, the type of construction indicates that the hoist was intended for medium duty, probably with an army in the field. The pulleys are arranged in tandem in the masts and in an iron suspended block. The arrangement is odd by modern standards, but normal for fifteenth-century hoists.[1] Iron is used for reinforcement collars, the windlass bearings and the hook. The sheaves are wooden.

[1] Seselschreiber (Cgm 973), fol. 65r shows a very similar hoist which he might have copied from our manuscript.

Folio 26v *supra*
(drawing 27r *supra*)

TRANSCRIPTION AND TRANSLATION

Item: cze machen ein brugk[1] *da man uber wasser graben get.*
So sal man dy seyllen[2] *dreyer schuch*[3] *prayt machen*
und ein halben hoch oder dick. Und dy palchen dy darauff gehoren
der sol yedlicher ix schuch lanck sein, und hollmen[4] *gehefft*
mitten auff dye seyllen.[5] *Und dy hol und dy*[6] *palchen*[7] *und dy*
seyllen untter gehoren auff[8] *daz wasser. Und an dem end*[9] *von*
der seyllen[10] *stett*[11] *sein hanthab, und wann mans*[12] *lengeren*[13]
wil, so sol man dy glydern in einander[14] *stossenn.*

To make a bridge to cross water-filled ditches. One should make the ropes[a] three feet apart and one half high[b] or thick. And the boards that go on them should each be nine feet long and have holes by which they are fastened to the middle of the ropes.[c] And the holes and the beams and the ropes beneath belong over the water. And at the end of the ropes there is [the bridge's] hand grip, and if one wishes to lengthen it, one should push the limbs together.[d]

1 Text: final *k* incomplete.
2 W: *das scheiben.*
3 Text: *schuch.*
4 Text: *holl men;* W: *ynen holl.*
5 W: *und setzen mitten auff die scheiben.*
6 Text: *den.*
7 W: *und die hoel von den palchenn.*
8 W: *scheiben soelen under sich gehoeren auff.*
9 Text: *ennd(?);* W: *und ain end.*
10 W: *scheiben.*
11 Text: *Stett.*
12 *Sic;* presumably a compression of *man sy* (i.e. the bridge).
13 W: *zuchen.*
14 Text: *ein ander.*

a A's drawing shows no ropes. W redraws the ring-like structures as if they were made of rope, but calls them *scheiben,* »disks« or »rings.«
b This must mean one-half foot.
c Following W here.
d Again, this feature is absent from the drawing. A probably had in mind some device to increase or decrease rope tension. Cf. W: *zuchen,* »draw tight.«

Folio 27r *supra*
(text 26v *supra*)

COMMENTARY

A bridge: the drawing is at considerable variance with the text and may be corrupt. The text seems to describe briefly a temporary suspension bridge whose pathway is slung along ropes. Bridges of this general type were common items in the repertoires of late medieval engineers.[1] Interpreting the drawing in the light of its supposedly descriptive text is very difficult.[2] Author A depicts two upright anchors with vertically-projecting eyebolts. A rigid rod passes through these eyes. Beneath the rod we see three horizontal planks which are mortised into one another to form a pathway. The planks are fastened together and joined to the horizontal rod by two large rings which have vertical projections ending in eyes. The eyes fit around the rod while the large rings pass through the planks. Note that each large ring intersects two planks, on either side of the mortise joint. This arrangement would improve the structural strength of the bridge, while eliminating the need for additional supports. The stark discrepancy between the drawing and the text seems to have confused even the copyist of Weimar, Hs. fol. 328, fol. 270 r, who emended A's *seyllen* to *scheiben* (following the drawing), but who then drew the rings as if they were made of rope. This level of confusion suggests that author A confounded two distinct bridge designs, one of which he described in his text, the other of which he depicted.

[1] Similar examples are found in Guido da Vigevano, fol. 48 r and ÖNB 3062, fols. 220 r and 225 v.

[2] The text could refer equally well to the suspended section of the pontoon bridge shown on fol. 12 r.

FOLIO 26v *infra*
(drawing 27 r *infra*)

TRANSCRIPTION AND TRANSLATION

Item: Cze machen[1] *dyssenn wagen*[2] *daz sol sein halb taÿl sein*
als ein schiff geclamert[3] *und glider auff gien als ein*
drisch,[4] *und als lang und als prayt als ein ander*
wagenn, und in wending[5] *ein druchlein*[6] *nach der leng*
dez wagens. Da ligenn sein steygk[7] *leder in und sein rueder.*
Und wann es in das wasser chumpt so[8] *ist es ein schyff*
und ein pruck, und auff dem lannd so ist es ein wagen
das da durffen czway pfärt czyechen. Und das ist ein
gütt vertigew[9] *chunst*[10] *etc.*

To make this wagon: it should be planked like a ship in its lower half and its sections unfold like a flail. It should be as long and as wide as any other wagon. Inside is a small chest[a] running the length of the wagon wherein its climbing ladder and its oars are kept. When it comes into the water it is a ship and a bridge and on land it is a wagon that two horses can pull. This is a good, useful device.

[1] Text: *czemachen.*

[2] Text: *Wagen.*

[3] Slash mark follows; signs of attempted erasure.

[4] I.e., *drischel,* a hinged threshing flail.

[5] Text: *in weniding.*

[6] W: *drelin.*

[7] W: *stuck.*

[8] Text: *s* blurred.

[9] Cf. fols. 5 v, 15 r, 20 v, and 21 r.

[10] Text: initial *d* converted to capital *C*. W: *ain ware rechten kunst.*

[a] *Druchlein* = a chest or container. Cf. Schmeller, I, 659, s. v. *Trühlein.*

Folio 27r *infra*
(text 26r *infra*)

COMMENTARY

An amphibious wagon-bridge: it can be linked to shore or to another similar wagon to form a bridge. The wagon itself is constructed like the hull of a boat and has two leaves which unfold at either end. The text indicates that the wagon is an »all-purpose« vehicle, since it has oars and can be rowed when on the water. It is unlikely that such a device was ever constructed, but there are several drawings of them from the fifteenth century engineering treatises.[1]

[1] Weimar Hs. fol. 328, fol. 137v reproduces this drawing and text.

Folio 27v

COMMENTARY

A framework: I am uncertain of its exact purpose. The drawing shows five poles, each pierced by a hole near its bottom, atop which a wooden roof stands. Fastened to the poles are two strips of wood running parallel to one another. A reddish-brown material (leather?) is stretched between these strips. The size of the poles and the curvature of the strips indicate that the entire framework curves away from the viewer. Possibly the drawing is intended to represent a portable shelter for use in an encampment. There is no explanatory text and I have seen no other drawing like this in any other manuscript from this period.[1]

[1] Thorndike, 17 calls this a pontoon bridge. I can see no evidence for his view.

Folio 28r[1]

COMMENTARY

Two drawings: 1. a horse carrying a flaming container; 2. a flaming barrel on a cart. Both drawings illustrate the ancient tactic of using animals or carts to carry burning material into enemy ranks.[2] This method remained common in fifteenth-century treatises; cats, dogs and birds were the usual animals.[3] Kyeser, who also illustrates this practice, gives a recipe for the flammable material which includes sulfur, resin and pitch.[4] Material of similar composition is doubtless used here.[5]

[1] Berthelot, 461, fig. 16; Canestrini, 48, fig. V reproduce this folio.
[2] Cf. Judges, 15:3–4, the tale of Sampson burning the fields of the Philistines by means of firebrands attached to foxes.

[3] Taccola (Clm 192,II), fol. 54v [57v] (cf. Canestrini, 50, fig. VI). The practice persisted into the sixteenth century, cf. ÖNB, Sn. 12769 (c. 1600), fol. 62v, Nuremberg, Germanisches Nationalmuseum, Hs. 27722 (late sixteenth century) fol. 102v.

[4] Quarg, *Kyeser*, I: 105v; II: 79.

[5] Not »Greek fire« as Canestrini, 48 maintains.

Folio 28v[1]

COMMENTARY

Three cannon on mounts: each figure shows a different sort of mount. The upper figure shows a cannon on a stationary mount. Here a second platform bears the weight of the cannon itself and rests on the lower platform. It apparently pivots from its front and is adjusted by means of the elevating arc at the rear.

The middle figure shows a different stationary mount with a cannon. Here the platform bearing the cannon's weight is suspended near its midpoint, and again is raised by means of an elevating arc.

The lower figure shows a cannon mounted on a wheeled carriage.[2] Note that, as in the previous figures, a suspended platform bears the weight of the cannon and that it is adjusted exactly as the others are. The elevating arc at the rear does not show holes or pegs to support the tail of the platform, but this is doubtless author A's oversight, not an intentional feature.

Coloration indicates that all parts of the mounts are wooden, while the cannon are made of iron.

[1] Gille, *Engineers*, 68 reproduces the upper and lower figures; Canestrini, 222, fig. LXXVI reproduces the middle one.

[2] Gille, *Engineers*, 68 refers to the lower figure as the first drawing of a cannon on a carriage; this is not correct, although such carriages were somewhat novel in the 1470's.

SECTION B

Folio 29r

COMMENTARY

A fixed tower: the figure in the nest at the top is shown firing an arquebus. The tower is permanently mounted in a masonry ring and can be climbed by means of the rungs fixed in the mast. Author B must have intended that it be part of a permanent fortification. Portable versions of this same type of tower appear frequently in technical treatises of the fifteenth century.

Folio 29v

COMMENTARY

A crossbow spanner: it is mechanically identical to the one sketched on fol. 18r. Here the hook to hold the bow's stirrup and the hook on the strap fixed to the lever are clearly shown. The machine is portable, being mounted on two iron-tired wheels. Iron is also used for the hooks, the reinforcement along the front of the spanner's platform, and the pin on which the lever pivots.

Folio 30r

COMMENTARY

Three illustrations: 1. a crossbow; 2. a belt with a rope loop to span a crossbow; 3. an A-frame, probably intended to serve as an extension for the crossbow to enable it to shoot larger arrows.

The first illustration shows a simple crossbow. The bow is leather covered, but is doubtless made of iron, as is the stirrup visible on its front. The belt with the rope loop would be worn around the bowman's waist. To span the bow he would place his foot into the stirrup and pass the loop around the bowstring, holding it in place with a gloved finger. By straightening up and thrusting down with his leg, he could pull the bowstring back until it caught on the trigger lock.[1] The belt is made of leather; the buckle and rope have not been tinted in this drawing.

The A-frame is difficult to explicate, but a similar one appears in a contemporary manuscript[2] attached to the front of a crossbow. There it serves to support a very large arrow, the butt of which touches the drawn bowstring. In our drawing, the frame, which is made of wood, has two cylindrical or tubular objects attached to the longer arms, while on the lower left a peg appears to fasten one of these tubes to the arms. I suspect that the frame is intended to serve some purpose with respect to crossbows, possibly as a sighting device.

[1] Cf. Quarg, *Kyeser*, I:77r; II:49; Wagner, 53 and Part IV, plt. 8; and Demmin, 902–04 for illustrations of contemporary belts and hooks used for his purpose. Karlsruhe Durlach ll, fol. 72r shows a similar loop arrangement.
[2] Vienna, Waffensammlung, Ms. P 5135, fol. 2r. (I can find no descriptive text relating to this.)

Folio 30v

COMMENTARY

A triangle: it seems to be a surveyor's or mason's tool for laying out a right angle. The triangle is made of wood, and is held together by bolts at each apex. Only those on the lower left in this drawing are properly colored; the rest are tinted to resemble wood. A rope stretched over two pegs is seen parallel to two of the triangle's sides. The illustration shows how a right triangle would be used to lay out a right angle during the construction of a building.

Folio 31r

COMMENTARY

A pictorial view of a castle: there is no mechanical or military device depicted in the illustration. The coloration indicates that the walls of the buildings are made of stone and brick, while the roofs are red tile or blue-colored metal (iron?). This is one of several illustrations of edifices in section B (cf. fols. 31v, 34v and 45r) that show no devices or machines. This probably indicates that author B had some architectural interests. He may have drawn these buildings from life, but I have not been able to identify any of them. Such drawings appear occasionally in technological manuscripts from the period.[1]

[1] Cf., e.g., Cpg 126, fols. 35r and 36r.

Folio 31v[1]

COMMENTARY

Two illustrations: 1. a tower, 2. a man walking tight-rope on a spar. The two drawings are not related. The tower is one of several edifices shown in section B (cf. fols. 31r, 34v ad

45r). The man walking the spar is dressed in a brightly colored red and blue costume and may represent a court jester or acrobat performing tricks. The spar is wooden and rests on a triangular frame at one end. The same scene with only minor background changes appears in Weimar, Hs. fol. 328, fol. 134v.

[1] Thorndike, 17 describes this folio.

Folio 32r[1]

COMMENTARY

A siege tower: it is designed to be used against walled areas. The four story structure is mounted on four wheels[2] and has iron-tipped poles hinged to its top for additional support. The top three stories have short wooden walls to protect the attacking force. The tower rolls forward edge-first to present oblique surfaces to defending fire. The total height of the structure is about twelve to thirteen meters.[3] The wheels are rimmed with iron. No means of propulsion is shown, but it was probably pushed into position by men and animals.
A similar, but not identical tower probably drawn from our manuscript, appears in Weimar, Hs. fol. 328, fol. 134r.

[1] Berthelot, 461, fig. 17; Canestrini, 46, fig. XXV reproduce this folio.
[2] Not six as Berthelot, 461 claims.
[3] Based on comparison with similar towers in contemporary manuscripts; cf. Chapter IV.

Folios 32v–33r[1]

COMMENTARY

A trebuchet: it is shown ready to hurl a dead animal. The trebuchet is powered by a large suspended chest filled with stones. In the sling lies a dead animal, presumably a horse, about to be cast over the walls of a besieged area.[2] This is one of the most carefully drawn and colored illustrations in this section. Author B has, however, neglected to show a trigger mechanism and a release for one rope of the sling. Note that the main beam is not one piece of wood, but it is made of two segments interlocked along the longitudinal axis to prevent the beam from bending. Iron is used extensively for cranks, ratchets, pivot bar, rings, reinforcement collars and chains lining the counterweight to the main arm.
There can be little doubt that this drawing is influenced by one in the *Bellifortis*.[3] It is not a copy of Kyeser's illustration, but it has many general similarities, especially in the proportions of the various parts. This is especially important because Kyeser gives the dimensions of these parts. Assuming that our trebuchet is roughly the same size as Kyeser's, it would be about fifteen meters long by about eight meters wide; the main beam would be

about eighteen meters long (fifteen meters for the longer arm and three for the shorter). The machine would be about seven meters tall. The size of the container for rocks is not given, but one can estimate two or three cubic meters at a minimum. Densely packed with rocks, the counterweight would weigh about two or three metric tons.[4] This weight would be barely adequate to fling the horse and should be regarded as a minimal estimate. It does, however, show quite clearly why iron was used so extensively and why the beam is made of interlocking segments.

[1] Gille, »Manuscrit,« 81–82, fig. 3 reproduces this figure.

[2] *Ibid.*, claims the animal is »sans doute ... un cheval porte-feu.« (Cf. also the caption to fig. 3.) There is no reason to suppose that the beast is filled with »un liquid enflammé.« Thorndike, 17 calls this »a picture of some animal lifting what seems to be a box of cannon balls by what appears to be a tread-mill.«

[3] Quarg, *Kyeser*, I: 30r; II: 24. This drawing is also quite carefully done in the Göttingen manuscript and was included in many of the copies of Kyeser's work.

[4] This is not excessive; in modern experiments with trebuchets a counterweight of about nine metric tons was used; cf. White, 103.

Folio 33v

COMMENTARY

A siege engine: the enclosed cart bears a rampway above it. The lower part is shaped like a pyramid and runs on twelve (eleven?) small wheels. Two poles extend upwards from the center to serve as supports for the rampway; it can be elevated to the desired angle by means of the rope at its rear. Note that a cannon is shown firing through a porthole in the cart's side. When an engine such as this reached the walls being stormed, the soldiers inside would swarm out and use the rampway to mount the wall. Kyeser's terminolgy for such a rampway is *coclear* [sic] (»spoon«) and the cart itself would be a *mus* or a *gattus* (»mouse« or »cat«).[1] Iron is used for reinforcing throughout the engine and for tires on the wheels. The method of propulsion is not shown, but it was probably pushed from within by the soldiers who manned it.

[1] Quarg, *Kyeser*, I: 16v; II: 31 shows the ancestor of this type of engine.

Folio 34r

COMMENTARY

An arrangement of straps with buckles: the straps seem designed to encircle a man's body. The lower two straps would pass around the lower torso and attach to the upper strap. The upper one would encircle the upper torso, while the curved portion would pass over the shoulders and behind the neck. The loop or ball at the end of this strap seems intended to fasten to something not shown in this drawing. Two metal right-angled hooks are visible at the intersections of the vertical and horizontal elements of the two lower straps. These would be on either side of a man's waist when the apparatus was being worn. I suspect that this is a type of girdle to aid in climbing or to help hold a mounted knight on his horse.[1]

[1] A somewhat similar square arrangement of straps is found in ÖNB 3062, fol. 34v. There it is one of several devices to assist in climbing. Karlsruhe, Durlach 11, fol. 16r shows a different arrangement of straps and hooks, and the German legend states that it is to help a man remain fast in the saddle. This drawing is reproduced in Weimar, Hs. fol. 328, fol. 134r, but without any explanatory note.

Folio 34v

COMMENTARY

A tower with two Copenhagen ladders hooked to it: the overall purpose of the drawing is unclear. The tower is of red brick with a crenelated masonry top. It is shown in isolation from any other fortification. The two ladders are each made of a single piece of wood and are equipped with iron hooks and feet. It is difficult to imagine why author B included this drawing, though Weimar, Hs. fol. 328, fol. 134r includes it without comment. Ladders of this type were common at this time (cf. Chapter IV) and their military use would have been clear to any soldier. This is one of several drawings in section B depicting edifices (cf. fols. 31r, 31v and 45r).

Folio 35r

COMMENTARY

A water system: the drawing shows a *noria* feeding water to a fountain through a line of pipes. The *noria* is clearly drawn but badly designed. Its catch-trough is much too low for maximum effectiveness (the position is reminiscent of the Roman *tympanum* or water-raising drum) and the aqueduct leading into the building housing the *noria* has no outflow channel. The pipes are made of wood, but the manner in which they are to be joined is not indicated. The short tubes represent the spigots for the fountain and are also made of wood. In spite of its errors, this drawing does show in a general, schematic way the sort of system many fifteenth-century cities used to supply water to their public fountains.

Folios 35v–36r[1]

COMMENTARY

An earth-hauling railway: it is shown at work in building fortifications. The drawing is complex. At the top there is a brick-colored crenelated wall in front of which the men and machine are working. Between the wall and the men is a trough running almost the width of the scene. It carries water to a square container serving as the counterweight on a horizontal beam. The container is reinforced with iron straps and is attached to the horizontal beam with strong iron chains and rings. On the same end of the beam there is a double rope to be used to raise the opposite end of the beam. The beam is balanced on a vertical post resting on a chassis with six iron-tired wheels. It runs in a frame resting on the ground and it can be moved back and forth by means of a screw turned by an X-handle (visible beneath the counterweight). The beam is made of interlocking wood pieces (like the beam on fols. 32v–33r) and is reinforced by iron bands. On its right end there is an iron ring with iron chains hanging from it which are attached to portable tracks. These tracks are U-shaped to fit the wheels of the cart at the bottom of the drawing and are lined with iron on the surfaces of the U. At the bottom of the track there are handles. Two workers are shown removing earth from a cart at the head of the tracks. The cart rides the tracks up and down the inclined earthwork under construction. It has four iron-tired wheels and is pushed on the tracks by means of the long pole extending behind it. The bend in the tracks in the lower part of the drawing indicates the transition from incline to level at the foot of the fortification. Finally, at the bottom left of the illustration there is a second cart (an afterthought by author B) with a top, an opening in one end and a hinged door in its side. This closed cart is probably an alternate design to keep the load from falling out as the cart ascends the steep incline.

The machine would operate in the following fashion: the cart would be loaded with earth or rubble and pushed up the tracks where the workers unload it and use the material as needed to make an earthwork fortification. The tracks are moved as the work progresses by means of the beam and chassis shown at the top of the drawing. The water flowing into the counterweight balances the weight of the tracks. By pulling on the ropes next to the counterweight, workers could lift the tracks at the top, while others could carry them with the handles on the bottom. By turning the X-handled screw the chassis and the beam could be moved to a new position, where the tracks would be set down and work resumed.[2]

It is difficult to know if any device exactly like this one was ever built. It would seem more reasonable to use the lifting power of the beam as a crane to draw the carts up the track rather than to move the track itself. This illustration may not be an exact drawing of a real machine,[3] but I strongly suspect that some rail-carts were in use at this time (see Chapter III). Similar earth-moving construction machines of better design appear in other treatises.[4] This illustration and the drawing by Formschneider from which it is copied are the earliest known representations of the use of rails to transport materials.

[1] Gille, »Manuscrit,« 80–81, fig. 2 reproduces this folio.

[2] Gille, »Manuscrit,« 80 claims that this is a mining cart running on rails; he does not repeat that claim in *Engineers*. Feldhaus, *TdAdM*, 349 correctly assessed the operation of this machine.

[3] Author B's drawing is merely a copy of one by Johann Formschneider; cf. Cgm 734, fols. 76v–77r and Munich, Deutsches Museum, Ms. 1949/258, fols. 76v–77r.

[4] More elaborate types are illustrated in Weimar, Ms. fol. 328, fols. 3v, 7v–8r. From the seventeenth century Lorini, 220 shows a rail-cart being pulled up an incline by a windlass.

FOLIO 36v[1]

COMMENTARY

A boring device: it is used to widen already-drilled wells. The drill with its wide bit is seen in the center of a tall vertical frame. As it is turned by the horizontal bars in the middle it is forced down by the pressure of a heavy weight mounted directly above the drill. This is stabilized by the horizontal beams extending to the vertical frame on either side. The weight and the bit are both iron. Note that the bit is not spiral-shaped, so it cannot remove earth from the hole it widens, but only loosen it. The tubular opening into which the bit descends is earth-colored, indicating that it is a well, not a cannon or mortar. The drawing would be clearer had author B shown the mouth of the hole being drilled in the same plane as the ground supporting the framework.

[1] Beck, 280, fig. 328; Gille, »Manuscrit,« 84, fig. 8 and *Engineers*, 69; Uccelli, 60, fig. 180; Feldhaus, *Technik*, 120 reproduce this folio.

FOLIO 37r[1]

COMMENTARY

A screw-lift cannon hoist: it functions exactly as the example on fol. 9r does. This version is better drawn and designed. It is equipped with an iron sleeve lining the hole in the platform through which the screw passes, which would reduce the effects of stress at that point. Also, the nut itself has an insert sleeve (of harder wood?) where it joins the screw for the same reason. This lift is portable and can be dismantled. The legs are mortised into the platform and are fastened together with iron hooks and eyes. Iron is also used for the brackets linking the handles with the nut, but the handles themselves are wooden.

[1] Gille, »Manuscrit.« 85, fig. 8 reproduces this folio.

FOLIO 37v[1]

COMMENTARY

A pile driver: the design is a simple one but it is not commonly seen in contemporary manuscripts. A four-footed frame holds a single pulley and a rope passing over the pulley is fixed to the driver's head. The rope has four separate ropes tied to its free end so that a crew of men can easily raise the driver-head. It requires no windlass or release mechanism. (Cf. Chapter III.)

[1] Berthelot, 464, fig. 18 and van Tyghem, 289, plt. 103 reproduce this folio.

Folio 38r[1]

COMMENTARY

A slewing crane: it is more fully developed than those in section A. A complete frame is shown holding a vertical mast topped by a short, braced mast set at an angle near its top. Beneath this the frame holds a vertical winding drum turned by a capstan. The simple pulley system gives no mechanical advantage. The crane is shown with a bucket instead of a hook, indicating that it was intended to lift loose material, e.g., mortar. Note that there is an iron pin bearing in the base of the vertical mast on which it pivots. The drawing of the ropes and pulleys is careless; at the upper and lower pulley the rope is shown fouled (slipped from the sheave) and the rope leading to the bucket is made to pass between two horizontal structural members.

[1] Beck, 273, fig. 311; Gille, Singer II, 656, fig. 600 B reproduce this folio.

Folio 38v[1]

COMMENTARY

A cannon mounted on a pivot firing through a shield: the shield pivots along with the cannon. This arrangement appears awkward and, unless the piece illustrated is of small caliber, would tend to be topped by the force of its own recoil.[2] It also seems incapable of elevation. The cannon itself is a breech-loader. A powder chamber (not shown) would fit into the excised portion of the barrel and would be held in place by means of clamps. See fols. 39v, 43r, 43v, 44v, 46r and 48r for similar breech loaders. Coloration indicates that the cannon and the edging of the shield are made of iron.

[1] Berthelot, 461 and fig. 19 reproduces this folio.
[2] Taccola (Clm 197, II), fol. 38r [33r] shows a similar cannon with shield; his is mounted on wheels, however.

Folio 39r

COMMENTARY

A siege engine: the shield at the front also serves as a bridge. The main part of the vehicle is a ten-wheeled flat platform. The long poles attached to its front are to brace the bridge when it spanned a gap or moat. The rope and pulley would be of no assistance in lowering the bridge. They are intended for use in raising it if the attack should fail and retreat becomes necessary. The bridge was extended into place by brute strength. Iron is employed extensively in this machine, especially as sheathing on the support poles and as reinforcements with grappling hooks on the bridge. This engine seems to be a synthesis of two of Kyeser's designs.[1]

[1] Quarg, *Kyeser*, I:31r; II:25 shows a *gattus* with a pivoting raised platform used as a shield and I:87v; II:55 shows a collapsable bridge which extends out when needed. A machine similar to fol. 39r is shown in Weimar, Hs. fol. 328, fol. 134v.

Folio 39v

COMMENTARY

A siege engine: it is essentially a shielded cannon carriage, pushed by men. Two armored panels join to form a wedge in front of two small cannon, which fire through ports cut in the armor. A long heavy beam in the center is equipped with handles to enable a crew of men to push the vehicle. Iron is used extensively for spiked edging on the armor panels and as tires on the four small wheels. This type of engine, as well as the one shown on fol. 43v, seems to have evolved from Kyeser's two-wheeled carts with spikes and scythes.[1]

[1] Quarg, *Kyeser*, I: 19v–23r, 24v–25r, 73v; II: 19–21 and 46 shows such carts.

Folio 40r[1]

COMMENTARY

A large crossbow on a four-legged trestle: the bow is shown spanned and loaded with three bolts which it hurls simultaneously. The spanning mechanism is a screw turned by an X-handle. The screw should carry a nut to which the leather strap with the spanning hook would be attached, but this is lacking in the drawing. Instead the strap appears to be attached to one part of the trestle. There are two possible interpretations of this drawing. First, it may represent a fairly conventional crossbow resting on a horizontal spanning frame from which it would be removed in order to aim and shoot. Second, it could be a misguided attempt to envision a siege crossbow mounted on a four legged stand. If the latter is correct, then B's machine is entirely impractical, because it could not be aimed at all. The former interpretation accounts for the presence of a conventional crossbow trigger and an equally conventional spanning stirrup on the bow (no such stirrup would be necessary if the bow were mounted on the trestle). It is nevertheless difficult to imagine how a bow powerful enough to discharge three missiles at once could simultaneously be light enough to hold and aim by hand. Thus, on either interpretation, author B's machine encounters severe practical difficulties. The similarities between this drawing and others from the fifteenth century indicate that author B probably intended to represent a siege crossbow (see Chapter V). Leather is used for the strap and to sheath the bow, which is doubtless made of steel; all other parts of the weapon are made of wood, including the screw.

[1] Berthelot, 461, fig. 20; Canestrini, 70, fig. XII reproduce this folio.

Folio 40v

COMMENTARY

A »booby trap«: a clearer redrawing in Weimar, Hs. fol. 328, fol. 135r makes its action more apparent. The spanned crossbow is placed so that it points at a spot where enemy soldiers are likely to pass. The crossbow's trigger is actuated by a rope loop that is in turn attached to the long handle of the spade-like piece near the bottom of the drawing. This constitutes the trigger mechanism of the entire trap. If an unwary victim should tread on the broad end (as depicted), his weight would trip the crossbow trigger, releasing a bolt in his direction. Fol. 40r does not make clear exactly how the triggering function operates, but judging from the Weimar version, author B apparently intended that the rope loop would exert downward pressure on the crossbow's rear stock, forcing the long trigger against the table and releasing the bowstring. This application of the crossbow is reminiscent of the primitive, self-acting animal traps from which all crossbows are presumably descended. The bow is leather-covered and is doubtless made of steel, as is the trigger. The rest of the device is wooden.

Folio 41r[1]

COMMENTARY

A wooden fortification: this may have been intended as a temporary fortification where stone edifices could not be built. The structure resembles that of a warship with three decks and masts having crow's nests. Cannon are shown protruding through gun-ports on the three decks and each deck appears to be steeped back slightly. The crow's nests are accessible by ladders and would serve as observation and firing platforms. It is uncertain whether such structures were ever built.

[1] Berthelot, 467, fig. 21, Canestrini, 108, fig. XXX reproduce this folio. Jähns, 428 describes it in some detail.

Folio 41v[1]

COMMENTARY

A siege engine: it is open at the top and back for easy exit. It runs on eight wheels mounted at the sides with one large, possibly steerable wheel at the front. A platform pivoted from two upright poles at the center of the vehicle serves both as a shield and as a rampway. The platform can be turned completely over by means of the rope at the front and has cross-pieces on the bottom to improve footing. Iron is used less extensively than on some of the other siege engines but serves to reinforce the edges of the platform and as spiked armor on the front and as tires. A similar siege engine appears in the *Bellifortis*.[2]

[1] Gille, »Manuscrit.« 83, fig. 5 reproduces this folio, misidentifying it as fol. 42r.
[2] Quarg, *Kyeser*, I : 31r and 32r; II : 25.

Folio 42r[1]

COMMENTARY

A grain mill: it is driven by a single-throw crankshaft and a connecting-rod. The millstones are shown from above, and the crankshaft from the side. The shaft is equipped with a set of four iron flyweights linked to arms extending from the shaft. Since these weights are connected to the arms by a single iron link, they will extend outward as the shaft's speed increases and serve to a small degree to regulate the machine's speed. This effect would have been more pronounced had author B used longer lengths of chain. The crankshaft and the connecting-rod are made of wood, although the eye of the rod looks as if it might be metal; it is colored to represent wood, however. The shaft is equipped with iron pivot bearings at each end. Thus there is no connection between the crankshaft and the millstones it is supposed to drive.

[1] Beck, 275–76, fig. 315; Gille, *Engineers*, 72 reproduce this folio.

Folio 42v

COMMENTARY

A cannon hoist: it is less practical than most of section A's and is mechanically misunderstood. Four masts support a horizontal cross-beam into which four sheaves are recessed. Unlike section A's hoists, this is powered by a crank with a ratchet. The suspended block is also different, being made of wood and capped with an iron retainer and pierced by iron sheave axles. Author B has misunderstood the pulley system. With four upper and only two lower sheaves it is impossible for the hoist to function as drawn. Faced with a surplus of pulleys on top, B has unwittingly drawn two discontinuous ropes, and neither has the necessary anchor.

Folio 43r[1]

COMMENTARY

Two cannon mounted on carriages: the carriages can be linked together to be drawn by a single team. The traces can be removed and the carriages separated by removing the coupling pins. Since the carriages as drawn constitute a four-wheeled vehicle, and since no means of turning the wheels is shown, this would be difficult to steer properly.[2] The carriages are solid platforms with no elevating device visible. The cannon are breech loaders, similar to those found on fols. 38v, 39v, 43v, 44v, 46r and 48r. Coloration indicates iron tires, iron armor on the front of the carriages, and iron coupling pins.

[1] Berthelot, fig. 21 reproduces this folio.

[2] Carriages designed to link together are shown in the *Hausbuch* (Bossert and Storck, figs. 58 and 68), but they are all equipped with turning trains.

Folio 43v[1]

COMMENTARY

A siege engine: it is similar to the one shown on fol. 39v. Here again two armored panels join to form a wedge behind which a cannon fires through a porthole. This version rests on two wheels and has two long poles extending behind it. It would be pushed rather like a wheel-barrow. The spiked panel extending in front of the wedge probably served to counter-balance the weight of the cannon. Spiked iron strips reinforce all the edges and iron tires are fitted on the wheels.

[1] Berthelot, 469, fig. 23 reproduces this folio.

Folio 44r[1]

COMMENTARY

A boat: it is obviously intended as an armored craft to transport troops for short distances on inland waterways. The boat is made of wood and has high V-shaped shields on its bow and stern. Note that it carries a cannon in its bow that can fire through a port in the shielding. Iron is used to reinforce these shields. The low sides would permit easy entry and exit from the boat. Since no means of propelling it is shown, one assumes that the soldiers being transported paddled it themselves. The craft is equipped with a side-steering rudder, which B has drawn on the wrong side. Similar types of simple boats appear in some of the fifteenth-century engineering manuscripts.[2]

[1] Berthelot, 470, fig. 24 reproduces this folio.

[2] Cf., e.g., Cgm 600, fol. 16v.

Folio 44v[1]

COMMENTARY

Boat with a cannon mounted in its bow: the boat is drawn quite simply, even schematically, while the cannon is shown disproportionately large. The illustration is meant to suggest the mounting of cannon on boats and to suggest a location and a type of mount suitable for this. It is not a detailed drawing of a specific boat. The mount shown does not seem capable of elevation. The cannon is a breech loader, similar to those found on fols. 38v, 39v, 43r, 43v, 46r and 48r.

[1] Berthelot, fig. 25 reproduces this folio.

Folio 45r

COMMENTARY

A tower with water pipes and a domed building: the purpose of the drawing is not clear. The tower is made of masonry and has a red tile roof. It is shown with seven wooden water pipes entering it at its bottom. This is one of several illustrations of edifices in section B (see fols. 31r, 31v and 34v). The drawing may be intended to represent the way in which a tower or other building could be supplied with water, but I cannot explain why author B drew so many pipes.

Folio 45v

COMMENTARY

Two drawings: a Nuremberg shears and two hollow wooden tubes. The shears are shown holding a cube of material by a rope. This is an integral part of the drawing, but the dots on the cubes making it into a die may be a later addition. The tubes to the left vaguely resemble cannon, but they are made of wood, and banded with iron at their mouths, on the small ends and at the junction of the small and large segments. A white band with an X-design runs around the larger portions of both tubes. On the small end of the lower one there is a ring to which a rope is attached. The rope leads to a metal peg or nail. I do not know what these are or what their relation to the Nuremberg shears could be.

Folio 46r[1]

COMMENTARY

Four cannon mounted on two carriages:[2] the carriages are linked together like those shown on fol. 43r, and the same problems noted there occur in this example, i.e., the lack of turning trains and elevating devices. The cannon are breech-loaders, similar to those shown on fols. 38r, 38v, 43r, 43v and 48r. Coloration indicates iron is used for tires, coupling pins and armor.

[1] Canestrini, 270, fig. XCVIII reproduces this folio.
[2] Thorndike, 17 refers to this as a »cannon with four wheels.«

Folio 46v

COMMENTARY

A wagon: it is heavily constructed and has sides that fold down. The general appearance of the wagon is that of a vehicle designed for combat. It requires six wheels to carry its weight. Note that B has not included any steering train to cope with the problems created by the extra set of wheels. Iron is used for reinforcement on the wheels and on the prongs of the poles carried on the wagon's side. The back side is shown hinged to the floor so that it would fold down if necessary, but author B has forgotten to leave enough room for the folded side to clear the wheels. Although this wagon is poorly designed and could never have been used, heavy wagons were employed as »tanks« in many armies in the late fifteenth century.[1]

[1] Cf. Chapter V. Thorndike, 17 calls this a cannon with six wheels.

Folio 47r[1]

COMMENTARY

A windmill: only the exterior view is shown, but in considerable detail. The way in which the mill housing rests on a single pole is readily visible. The post is braced by diagonal struts, two of which author B has omitted. The mill is located on an artificial mound made of brick. There is no tail pole shown with which to turn the mill, but the ladder extending from the door could be used in its place. There are two unusual features about the mill. The wind-shaft is shown extending through the back wall of the mill housing. Since the wind-shafts must bear the thrust of the wind pressing on the sails, they were usually mounted on heavy bearings within the mill. Author B's design would seem to create unnecessary problems in mounting the wind-shaft to bear such stress. Above the wind-shaft we see a sack hoist consisting of a rotating shaft around which a rope winds to lift unmilled grain to the millstones. Although some scholars have claimed that this shaft is turned by the wind-shaft,[2] it is more likely that it was powered by the miller himself through a rope. The clutch required to take power from the rotating wind-shaft did not exist in the late fifteenth century. Even sixteenth-century works showing windmill interiors depict such sack hoists as manually powered.[3]

[1] Beck, 278, fig. 320; Feldhaus, *Technik*, 722, *TdAdM*, 349 and *DM*, 241; Uccelli, 8, fig. 20; Gleisberg, 50 reproduce this folio.
[2] Feldhaus, *DM*, 241; Uccelli, 8.
[3] Ramelli, fig. 133, for example.

Folio 47v

COMMENTARY

A gearing study: the mechanism is not part of any machine to do useful work. It shows right-angle gearing turned by a crank equipped with flyweights. The crank and the end disks of the lantern pinion are made of iron and the pinion has iron pivot bearings on either end. The crown gear and the flyweights are made of wood. The gearing is held in a three-legged frame which lacks the necessary bracing for the middle member. The drawing is done in poor perspective.

Folio 48r

COMMENTARY

A siege engine: it consists of a triangular frame studded with spikes on which two cannon are mounted. The engine lacks any armor and the cannon are set to shoot across one another's lines of fire. A massive pole with handles runs along the center line so that a crew of men might push the vehicle. The drawing is plagued with perspective problems. The hinter rear wheel does not look as if it would touch ground and only one front wheel is shown. This type is descended from some of Kyeser's triangular siege engines.[1]

[1] Quarg, *Kyeser*, I: 13v–14r; II: 17 for example.

Folio 48v

COMMENTARY

A siege engine: it runs in a trough which serves as a track. The engine is a shelter, shaped like a prism tapering to a point at one end. It is mounted on only one pole at the foot of which three wheels can be seen. These wheels run in a U-shaped trough so that the shelter could roll back and forth over a short distance like a cart on rails. The angle of the shelter can be controlled by means of a rope at the back. The purpose of this device is not clear; perhaps it was intended to shield a large siege cannon during loading and would be drawn back when the cannon was ready to fire. The trough is too short for this to have been a mobile engine. The mounting of such a machine in a trough is an interesting example of B's desire to apply the principle of railed vehicles in different settings (cf. fols. 35v–36r).

BIBLIOGRAPHY

Agricola, Georgius. *De re metallica.* Trans. Herbert Clark Hoover and Lou Henry Hoover. London: The Mining Magazine, 1912.

Alfonsus, Petrus. *De disciplina clericalis.* Ed. Angel Gonzalez Palencia. Madrid: Editorial Maestre, 1948.

Anderson, R. H. »The Technical Ancestry of Grain-Milling Devices,« *Mechanical Engineering,* LVII (1935), pp. 611 ff.

Anderson, William. *Castles of Europe: From Charlemagne to the Renaissance.* New York: Random House, 1970.

Andreae, S. J. F. »Embanking and Drainage Authorities in the Netherlands during the Middle Ages,« *Speculum,* XXVII (1952), pp. 158–67.

Andreas von Regensburg. *Sämtliche Werke.* Ed. Georg Leidinger. (»Quellen und Untersuchungen zur bayerischen und deutschen Geschichte,« N. S. I) München: Rieger, 1917.

Anonymous. *Büchsenmeisterei: Von Geschoss, Büchsen, Pulver, Salpeter und Feuerwerk, etc.* (i. e. a variant of the *Feuerwerkbuch*). Strassburg: Christian Engenolphen, 1529.

Antier, Jean-Jacques. *Histoire mondiale du sous-marine.* Paris: Laffont, 1968.

Arnold, Gerhard. *Bilder aus der Geschichte der Kraftmaschinen: Von der Antike bis zum Beginn des 20. Jahrhunderts.* München and Berlin: Heinz Moos Verlag, 1968.

Atkinson, Frank. »The Horse as a Source of Rotary Power,« *Transactions of the Newcomen Society,* XXXIII (1960–61), pp. 31–55.

Baader, Joseph. »Nürnbergs Stadtviertel im Mittelalter,« *Zweiunddreissigster Jahresbericht des historischen Vereins in Mittelfranken.* Ansbach: 1864, pp. 52–84.

Baarmann, O. »Die Entwicklung der Geschützlafette bis zum Beginn des 16. Jahrhunderts und ihre Beziehungen zu der des Gewehrschaftes,« in *Beiträge zur Geschichte der Handfeuerwaffen: Festschrift zum achtzigsten Geburtstag von Moritz Thierbach.* Dresden: Verlag und Druck von Wilhelm Baensch, 1905, pp. 54–86.

Bacon, Roger, *Fr. Rogeri Bacon Opera quaedam hactenus inedita.* Ed. John Sherren Brewster. London: Longman, Green, 1859.

Bathe, G. *Horizontal Windmills.* Philadelphia: Private Press, 1948.

Battison, Edwin A. »Stone Cutting and Polishing Lathe by Jacques Besson,« *Technology and Culture,* VII (1966), pp. 202–05.

Bayerische Akademie der Wissenschaften, Historische Kommission. *Die Chroniken der deutschen Städte.* Leipzig, etc.: Hirzel, etc., 1862–1931. Vols. I, II, VII.

Beck, James H. »The Historical ›Taccola‹ and the Emperor Sigismund in Siena,« *The Art Bulletin,* L (1968), pp. 309–20.

Beck, James H. (ed.). *Liber tertius de ingeneis ac edifitiis non usitatis di Mariano di Jacopo detto il Taccola.* Milan: Polifilo, 1969.

Beck, Theodor. *Beiträge zur Geschichte des Maschinenbaues.* Berlin: 1899; reprinted Hildesheim: Olms, 1969.

Beckmann, Johann. *Beyträge zur Geschichte der Erfindungen.* Leipzig: Kummer, 1788.

Bedini, Silvio A. and Maddison, Francis R. *Mechanical Universe–The Astrarium of Giovanni de'Dondi.* (»Transactions of the American Philosophical Society,« N.S. Vol. LVI, part 5), Philadelphia: American Philosophical Society, 1966.

Beeler, John. *Warfare in Feudal Europe 730–1200.* Ithaca, N.Y.: Cornell, 1971.

Benecke, Georg Friedrich, Müller, Wilhelm and Zarncke, Friedrich. *Mittelhochdeutsches Wörterbuch.* 4 vols.; Leipzig: S. Hirzel Verlag, 1854–66.

Benedum, Jost and Michler, Markwart. »Zu den Schrauben-gewinden antiker Specula: Ein Diskussionsbeitrag zu A. Mutz, ›Römische Bronzegewinde‹,« *Technikgeschichte,* XXXVII (1970), pp. 353–57.

Bennett, Richard and Elton, John. *History of Corn Milling.* Vol. II: *Watermills and Windmills.* London: Simpkin, Marshall, 1899.

Berthelot, Marcellin. »Les manuscrits de Léonard de Vinci et les machines de guerre,« *Journal des savants,* 3me sér. LXVII (1902), pp. 116–20.

Berthelot, Marcellin. »Pour l'histoire des art mécaniques et de l'artillerie vers la fin du moyen age,« *Annales de chemie et de physique,* 6me sér., XXIV (Decembre, 1891), pp. 433–521.

Besson, Jacques. *Theatrum instrumentorum et machinarum.* 2nd ed.; Lyons: Bartholemy Vincent, 1578. Editio princeps: Orleans (?): 1569 (?) under the title *Instrumentorum et machinarum liber primus.*

Bezold, Friedrich von. *König Sigismund und die Reichskriege gegen die Hussiten bis zum Ausgang des dritten Kreuzzugs.* Munich: Ackermann, 1872–77.

Bibliothèque Nationale, Départment des Manuscrits. *Psautier de Saint Louis* (i.e. facsimile edition of Ms. latin 10525). Paris: Catala Frères, n. d. (c. 1900).

Biringuccio, Vannoccio. *De la pirotechnia libri X.* Venezia: Curtio Navo et Fratelli, 1540.

Biringuccio, Vannoccio. *The Pirotechnia of Vannoccio Biringuccio.* Trans. Cyril Stanley Smith and Martha Teach Gnudi. 2nd ed.: New York: Basic Books, 1959.

Birkenmajer, Alexander. »Zur Lebensgeschichte und wissenschaftlichen Tätigkeit von Giovanni Fontana (1395?–1455?),« *Isis,* XVII (1932), 34–53.

Bittner, L. *Gesamtinventar des Wiener Haus-, Hof- und Staatsarchivs.* Vol. I; Wien: Verlag Adolf Holzhausens Nachfolger, 1936.

Blaine, Bradford B. »The Application of Water-Power to Industry During the Middle Ages.« Unpublished Ph.D. Dissertation, Department of History, University of California, Los Angeles, 1966.

Bloch, Marc. »Les inventions medievales,« *Annales, Economies, Sociétés, Civilisations,* VII (1935), pp. 634 ff. and VIII (1936), pp. 513 ff.

Bloch, Marc. »Avenement et conquetes du moulin à eau,« *Annales, Economies, Sociétés, Civilisations,* VII (1935), pp. 538 ff.

Boeheim, Wendelin. »Studie über die Entwicklung des Geschützwesens in Deutschland,« *Zeitschrift für historische Waffenkunde,* I (1899), pp. 57 ff.

Bossert, Helmuth Theodor and Storck, Willy F. (eds.). *Das mittelalterliche Hausbuch.* Leipzig: E. A. Seemann, 1912.

Bowles, Edmund A. »On the Origins of the Keyboard Mechanism in the Late Middle Ages,« *Technology and Culture,* VII (1966), pp. 152–62.

Boyer, Marjorie Nice. »Bridges and Mill Sites in Medieval France,« *Actes du XIIe Congrès International d'Histoire des Sciences: Paris, 1968.* Paris: Blanchard, 1971. Vol. X, Part B, pp. 13–17.

Boyer, Marjorie Nice. »Medieval Pivoted Axles,« *Technology and Culture,* I (1960), pp. 128–38.

Boyer, Marjorie Nice. »Medieval Suspended Carriages,« *Speculum,* XXXIV (1959), pp. 359 ff.

Brandt, Paul. *Schaffende Arbeit und bildende Kunst.* Vol. I: *Im Altertum und Mittelalter.* Leipzig: Alfred Kröner Verlag, 1927.

Brett, G. »A Byzantine Water Mill,« *Antiquity*, XIII (1939), pp. 354ff.

Briquet, Charles Moïse. *Les Filigranes: Dictionaire historique des marques du papier des leur apparition vers 1281 jusqu'en 1600.* 4 vols., 2nd ed.; Leipzig: Verlag von Karl W. Hiersemann, 1923.

Brockman, Eric. *The Two Sieges of Rhodes: 1480–1522.* London: John Murray, 1969.

Bruhn, Wolfgang and Tilke, Max. *Das Kostümwerk.* Berlin: Wasmuth, 1941.

Bühler, Curt F. »Watermarks and the Dates of Fifteenth-Century Books,« *Studies in Bibliography*, IX (1957), pp. 217–24.

Buffet, Bernard and Evrard, René. *L'eau potable à traverse les âges.* 2nd ed.; Liège: Editions Soledi, 1951.

Calder, Ritchie. *The Evolution of the Machine.* New York: American Heritage, 1968.

Canestrini, Giovanni. *Arte militare meccanica medievale.* Milano: Toninelli, n. d. (1945).

Carman, W. Y. *A History of Firearms.* New York: 1955.

Cary, George. *The Medieval Alexander.* Ed. D. J. A. Ross. Cambridge: Cambridge University Press, 1956.

Caus, Salomon de. *Von gewaltsamen Bewegungen.* Frankfurt am Main: Abraham Pacquart, n. d. (1615?).

Cipolla, Carlo. *Clocks and Culture: 1300–1700.* London: Collins, 1967.

Cipolla, Carlo. *Guns, Sails and Empires.* New York: Pantheon Books, 1965.

Clagett, Marshall. »The Life and Works of Giovanni Fontana,« *Annali dell'Istituto e Museo di Storia della Scienza di Firenze*, I (1976), pp. 5–28.

Clephan, Robert Coltman. *The Defensive Armour and the Weapons and Engines of War of Mediaeval Times and of the »Renaissance.«* London: Walter Scott, 1900.

Commena, Anna. *The Alexiad.* Trans. Elizabeth A. S. Dawes. London: K. Paul, Trench, Trubner, 1928.

Conant, Kenneth J. »The After-Life of Vitruvius in the Middle Ages,« *Journal of the Society of Architectural Historians*, XXVII (1968), pp. 30–38.

de la Croix, Horst. »The Literature on Fortifications in Renaissance Italy,« *Technology and Culture*, IV (1963), pp. 30–50.

Crombie, A. C. *Medieval and Early Modern Science.* Garden City, N. Y.: Doubleday, 1959.

Crous, Ernst and Kirchner, Joachim. *Die gotischen Schriftarten.* Leipzig: Klinkhardt & Biermann, 1928.

Curwen, E. Cecil. »The Problem of Early Water-Mills,« *Antiquity*, XVII (1944), pp. 130–46.

Daumas, Maurice (general ed.). *Histoire général des techniques.* Vol. I: *Les origines de la civilisation technique.* Paris: Presses Universitaires de France, 1962. English edition: *A History of Technology.* New York: 1969.

Delbrück, Hans. *Geschichte der Kriegskunst im Rahmen der politischen Geschichte.* 4 vols., 2nd ed., 1923; reprinted Berlin: Walter de Gruyter, 1964.

Demmin, August. *Die Kriegswaffen in ihren geschichtlichen Entwicklungen.* 4th ed.; Leipzig: P. Friesenhahn, 1893.

Denk, F. »Zwei mittelalterliche Dokumente zur Fluggeschichte und ihre Bedeutung,« *Sitzungsberichte der Phys.-med. Sozietät zu Erlangen*, LXXI (1939), pp. 353–68.

Dibner, Bern. *Leonardo da Vinci, Military Engineer.* New York: Burndy Library, 1946.

Diels, H. and Schramm, E. *Herons Belopoiika.* (»Abhandlungen der königlichen preussischen Akademie der Wissenschaften, philosophisch-historische Klasse,« No. 2). Berlin: 1918.

Downey, Fairfax. *Cannonade: Great Artillery Actions of History.* Garden City, N. Y.: Doubleday, 1966.

Drachmann, A.G. *Ancient Oil Mills and Presses.* (»Det Kgl. Danske Videnskabernes Selskab. Archaeologisk-kunsthistoriske Meddelelser,« Vol. I, No. 1). Copenhagen: Levin & Munksgaard, 1932.

Drachmann, A.G. *Ktesibios, Philon, Heron.* Copenhagen: 1948.

Drachmann, A.G. *The Mechanical Technology of Greek and Roman Antiquity.* Madison: University of Wisconsin Press, 1963.

Drescher, Karl. »Johann Hartlieb: Über sein Leben und seine schriftstellerische Tätigkeit,« *Euphorion,* XXV (1924), pp. 225–41, 354–70, 569–90 and XXVI (1925), pp. 341–67, 481–564.

Dugas, René. *A History of Mechanics.* Trans. J.R. Maddox. Neuchatel: Neuchatel: Griffon, 1955.

Durdík, Jan. *Hussitisches Heerwesen.* Trans. Eberhard Wolfgramm. Berlin: Deutscher Militärverlag, 1961.

Ebendorfer von Hasselsbach, Thomas. *Chronicon austriacum.* Ed. Alfons Lhotsky. (»Monumenta germaniae historica: Scriptores rerum germanicarum,« N.S., Vol. XIII). Berlin: Weidmannsche Verlagsbuchhandlung, 1967.

Eco, Umberto and Zorzoli, Giovanni G. *Storia figurata delle invenzioni.* Milan: Bompiani, 1968.

Ehlers, G. »Die Wasserversorgung der deutschen Städte im Mittelalter,« *Technikgeschichte,* XXV (1936), pp. 13–25.

Eis, Gerhard. *Mittelalterliche Fachliteratur.* Stuttgart: J.B. Metzlersche Verlagsbuchhandlung, 1962.

Eis, Gerhard. *Historische Laut- und Formenlehre des Mittelhochdeutschen.* Heidelberg: Carl Winter Universitätsverlag, 1950.

Essenwein, August von. »Eigenthümliche Wagen, Schiffe und Schlitten vom 15.–18. Jahrhundert,« *Anzeiger für Kunde der deutschen Vorzeit,* XXX (März-April, 1883), pp. 57–79.

Essenwein, August von. »Mittelalterliche Taucher und Schwimmapparate,« *Anzeiger für Kunde der deutschen Vorzeit,* N.F. XVIII (1871), pp. 257 ff.

Essenwein, August von. *Quellen zur Geschichte der Feuerwaffen.* 2 vols., 1877; reprinted Graz: Akademische Druck und Verlagsanstalt, 1969.

Ewebank, Thomas. *A Descriptive and Historical Account of Hydraulic and Other Machines for Raising Water, Ancient and Modern.* 3rd ed.; New York: Greeley and McElrath, 1849.

Feldhaus, Franz Maria. *Die Geschichte des Schleifmittels.* Hannover-Hainholz: Vereinigte Schmirgel- und Maschinen-Fabriken, 1919.

Feldhaus, Franz Maria. *Geschichte des technischen Zeichnens.* 3rd ed., rev.; Wilhelmshaven: Franz Kuhlmann, 1967.

Feldhaus, Franz Maria. *Die Maschine im Leben der Völker.* Basel and Stuttgart; Verlag Birkhäuser, 1954.

Feldhaus, Franz Maria. *Die Technik der Antike und des Mittelalters.* Wildpark-Potsdam: Akademische Verlagsgesellschaft Athenaion, 1931.

Feldhaus, Franz Maria. *Die Technik der Vorzeit, der geschichtlichen Zeit und der Naturvölker.* 2nd ed., 1914: reprinted Heinz Moos Verlag, 1965.

Ffoulkes, Charles. *Arms and Armament: An Historical Survey of the Weapons of the British Army.* London: George G. Harrap, 1945.

Fischer, Hermann. »Beiträge zur Geschichte der Werkzeugmaschinen: Bohrer und Bohrmaschinen für Holz und Metall,« *Beiträge zur Geschichte der Technik und Industrie,* IV (1912), pp. 274–308.

Fischer, Hermann. »Beiträge zur Geschichte der Werkzeugmaschinen: Schmiedmaschinen,« *Beiträge zur Geschichte der Technik und Industrie,* VI (1914–15), pp. 1–34.

Fischer, Karl H. and Walther, Leo. *Die Wasserversorgung der Stadt Nürnberg von der reichstädtischen Zeit bis zur Gegenwart: Festschrift zur Eröffnung der Wasserleitung Rama.* Nürnberg: Stadtmagistrat, 1912.

Forti, Umberto. *Storia della Tecnica dal Medioevo al Rinascimiento.* Florence: Sansoni, 1957.

Fowler, Kenneth. *The Age of the Plantaganets and Valois: The Struggle for Supremacy.* New York: Putnam's, 1967.

Francesco di Giorgio Martini. *Trattati di architettura, ingegneria e arte militare.* Ed. Corrado Maltese. Milano: Edizioni il Polifilio, 1967.

Frauenholz, Eugen von (ed.). *Lazarus von Schwendi: Der erste deutsche Verkünder der allgemeinen Wehrpflicht.* Hamburg: Hanseatische Verlagsanstalt, 1939.

Freese, Stanley. *Windmills and Millwrighting.* Cambridge: Cambridge University Press, 1957.

Gasparinetti, A. F. »Paper, Papermakers and Paper-Mills of Fabriano,« (Italian essay of 1938, revised 1953) in *Zonghi's Watermarks.* (»Monumenta chartae papyraceae historiam illustrantia,« Vol. III). Hilversum, Holland: Paper Publications Society, 1953, pp. 63–82.

Gille, Bertrand. *Engineers of the Renaissance.* Trans. anonymous. Cambridge, Massachusetts: M. I. T. Press, 1966.

Gille, Bertrand. »Etudes sur les manuscrits d'ingenieurs du XVe siècle: Le manuscrit dit de la guerre hussite,« *Techniques et civilisations,* V (1956), pp. 77–86.

Gille, Bertrand. »Léonard de Vinci et son temps,« *Techniques et civilisations,* II (1952), pp. 69–84.

Gille, Bertrand. »La naissance du système bielle-manivelle,« *Techniques et civilisations,* II (1952), pp. 42–44.

Gleisberg, Hermann. *Technikgeschichte der Getreidemühle.* (»Abhandlungen und Berichte des deutschen Museums,« Vol. XXIV, Heft 3). München: R. Oldenbourg, 1956.

Gleisberg, Hermann. *Triebwerke in Getreidemühlen.* (»Technikgeschichte in Einzeldarstellungen,« Vol. XV). Düsseldorf: V. D. I. Verlag, 1970.

Götze, Alfred A. W. *Frühneuhochdeutsches Glossar.* 5th ed.; Berlin: Walter de Gruyter, 1956.

Grimm, Jacob and Grimm, Wilhelm. *Deutsches Wörterbuch.* 16 vols.; Leipzig: S. Hirzel Verlag, 1854–1954.

Grosz, August and Thomas, Bruno. *Katalog der Waffensammlung in der neuen Burg.* Wien: Kunsthistorisches Museum, 1936.

Guttmann, Oscar. *Monumenta pulveris pyrii.* London: Artist's Press, 1906.

Haage, Bernard, »Zum ›Kunstbuch‹ des Hans Hassenwein in Hermannstadt,« *Sudhoffs Archiv,* LI (1967), pp. 268–69.

Hagenmeyer, Christa. »Kriegswissenschaftliche Texte des ausgehenden 15. Jahrhunderts: Schermers Basteibau, Wagenburgordnung, Feuerwerksrezepte,« *Leuvense Bijdragen: Tidjschrift voor moderne Filologie,* LVI (1967), pp. 169–97.

Hale, J. R. »Gunpower and the Renaissance: An Essay in the History of Ideas,« in C. H. Carter (ed.). *From the Renaissance to the Counter-Reformation.* New York: 1965.

Hall, A. Rupert. *Ballistics in the Seventeenth Century.* Cambridge: Cambridge University Press, 1952.

Hall, A. Rupert. »The Military Inventions of Guido da Vigevano,« *Actes du VIIIe Congrès International d'Histoire des Sciences: Florence, 1956.* Florence: Gruppo Italiano di Storia delle Scienze, 1958. Vol. III, pp. 966–69.

Hall, A. Rupert. »More on Medieval Pivoted Axles,« *Technology and Culture,* II (1961), pp. 17–22.

Hall, Bert S. and West, Delno C. *On Pre-Modern Technology and Science: A Volume of Studies in Honor of Lynn White, jr.* (»Humana Civilitas,« Vol. I; Center for Medieval and Renaissance Studies, University of California, Los Angeles). Malibu, California: Undena Publications, 1976.

Hall, Bert S. »Crossbows and Crosswords,« *Isis,* LXIV (1973), pp. 527–33.

Halm, Karl von and Laubmann, Georg. *Catalogus codicium latinorum bibliothecae regiae monacensis.* 2nd ed.; München: Bibliothecae Regiae, 1892.

Hampe, Theodor. »Ausrüstung einer Wagenburg im 15. Jahrhundert,« *Mitteilungen aus dem germanischen Nationalmuseum*, (1897), pp. 79–80.

Harper, G. A. »Controlling Power,« *Chartered Mechanical Engineer*, XIV (1967), pp. 120–25.

Hart, Clive. »Medieval Kites and Windsocks,« *The Aeronautical Journal*, LXXIII, No. 708 (December, 1969), pp. 1019–26.

Hartig, Otto. »Christof Seselschreiber und sein Buch über Büchsenmeisterei und technische Erfindungen,« *Kultur des Handwerks*, Heft 9 (July, 1927), pp. 278–84.

Hartig, Otto. *Die Gründung der Münchner Hofbibliothek.* (»Abhandlungen der königlichen bayerischen Akademie der Wissenschaften, philosophische, philologische und historische Klasse,« Vol. XXVIII, Abhandlung 3). München: 1917.

Hassenstein, Wilhelm. *Das Feuerwerkbuch von 1420.* München: Verlag der deutschen Technik, n. d. (1941–43).

Heimpel, Hermann. Review of *Conrad Kyeser aus Eichstätt, Bellifortis*, ed. and trans. Götz Quarg. *Göttingische gelehrte Anzeigen*, CCXXIII, Heft 1/2 (1971), pp. 115–48.

Heinrichs, Wilhelm. »Der Werdegang der Feuerwehrleiter,« *Beiträge zur Geschichte der Technik und Industrie*, XVIII (1928), pp. 118–28.

Held, Robert. *The Age of Firearms.* New York: Harper and Row, 1957.

Heller, Edmund K. »Ludwig Hohenwang's ›Von der Ritterschaft:‹ An Evaluation, and a Survey of His Military and Naval Terms,« *University of California Publications in Modern Philology*, XXXVI, No. 6 (1952), pp. 173–83.

Herberger, Theodor. *Augsburg und seine frühere Industrie.* Augsburg: Albrecht Volkhart'sche Buchdruckerei, 1852.

Herold, Hans. »The Extraction of Saltpeter in Bygone Ages,« *Sandoz Bulletin*, XIII (1969), pp. 25–40.

Herrade of Landsperg. *Hortus deliciarum par l'abbesse Herrade de Landsperg.* Ed. A. Straub and G. Keller. 11 fascicles; Strasbourg: Société pour la conservation des monuments historiques d'Alsace, 1879–99.

Heymann, Frederick G. *George of Bohemia, King of Heretics.* Princeton: Princeton University Press, 1965.

Heymann, Frederick G. *John Žižka and the Hussite Revolution.* Princeton: Princeton University Press, 1955.

Hill, Donald R. »Trebuchets,« *Viator*, IV (1973), pp. 99–114.

Höfler, K. (ed.). *Geschichtsschreiber der hussitischen Bewegung in Böhmen.* (»Fontes rerum austriacarum,« Abteilung I, Vols. II, VI and VII). Wein: Kaiserliche-Königliche Hof- und Staatsdruckerei, 1856.

Holder, Alfred. *Die Durlacher und Rastatter Handschriften.* (»Die Handschriften der grossherzoglichen badischen Hof- und Landesbibliothek in Karlsruhe,« Abteilung III). Karlsruhe: Verlag von Ch. Th. Groos, 1895.

Hornung, Wolfgang. »Vom Steigbaum zur Drehleiter,« *Technikgeschichte*, XXXVI (1969), pp. 177–205.

Horwitz, Hugo T. »Technische Darstellungen aus alten Miniaturwerken,« *Beiträge zur Geschichte der Technik und Industrie*, X (1920), pp. 175–78.

Horwitz, Hugo T. »Zur Geschichte des Schaufelradantriebes,« *Zeitschrift des österreichischen Ingenieur- und Architekten-Vereins*, LXXXII (1930), pp. 309–13 and 356–60.

Husa, Václav, Petráň, Joseph and Šubrtova, Alena. *Homo Faber.* Praha: Artia, 1967.

Huuri, Kalervo. *Zur Geschichte des mittelalterlichen Geschützwesens aus orientalischen Quellen.* Helsinki: Finnische Literaturgesellschaft, 1941.

Ising, Gerhard. *Zur Wortgeographie spätmittelalterlicher deutscher Schriftdialekte.* 2 vols.; Berlin: Akademie Verlag, 1968.

Jacobi, H. »Römische Getreidemühlen,« *Saalburg Jahrbuch,* III (1912), pp. 75–95.

Jähns, Max. *Geschichte der Kriegswissenschaften, vornehmlich in Deutschland.* Vol. I: *Altertum, Mittelalter, XV. und XVI. Jahrhundert.* München and Leipzig: K. Oldenbourg, 1889.

Jegel, August. *Alt-Nürnberger Handwerksrecht und seine Beziehungen zu anderen.* Neustadt an der Aisch: Verlagsdruckerei Ph. C. W. Schmidt, 1965.

Jespersen, Anders. *A Preliminary Analysis of the Development of Gearing in Watermills in Western Europe.* Virum, Denmark: by the author, 1953.

Johannis de Capua. *Directorium vitae humanae.* Ed. Joseph Derenbourg. Paris: F. Vieweg, 1887.

Kaminsky, Howard. *A History of the Hussite Revolution.* Berkeley and Los Angeles: University of California Press, 1967.

Kammerer, O. »Die Entwicklung der Zahnräder,« *Beiträge zur Geschichte der Technik und Industrie,* IV (1912), pp. 242–73.

Keller, Alexander G. »Mechanical Linkages,« *Chartered Mechanical Engineer,* XIV (1967), pp. 322–27.

Keller, Alexander G. »The Missing of Jacques Besson: Inventor of Machines, Distiller of Oils, and Huguenot Pastor,« *Technology and Culture,* XIV (1973), pp. 28–39.

Keller, Alexander G. *A Theater of Machines.* New York: MacMillan, 1965.

Kellermann, Rudolf and Treue, Wilhelm. *Die Kulturgeschichte der Schraube.* 2nd ed., rev.; München: F. Bruckmann, 1962.

Kiechle, Franz. »Zur Verwendung der Schraube in der Antike,« *Technikgeschichte,* XXXIV (1967), pp. 14–22.

Kirchner, Joachim. *Scriptura gothica libraria.* München: R. Oldenbourg, 1966.

Klinckowstroem, Carl, Graf von. »Zur Geschichte des Göpels,« *Technikgeschichte,* XXXIV (1967), pp. 29–35.

Kluge, Friedrich. *Etymologisches Wörterbuch der deutschen Sprache.* Berlin: 1963.

Köhler, G. *Die Entwicklung des Kriegswesens und der Kriegsführung in der Ritterzeit.* Breslau: 1887.

Kretzschmer, Fritz. *Bilddokumente römischer Technik.* 3rd. ed.; Düsseldorf: V.D.I. Verlag, 1967.

Krinsky, Carol Herselle. »Seventy-Eight Vitruvius Manuscripts,« *Journal of the Warburg and Courtauld Institutes,* XXX (1967), pp. 36–70.

Kyeser, Conrad. *Conrad Kyeser aus Eichstätt, Bellifortis.* Ed. and trans. Götz Quarg. Vol. I: *Facsimile-Ausgabe der Pergamenthandschrift, Cod. Ms. philos. 63 der Universitätsbibliothek, Göttingen.* Vol. II: *Umschrift und Uebersetzung nebst Erläuterungen.* Düsseldorf: Verlag des Vereins deutscher Ingenieure, 1967.

Labarre, E.J. *A Dictionary and Encyclopedia of Paper and Papermaking.* 2nd ed.; Amsterdam: Swets & Zeittinger, 1952.

Lafuente y Zamalloa, Modesto. *Historia general de España.* Vol. X; Barcelona: Montaner y Simon, 1888.

Lane, Frederick. »The Crossbow in the Nautical Revolution of the Middle Ages,« *Explorations in Economic History,* VII (1969–70), pp. 161–71.

Latham, J.D. and Paterson, W.F. *Saracen Archery: An English Version of a Mameluke Work on Archery (ca. A.D. 1368).* London: Holland Press, 1970.

Lawrence of Březová (Vauřince z Brezove). *Kronika husitska.* (»Fontes Rerum Bohemicarum« Vol. V). Prague: 1893.

le Cerf, G. and Labande, E.R. *Instruments de musique au XVe siècle: les traités d'Henri-Arnaut de Zwolle et de divers anonymes.* Paris: Picard, 1932.

Lee, Charles. »Some Railway Facts and Fallacies,« *Transactions of the Newcomen Society,* XXXIII (1960–61), pp. 1 ff.

Leonardo da Vinci. *Codex Atlanticus.* Ed. G. Piumati. 6 vols.; Milan: 1894–1904.

Leonardo da Vinci. *Fragments at Windsor Castle from the Codex Atlanticus.* Ed. Carlo Pedretti. London: Phaidon, 1957.

Leonardo da Vinci. *Madrid Codices.* Ed. and trans. Ladislao Reti. New York: McGraw-Hill, 1974.

Leonardo da Vinci. *I manoscritti di Leonardo da Vinci: Codice sul volo degli uccelli.* Ed. Teodore Sabachnikoff and Giovanni Piumati. Paris: Rouveyre, 1893.

Leonardo da Vinci. *I manoscritti e disegni di Leonardo da Vinci;* Vol. I *Codex Arundel 263.* Rome: Danesi, 1923.

Leonardo da Vinci. *The Notebooks of Leonardo da Vinci.* Ed. and trans. Edward MacCurdy. New York: Braziller, 1958.

Lewis, Michael J.T. *Early Wooden Railways.* London: Routledge and K. Paul, 1970.

Lexer, Matthias von. *Mittelhochdeutsches Handwörterbuch.* 3 vols.; Leipzig: S. Hirzel Verlag, 1872–78.

Lipuius, Justus. *Poliorceticwv [sic], sive de machinis, tormentis, telis libri quinque.* 3rd ed.; Amsterdam: 1605.

Lorini, Buonaiuto. *Le fortificationi.* Venice: Presso Francesco Rampazetto, 1609.

Lot, Ferdinand. *L'art militaire et les armeés au moyen âge.* Paris: 1949.

Lutze, Eberhardt. *Die Bilderhandschriften der Universitätsbibliothek Erlangen.* Erlangen: Universitätsbibliothek, 1936.

Macek, Josef. *The Hussite Movement in Bohemia.* 2nd ed.; Prague: Orbis, 1958.

MacGregor, John. »On the Paddle-Wheel and Screw Propellor from the Earliest Times,« *Journal of the Society of Arts,* VI, 282 (April 16, 1858), pp. 335–42.

Mahr, O. »Zeittafel zur Geschichte des Geschützwesens bis zum Weltkrieg.« *Beiträge zur Geschichte der Technik und Industrie,* XXVII (1938), pp. 107–115.

Marsden, Eric W. *Greek and Roman Artillery.* Vol. I: *Historical Development.* Vol. II: *Technical Treatises.* Oxford: Clarendon Press, 1969–71.

Marx, E. »Bericht über ein Dokument mittelalterlicher Technik,« *Beiträge zur Geschichte der Technik und Industrie,* XVI (1926), pp. 317–21.

Matschoss, Conrad. *Geschichte des Zahnrades.* Berlin: V.D.I. Verlag, 1940.

Maurizio, A. *Die Geschichte unserer Pflanzennahrung.* Berlin: Verlagsbuchhandlung Paul Parey, 1927.

Mayr, Otto. *Zur Frühgeschichte der technischen Regelungen.* München and Wien: R. Oldenbourg, 1969.

Mehring, Gebhard. »Des Münchner Gelehrten Hans Hartlieb Angriffs- und Verteidigungskunst,« *Das Bayerland,* XIII (1902), pp. 501–03 and 514–15.

Meller, Emil. »Aus der Geschichte des Getreidemühlenbaus,« *Beiträge zur Geschichte der Technik und Industrie,* XX (1930), pp. 111–18.

Mehringhausen, M. »Die häusliche Wasserverwendung und Abwasserwirtschaft im Mittelalter,« *Technikgeschichte,* XXV (1936), pp. 43–56.

Marcolongo, Roberto. *La dinamica di Leonardo.* Milan: Tambarini, 1929.

Marcolongo, Roberto. »La meccanica di Leonardo da Vinci,« *Atti della reale accademia delle scienze fisiche e matematiche.* Ser. nov. XIX, No. 2. Naples: 1933.

Moritz, L.A. *Grain Mills and Flour in Classical Antiquity.* Oxford: Clarendon Press, 1958.

Moser, Virgil. *Frühneuhochdeutsche Grammatik.* 2 vols.; Heidelberg: Carl Winter Universitätsbuchhandlung, 1929–51.

Müller, Heinrich. *Deutsche Bronzegeschützröhre: 1400–1750.* Berlin: Deutscher Militärverlag, 1968.

Müller, Max. *Johann Albrecht von Widmannstetter, 1506–1557: Sein Leben und Wirken.* Inaugural-Dissertation; München; Bamberg: Handles-Druckerei, 1907.

Muendel, John. »The Horizontal Mills of Medieval Pistoia,« *Technology and Culture*, XV (1974), pp. 194–225.

Murr, Christoph Gottlieb von. *Beschreibung der vornehmsten Merkwürdigkeiten in der H.R. Reichs freyen Stadt Nürnberg*. Nürnberg: Johann Eberhard Zeh. [sic], 1778.

Mutz, Alfred. »Römische Bronzegewinde,« *Technikgeschichte*, XXXVI (1969), pp. 161–67.

Napoléon III, Emperor of the French and Favé, Ildefonse. *Etudes sur le passé et l'avenir de l'artillerie*. Vol. III: *Histoire des progrès de l'artillerie*. Paris: Librairie Militaire, 1862.

Navarrete, Martin Fernandez de. *Coleccion de los viages y descubrimentos que hicieron por mar los Españoles desde fines del siglo XVI*. Vol. I, 1826; reprinted Buenos Aires: Editorial Guarania, 1945.

Needham, Joseph. *Clerks and Craftsmen in China and the West*. Cambridge: Cambridge University Press, 1970.

Needham, Joseph. »Implications of Chinese Textile and Cast-Iron Technology for Certain Basic Inventions in Mechanical Engineering,« *Actes du XIe Congrès International d'Histoire des Sciences: Warsaw-Cracow, 1965*. Warsaw: Ossolineum, 1968. Vol. VI, pp. 61–64.

Needham, Joseph. »L'unité de la science: L'apport indispensable de l'Asie,« *Archives internationales d'histoire des sciences*, XXVIII (1949), pp. 563–82.

Needham, Joseph and Wang Ling. *Science and Civilisation in China*. Vol. IV: *Physics and Physical Technology*; Part 2: *Mechanical Engineering*. Cambridge: Cambridge University Press, 1965.

Nelson, Alan H. »Six-wheeled Carts: An Underview,« *Technology and Culture*, XIII (1972), pp. 391–416.

Neubauer, Kurt. *Das Kriegsbuch des Philipp von Seldeneck vom Ausgang des 15. Jahrhunderts*. Ph. D. Dissertation; Heidelberg: by the author, 1963.

Nolthenius, A.T. »Les moulins à main au moyen-âge,« *Techniques et civilisations*, IV (1955), pp. 149–52.

Odložilík, Otakar. *The Hussite King: Bohemia in European Affairs 1440–1471*. New Brunswick, New Jersey: Rutgers University Press, 1965.

Oliver, R.P. »A Note on the *De rebus bellicis*,« *Classical Philology*, L (1955), pp. 113ff.

Oman, Charles W.C. *The Art of War in the Middle Ages*. 2nd ed.; rev.; Ithaca, N.Y.: Cornell University Press, 1953.

Oman, Charles W.C. *A History of the Art of War in the Middle Ages*. 2 vols., 2nd ed., rev.; London: Methuen, 1924.

Ongania, Ferdinand (ed.). *Early Venetian Printing Illustrated*. Venice, London, New York: Ongania, John C. Nimmo, Charles Scribner's Sons, 1895.

Palacky, Franz. *Geschichte von Böhmen*. Vol. III, Part 2; Prag: Kronenberger, 1851. Vol. IV, Part 2; Prag: Friedrich Tempsky, 1860.

Parry, V.J. Article »Bārūd [Gunpowder]; iv; The Ottoman Empire,« in *The Encyclopedia of Islam*. 2nd ed., Vol. I; Leiden: E.J. Brill, 1960, pp. 1060–66.

Parsons, William Barclay. *Engineers and Engineering in the Renaissance*. 1939; reprinted Cambridge, Massachusetts: M.I.T. Press, 1968.

Partington, J.R. *A History of Greek Fire and Gunpowder*. Cambridge: W. Heffer and Sons, 1960.

Payne-Gallwey, Ralph. *The Crossbow*. 1903; reprinted New York: Bramhall House, n. d. (1958).

Payne-Gallwey, Ralph. *The Projectile-Throwing Engines of the Ancients*. London: 1907; reprinted Totowa, New Jersey: Rowman and Littlefield, 1973.

Picatoste y Rodriguez, Felipe. *Apuentes para una biblioteca cientifíca Española del siglo XVI*. Madrid: Manuel Tello, 1891.

Piccard, Gerhard. *Die Kronen-Wasserzeichen* (»Veröffentlichungen der staatlichen Archivverwaltung Baden-Württemberg, Sonderreihe: Die Wasserzeichendartei Piccard im Hauptstaatsarchiv Stuttgart,« Findbuch I). Stuttgart: Verlag W. Kohlhammer, 1961.

Piccard, Gerhard. *Die Ochsenkopf-Wasserzeichen* (»Veröffentlichungen der staatlichen Archivverwaltung Baden-Württemberg, Sonderreihe: Die Wasserzeichenkartei Piccard im Hauptstaatsarchiv Stuttgart,« Findbuch II). 3 vols.; Stuttgart: Verlag W. Kohlhammer, 1966.

Pouderoyen, Adrien. »Les polders aux Pays-Bas, leur asséchement, leur drainage; Les moulins de pompage, d'industrie et les moulins à boue,« *Techniques et civilisations,* II (1953), pp. 142–46 and III (1954), pp. 16–32 and 39–62.

Prager, Frank D. »Fontana on Fountains: Venetian Hydraulics of 1418,« *Physis,* XIII (1971), pp. 341–60.

Prager, Frank D and Scaglia, Gustina. *Brunelleschi: Studies of His Technology and Inventions.* Cambridge, Massachusetts: M. I. T. Press, 1970.

Prager, Frank D. and Scaglia, Gustina. *Mariano Taccola and His Book* **De ingeneis**. Cambridge, Mass.: M. I. T. Press, 1972.

Price, Derek J. de Solla. »An Ancient Greek Computer,« *Scientific American,* CC, 40 (June, 1959), pp. 60–67.

Price, Derek J. de Solla. *Gears from the Greeks: The Antikythera Mechanism–A Calendar Computer from ca. 80 B.C.* (»Transactions of the American Philosophical Society,« N.S. Vol. LXIV, Part 7). Philadelphia: American Philosophical Society, 1974.

Promis, Carlo. »Gl'i ingegneri militari,« *Miscellanea di storia italiana,* XII (Turin: 1871), pp. 418–19.

Prowe, Leopold. *Nicolaus Coppernicus.* 2 vols.; Berlin: Weidmannsche Buchhandlung, 1883–84.

Quarg, Götz. »Der *Bellifortis* von Conrad Kyeser aus Eichstätt, 1405.« *Technikgeschichte,* XXXII (1965), pp. 293–324.

Ramelli, Agostino. *Le diverse et artificiose machine.* Paris: by the author, 1588.

Randall, John Herman. »The Place of Leonardo da Vinci in the Emergence of Modern Science,« *Journal of the History of Ideas,* XIV (1953), pp. 191–202.

Randall, Lillian M. C. *Images in the Margins of Gothic Manuscripts.* Berkeley and Los Angeles: University of California Press, 1966.

Rathgen, Bernard (ed. Johannsen, Otto). »Beiträge zur Geschichte des Geschützwesens; I: Zur Geschichte und Geschützfabrikation,« *Beiträge zur Geschichte der Technik und Industrie,* XVIII (1928), pp. 41–44.

Rathgen, Bernard. *Das Geschütz im Mittelalter.* Berlin: V. D. I. Verlag, 1928.

Reindl, C. »Die Entwicklung der Wasserkraftnutzung und der Wasserkraftmaschinen,« *Wasserkraft Jahrbuch* (1924), pp. 3–31.

Renner, Peter. *Das Kriegsbuch Herzog Philipps von Cleve.* Ph. D. Dissertation; Heidelberg: by the author, 1960.

Reti, Ladislao. »The Codex of Juanelo Turriano (1500–1585),« *Technology and Culture,* VIII (1967), pp. 53–66.

Reti, Ladislao. »The Double-Acting Principle in East and West,« *Technology and Culture,* XI (1970), pp. 178–200.

Reti, Ladislao. »Francesco di Giorgio Martini's Treatise on Engineering and Its Plagiarists,« *Technology and Culture,* IV (1963), pp. 287–98.

Reti, Ladislao. »The Horizontal Waterwheels of Juanelo Turriano (ca. 1565): A Prelude to Besacle,« *Actes du XIIe Congrès International d'Histoire des Sciences: Paris, 1968.* Paris: Blanchard, 1971. Vol. X, Part B, pp. 79–82.

Reti, Ladislao. »Leonardo and Ramelli,« *Technology and Culture,* XIII (1972), pp. 577–605.

Reti, Ladislao. »The Leonardo da Vinci Codices in the Biblioteca Nacional of Madrid,« *Technology and Culture*, VIII (1967), pp. 437–45.

Reti, Ladislao. »Leonardo on Bearings and Gears,« *Scientific American*, CCXXIV, No. 2 (February, 1971), pp. 100–10.

Reti, Ladislao. »A Postscript to the Filarete Discussion: On Horizontal Waterwheels and Smelter Blowers in the Writings of Leonardo da Vinci and Juanelo Turriano,« *Technology and Culture*, VI (1965), 428–41.

Reti, Ladislao (ed.). *The Unknown Leonardo*. New York: McGraw-Hill, 1974.

Reti, Ladislao and Dibner, Bern. *Leonardo da Vinci, Technologist*. Norwalk, Conn.: Burndy Library, 1969.

Ritter, Karl. »Die mittelalterliche Steinbüchse aus Schmiedeisen,« *Beiträge zur Geschichte der Technik und Industrie*, XXVII (1938), pp. 22–29.

Robins, F. W. *The Story of Water Supply*. London: Oxford University Press, 1946.

Rodakiewicz, Erla. »The *editio princeps* of Roberto Valturio's *De re militari* in Relation to the Dresden and Munich Manuscripts,« *Maso Finiguerra*, V (1940), pp. 15–82.

Romocki, S. J. von. *Geschichte der Explosivstoffe*. Teil I: *Geschichte der Sprengstoffchemie, der Sprengtechnik und des Torpedowesens*. Berlin: Robert Oppenheim, 1895.

Ronse, Alfred. *De Wondmolen*. Brugge: Brouwer, 1934.

Rose, Paul Lawrence. »The Taccola Manuscripts,« *Physis*, X (1968), pp. 337–46.

Ross, David J. A. *Alexander and the Faithless Lady: A Submarine Adventure*. (»Inaugural Lecture, Birkbeck College, University of London«). London: University of London, 1967.

Sarton, George. *Introduction to the History of Science*. Vol. III: *Science and Learning in the Fourteenth Century*. Baltimore: Williams and Wilkins, 1948.

Scaglia, Gustina. »Drawings of Brunelleschi's Mechanical Inventions for the Construction of the Cupola,« *Marsyas*, X (1960–61), pp. 45–68.

Scaglia, Gustina. »Drawings of Machines for Architecture from the Early Quattrocento in Italy,« *Journal of the Society of Architectural Historians*, XXV (1966), pp. 90–114.

Schefold, Max. »Das mittelalterliche Hausbuch als Dokument für die Geschichte der Technik,« *Beiträge zur Geschichte der Technik und Industrie*, XIX (1929), pp. 127–32.

Schiller, Karl C. and Lübben, Heinrich A. *Mittelniederdeutsches Wörterbuch*. 6 vols.; Bremen: J. Kühtmann, 1875–81.

Schmeller, Johann Andreas. *Bayerisches Wörterbuch*. Ed. G. Karl Fromann. 2nd ed., 2 vols., 1872–77; reprinted Leipzig: K. F. Köhler Verlag, 1939.

Schneider, Rudolf. *Die Artillerie des Mittealters: Nach Angaben der Zeitgenossen dargestellt*. Berlin: Weidmannsche Buchhandlung, 1910.

Schubert, H. »The First Cast-Iron Cannon Made in England,« *Journal of the Iron and Steel Institute*, CXLVI (1942), pp. 131 ff.

Schultz, Alwin. *Das häusliche Leben der europäischen Kulturvölker vom Mittelalter bis zur 2. Hälfte des 18. Jahrhunderts*. Munich: 1903.

Schultz, Alwin. *Das höfische Leben zur Zeit der Minnesinger*. Leipzig: 1889.

Shaw, Joseph W. »A Double-Sheaved Pulley Block from Kenchreai,« *Hesperia*, XXXVI (1967), pp. 389–401.

Shelby, Lon R. »Medieval Mason's Tools: The Level and Plumb Rule,« *Technology and Culture*, II (1961), pp. 127–30.

Shelby, Lon R. *John Rogers: Tudor Military Engineer*. Oxford: Oxford University Press, 1967.

Singer, Charles Joseph *et al.* (eds.). *A History of Technology*. 5 vols.; Oxford: Clarendon Press, 1954–58.

Smith, Cyril Stanley and Hawthorne, John G. (eds.). *Mappae Clavicula: A Little Key to the World of Medieval Techniques.* (»Transactions of the American Philosophical Society,« N.S. Vol. LXIV, Part 4). Philadelphia: American Philosophical Society, 1974.

Stammler, Wolfgang (ed.). *Deutsche Literatur des Mittelalters: Verfasserlexikon.* Berlin: de Gruyter, 1933–55.

Stammler, Wolfgang (ed.). *Deutsche Philologie im Aufriß.* Berlin: Schmidt, 1954–62.

Stetten, Paul von (der Jüngere). *Kunst, Gewerb und Handwerks Geschichte* [sic] *der Reichs-Stadt Augsburg.* 2 vols.; Augsburg: Conrad Heinrich Stage, 1779–88.

Stöcklein, Ansgar. *Leitbilder der Technik: Biblische Tradition und technischer Fortschritt.* Munich: Heinz Moos, 1969.

Storck, John and Teague, Walter Dorwin. *Flour for Man's Bread: A History of Milling.* Minneapolis: University of Minnesota Press, 1952.

Strauss, Felix F. »Mills Without Wheels' in the Sixteenth-Century Alps,« *Technology and Culture,* XII (1971), pp. 23–42.

Streidl, H. »Die Bucherei des Orietalisten Johann Albrecht Widmannstetter,« in H.J. Kissling and Alois Schmaus (eds.). *Serta monacensia: Franz Babinger zum 15. Januar 1951 als Festgruß dargebracht.* Leiden: Brill, 1952, pp. 200–44.

Taccola, Mariano di Jacopo. *De machinis.* Ed. Gustina Scaglia. Wiesbaden: Dr. Ludwig Reichert Verlag, 1971.

Taccola, Mariano di Jacopo. *Liber Tertius de ingeneis ac edifitiis non usitatis.* Ed. James H. Beck. Milano: Polifilo, 1969.

Taenzler, Walter. *Der Wortschatz des Maschinenbaues im 16., 17. und 18. Jahrhundert: Ein lexikalischer und sprachwissenschaftlicher Beitrag zur Kultur- und Volkskunde.* Inaugural-Dissertation; Bonn: by the author, 1955.

Taylor, F.R. Forbes. »The Winch from Well-Head to Goliath Crane,« *Chartered Mechanical Engineer,* IX (1962), pp. 200–09.

Theophilus Presbyter. *De diversis artibus.* Ed. and trans. C. R. Dodwell. London: Thomas Nelson and Sons, 1961.

Theophilus Presbyter. *On Divers Arts.* Ed. and trans. John G. Hawthorne and Cyril Stanley Smith. Chicago: University of Chicago Press, 1963.

Thompson, E.A. *A Roman Reformer and Inventor: Being a New Text of the Treatise* **De rebus bellicis**. Oxford: Clarendon Press, 1952.

Thorndike, Lynn. »Marianus Jacobus Taccola,« *Archives internationales d'histoire des sciences,* XXXIV (1955), pp. 7–26.

Todericu Doru. »Raketentechnik im 16. Jahrhundert: Bemerkungen zu einer in Sibiu (Hermannstadt) vorhanden Handschrift des Conrad Haas,« *Technikgeschichte,* XXXIV (1967), pp. 97–114.

Todericu Doru. *Preistoria rachetei moderne: Manuscrisul de la Sibiu (1400–1569).* Bucharest: Editura Academiei Republicii Socialiste Romania, 1969.

Toussaint, Fritz. *Lastenförderung durch fünf Jahrtausende.* Mainz: K. G. Krauskopf, 1965.

Tout, T.F. »Firearms in England in the Fourteenth Century,« *English Historical Review,* XXVI (1911), pp. 666–702.

Treue, Wilhelm. *Achse, Rad und Wagen.* München: F. Bruckmann, 1965.

Treue, Wilhelm *et al.* (eds.). *Das Hausbuch der Zwölfbrüderstiftung zu Nürnberg.* 2 vols.; München: F. Bruckmann, 1965.

Treue, Wilhelm. *Kulturgeschichte der Schraube von der Antike bis zum achtzehnten Jahrhundert.* München: F. Bruckmann, 1955.

Tucher, Endres. *Endres Tuchers Baumeisterbuch der Stadt Nürnberg.* Ed. Matthias von Lexer. Introduction and apparatus by Friedrich von Weeck. (»Bibliothek des litterarischen Vereins in Stuttgart,« Vol. LXIV). Stuttgart: 1862.

Tyghem, Frieda van. *Op en Om de Middeleeuwse Bouwerf: De Gereedschappen en Toestellen Gebruikt bij het Bouwen van de Vroege Middeleeuwen tot Omstreeks 1600.* (»Verhandelingen van de Koninklijke Vlaamse Academie voor Wetenschappen, Letteren en Schone Kunsten van Belgie; Klasse der Schone Kunsten.« Vol. XXVIII, Nr. 19, Parts 1 and 2.) Brussels: 1966.

Uccelli, Arturo. *Storia della tecnica dal medio evo ai nostri giorni.* Milano: U. Hoepli, n. d. (1944).

Unterkirchner, Franz. *Inventar der illuminierten Handschriften, Inkunabeln und Frühdrucke der österreichischen Nationalbibliothek.* Part I: *Die abendländischen Handschriften.* Wien: Georg Pachner Verlag, 1957.

Urbani, Domenico. *Segni di cartiere antiche.* Venezia: Naratovitch, 1870.

Usher, Abbot Payson. *A History of Mechanical Inventions.* 2nd ed., rev.; Cambridge, Massachusetts: Harvard University Press, 1954.

Valturio, Robert. *De re militari.* Verona: Boninus de Boninis, 1483. Editio princeps: Verona: 1472.

Veen, Johann van. *Dredge, Drain, Reclaim: The Art of a Nation.* 5th ed.; The Hague: Martinus Nijhoff, 1962.

Vegetius. *Flavi Vegeti Renati Epitoma rei militaris.* Ed. Carol Lang, Stuttgart: Teubner, 1967.

Veranzio, Fausto. *Machinae novae.* N. d. (1615?); reprinted München: Heinz Moos Verlag, 1965.

Verriest, Leo (ed.). *Le polyptique illustré dit »Veil rentier« de Messire Jehan de Pamele-Audenarde, vers 1275.* Bruxelles: 1950.

Villard de Honnecourt. *Sketchbook.* Ed. Theodore Bowie. 2nd ed.; New York: George Wittenborn, 1962.

Villard de Honnecourt. *Album de Villard de Honnecourt.* Ed. Alfred Darcel. Introduction and glossary by J. B. A. Lassus. Paris: Leonce Laget, 1968.

Vitruvius. *Vitruvius on Architecture.* Ed. and trans. F. Granger. (»The Leob Classical Library«). London: Heinemann, 1934.

Wagner, Eduard, Drobná, Zoroslava and Durdík, Jan. *Medieval Costume, Armour and Weapons.* Trans. Jean Layton. London: Andrew Dakers, n. d. (1957?).

Wailes, Rex. *The English Windmill.* London: Routledge and Kegan Paul, 1954.

Wailes, Rex. »Some Windmill Fallacies,« *Transactions of the Newcomen Society,* XXXII (1959–60), pp. 93–109.

Waldburg-Wolfegg, Johannes, Graf von. *Das mittelalterliche Hausbuch.* München: Prestel-Verlag, 1957.

Webb, Henry J. »The Science of Gunnery in Elizabethan England,« *Isis,* XLV (1954), pp. 10–21.

Wegner, Hans. *Beschreibendes Verzeichnis der deutschen Bilder-Handschriften des späten Mittelalters in der Heidelberger Universitäts-Bibliothek.* Leipzig: Verlagsbuchhandlung J. J. Weber, 1927.

Werlin, Joseph. »Ein Feuerwerkstext aus dem 16. Jahrhundert,« *Centaurus,* IX (1963), pp. 272–87.

White, Lynn, jr. »Jacopo Aconcio as an Engineer,« *American Historical Review,* LXXII (1967), pp. 425–44.

White, Lynn, jr. »Medical Astrologers and Late Medieval Engineering,« *Viator,* VI (1975), pp. 295–315.

White, Lynn, jr. *Medieval Technology and Social Change.* New York: Oxford University Press, 1962.

White, Lynn, jr. »Theophilus Redivivus,« *Technology and Culture,* V (1964), pp. 224–33.

Wilbur, C. Martin. »The History of the Crossbow,« *Annual Report of the Smithsonian Institution* (1936), pp. 427–38.

Wilson, Paul N. *Watermills with Horizontal Wheels.* (»Society for the Protection of Ancient Buildings,« Booklet No. 7). Kendal: Titus Wilson and Son, 1960.

Wolf, A. *A History of Science, Technology and Philosophy in the Sixteenth and Seventeenth Centuries.* 2nd ed.; London: Allen and Unwin, 1950.

Wolf, Herbert. *Studien zur deutschen Bergmannssprache in den Bergmannsliedern des 16.–20. Jahrhunderts vorwiegend nach mitteldeutschen Quellen.* Tübingen: Max Niemeyer Verlag, 1958.

Wolfram von Eschenbach. *Wolfram von Eschenbach, Werke.* Ed. Karl Lackmann. 6th ed.; Berlin and Leipzig: Walter de Gruyter, 1926.

Woodbury, Robert S. *Studies in the History of Machine Tools.* Cambridge, Mass.: M.I.T. Press, 1972.

Wulf, Max von. *Die hussitische Wagenburg.* Inaugural-Dissertation; Berlin: Gustav Schade, 1889.

Zeising, Heinrich. *Theatri machinarum: erster–fünfter Theill.* Leipzig: Grossen, 1612–14.

Zubov, V.P. *Leonardo da Vinci.* Trans. David H. Kraus. Cambridge, Mass.: Harvard, 1968.

INDEX CODICUM

Bessnitzer, Ulrich

Heidelberg, Univ.-Bibl.
  Cod. palat. germ. 502

Feuerwerkbuch

Berlin, Staatsbibl. Preussischer Kulturbesitz
  Ms. germ. fol. 94
  Ms. germ. qu. 128
  Ms. germ. qu. 621
  Ms. germ. qu. 867
  Ms. germ. qu. 1018
  Ms. germ. fol. 1117
  Ms. germ. qu. 1187
  Ms. germ. qu. 1188

Dresden, Sächs. Landesbibl.
  Hs. App. 463
  Hs. C 113

Frankfurt, Stadt- u. Univ.-Bibl.
  Ms. germ. qu. 14

Freiburg i. B., Univ.-Bibl.
  Hs. 262

Göttingen, Univ.-Bibl.
  Cod. phil. 64

Heidelberg, Univ.-Bibl.
  Cod. palat. germ. 502
  Cod. palat. germ. 562
  Cod. palat. germ. 787

Kassel, Landesbibl.
  Ms. math. qu. 14

Leipzig, Univ.-Bibl.
  Ms. 1597

München, Bayerische Staatsbibl.
  Cod. hebr. 235
  Cod. germ. 356
  Cod. germ. 399
  Cod. germ. 734
  Cod. germ. 4902

New York, New York Public Library
  Ms. Spencer 104

Nürnberg, Germanisches Nationalmuseum
  Hs. 1480
  Hs. 1481 a

Sibiu, Archives munic.
  Ms. Varia II 374

Strasbourg, Bibl. Nat. et Univ.
  Ms. 2259

Weimar, Nat. Forschungs- u. Gedenkstätten d. kl. Deutschen Lit.
  Ms. qu. 342

Wien, ÖNB
  Cod. 2952
  Cod. 2987
  Cod. 3062
  Cod. 3064
  Cod. 10,855
  Cod. 10,895
  Cod. 10,917
  Cod. 10,953
  Cod. Ser. nova 12,769

»Formschneider« Types

Frankfurt, Stadt- u. Univ.-Bibl.
  Ms. germ. qu. 14

München, Bayerische Staatsbibl.
  Cod. germ. 734

München, Deutsches Museum
  Ms. 1949/258

Nürnberg, Germanisches Nationalmuseum
  Hs. 719

Paris, Bibl. Nat.
  Ms. franc. 20006

Sibiu, Archives munic.
  Ms. Varia II 734

Weimar, Nat. Forschungs- u. Gedenkstätten d. kl. Deutschen Lit.
  Ms. qu. 342

Francesco di Giorgio

London, British Museum
Ms. Additional 34113
Dept. of Prints and Drawings, Ms. 197, 621

Giovanni da Fontana

München, Bayerische Staatsbibl.
Cod. icon. 242

Guido da Vigevano

Paris, Bibl. Nat.
Ms. latin 11015

Helm, Franz

Wien, ÖNB
Cod. 10,895
Cod. 10,917
Cod. 10,953
Cod. Ser. nova 12,769

Kyeser, Conrad (Bellifortis)

Besançon, Bibl. munic.
Ms. 1360

Berlin, Staatsbibl. Preussischer Kulturbesitz
Ms. germ. fol. 94
Ms. germ. qu. 621

Budapest, Bibl. Acad. Sci. Hung.
Latin cod. fol. 14

Chantilly, Musée Condé
Ms. 348 (633)

Colmar, Bibl. de la Ville
Ms. 491

Donaueschingen, Fürstl.-Fürstenberg. Hofbibl.
Ms. 860

Erlangen, Univ.-Bibl.
Ms. 1390

Frankfurt, Stadt- u. Univ.-Bibl.
Ms. germ. qu. 14
Ms. germ. qu. 15

Göttingen, Univ.-Bibl.
Cod. phil. 63
Cod. phil. 64

Heidelberg, Univ.-Bibl.
Cod. palat. germ. 787

Innsbruck, Ferdinandeum
Cod. 16.0.7.

Karlsruhe, Badische Landesbibl.
Ms. Durlach 11

Köln, Stadtarchiv
Ms. 232

München, Bayerische Staatsbibl.
Cod. hebr. 235
Cod. germ. 356

New York, New York Public Library
Ms. Spencer 58
Ms. Spencer 104

Paris, Bibl. Nat.
Ms. franc. 20006

Sibiu, Archives munic.
Ms. Varia II 374

Strasbourg, Bibl. Nat. et Univ.
Ms. 2259

Weimar, Nat. Forschungs- u. Gedenkstätten d. kl. Deutschen Lit.
Ms. fol. 328

Wien, ÖNB
Cod. 3062
Cod. 3068
Cod. 3069
Cod. 5278
Cod. 5518

Wien, Waffensammlung d. Kunsthist. Museums
Ms. P 5014
Ms. P 5135
Ms. 5342 A
Ms. P 6562

Zürich, Zentralbibl.
Ms. Rh. hist. 33 b

Monch, Phillibs

Heidelberg, Univ.-Bibl.
Cod. palat. germ. 126

Ordnance Inventories

Augsburg, Stadtarchiv
Ms. Schätze 137 f.

München, Bayerische Staatsbibl.
Cod. icon. 222

»Primitive« Types

Frankfurt, Stadt- u- Univ.-Bibl.
Ms. germ. qu. 14

München, Bayerische Staatsbibl.
Cod. germ. 600

Nürnberg, Germanisches Nationalmuseum
Ms. 25,801

Wien, ÖNB
Cod. 3069

Wien, Waffensammlung d. Kunsthist. Museums
Ms. P 5014

Zürich, Zentralbibl.
Ms. Rh. hist. 33b

Seselschreiber, Christof

München, Bayerische Staatsbibl.
Cod. germ. 973

NAME AND SUBJECT INDEX

Like any index, this is meant to facilitate the reader's task by referring him to information contained within the work. Because of this book's format, and because of the need to include as technical terms foreign words, nicknames, or slang terms, this index has several features which need to be understood in advance to avoid confusion. Most entries in the index are common English words, followed by simple page numbers, thus, **brace-and-bit: 99.** Some entries refer the reader to footnotes; in such cases the page number containing the footnote is given first, followed by the footnote number (designated by f.), thus, **Auberoche: 61, f.37**. In some instances, the reader is directed to information contained in transcriptions from or commentaries upon the individual folios of Clm 197, I; these entries contain *both* the relevant page number *and*, following in italics, the folio number to which reference is made, thus, **bellows: 151, 12r**. In a few cases, entries refer to information in footnotes to such transcriptions or commentaries; for these three numbers are given, the page number, the folio number (in italics), and the footnote number (designated by f.). Hence an entry such as **Valturio, Roberto, De re militari: ... 161, 17v supra, f.2** should lead the reader to that footnote number 2 on page 161 which is connected to the commentary on fol. 17v *supra*. As in the text, wherever possible modern English terms have been employed, but it has been necessary to retain some foreign words, as well as nicknames or slang terms for certain devices. Foreign words are indicated by italics. Nicknames or slang terms are set within German *Anführungszeichen* thus, » «. An entry such as **gastraphetes, see »belly-bow«.** clearly distinguishes for the reader the Greek term and its informal English equivalent. However, the reader should note that foreign words which are used as nicknames or slang are designated only by » «, not by italics. Finally, cross-references between individual folios of Clm 197, I are given in a separate index on pages 220–221.

INDEX OF FOLIOS CITED

This index refers the reader to the location of each reference to each folio of *Codex latinus monacensis* 197, part 1 that appears in this work. The appearance of folio numbers in this index means that the folio cited is mentioned in the commentary to a particular folio. One can thus find, for example, a reference to fol. 1r on page 143 in the commentary on fol. 4v. This index therefore provides a rough cross-tabulation of folios having some elements of similarity.

CODEX LATINUS MONACENSIS 197, 1

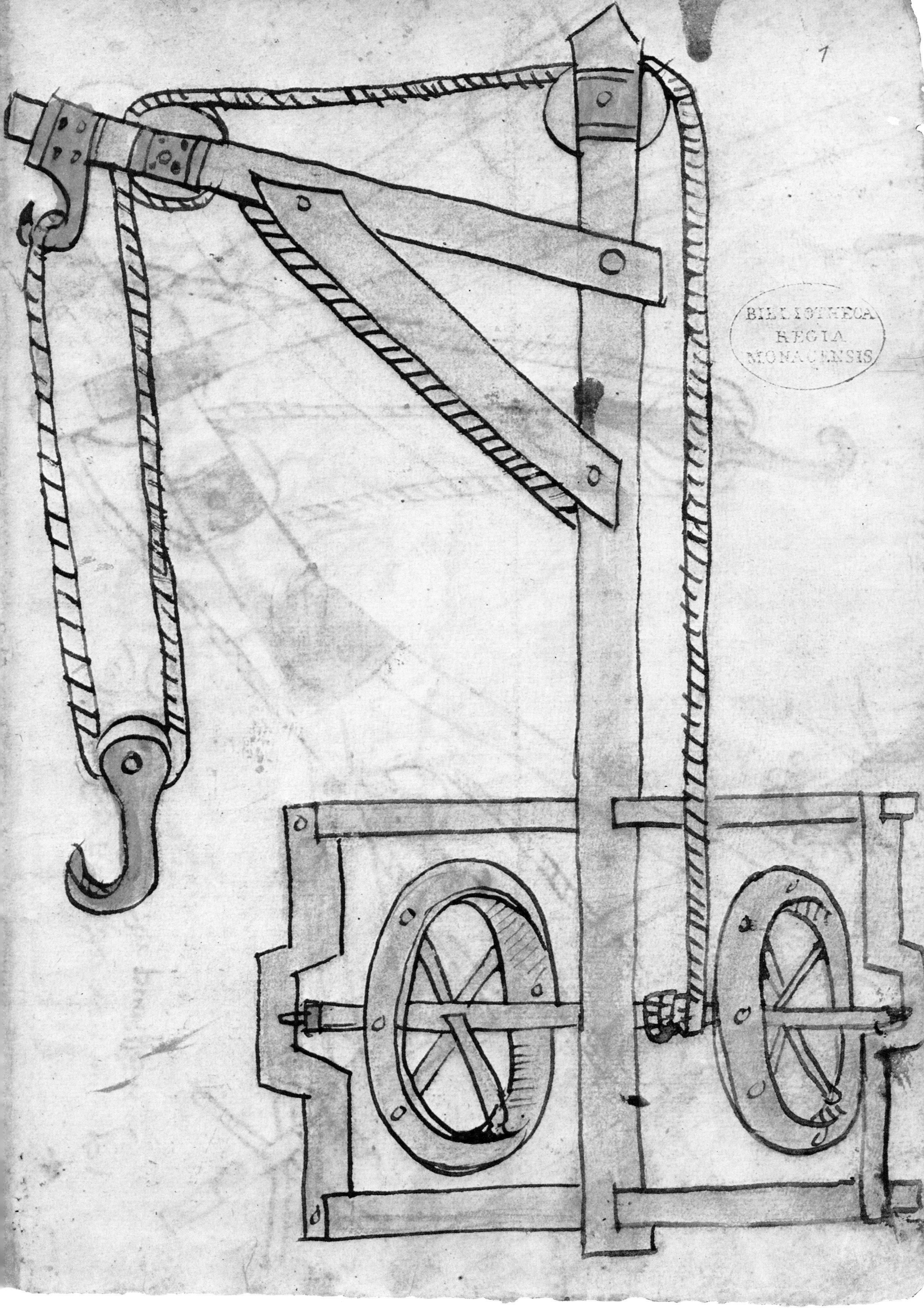

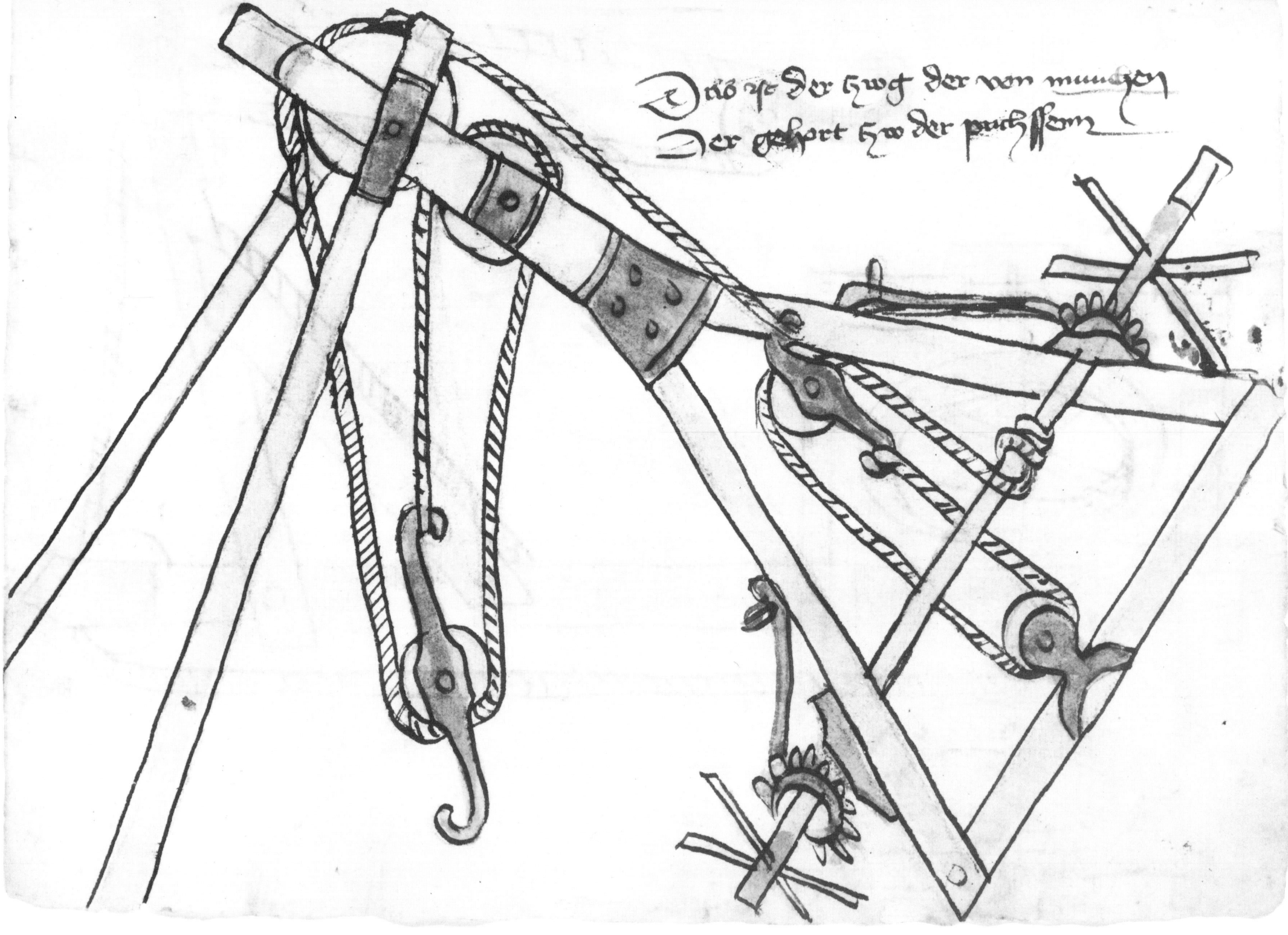

Das ist der zwg der von munchen
Der gehort zw der puchssen

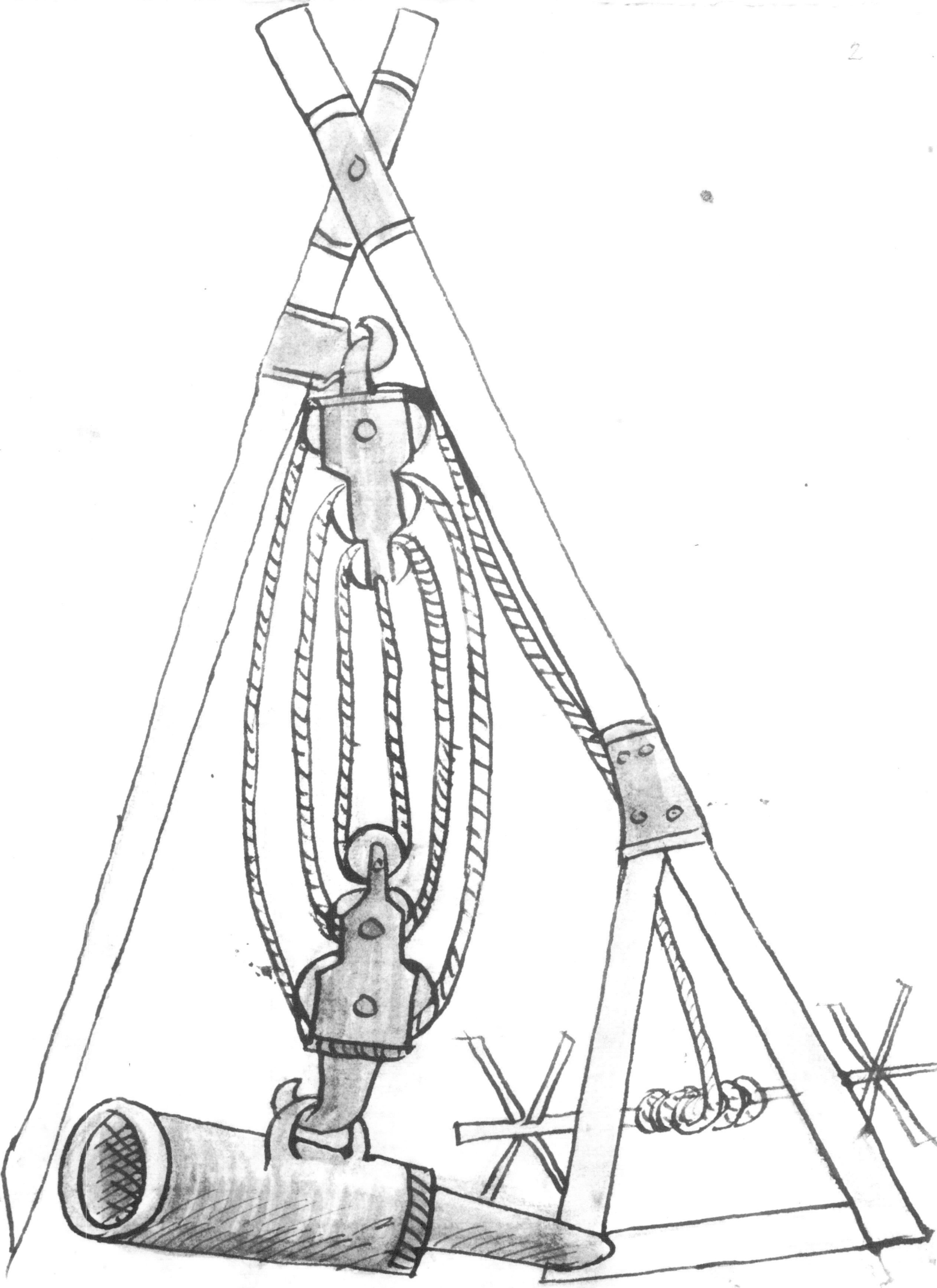

Item daz ist ein zug mit zwain kasten vnd der ain auff
der ander ab zu nurenberg auch dez grabn den treyben
zwen pfard

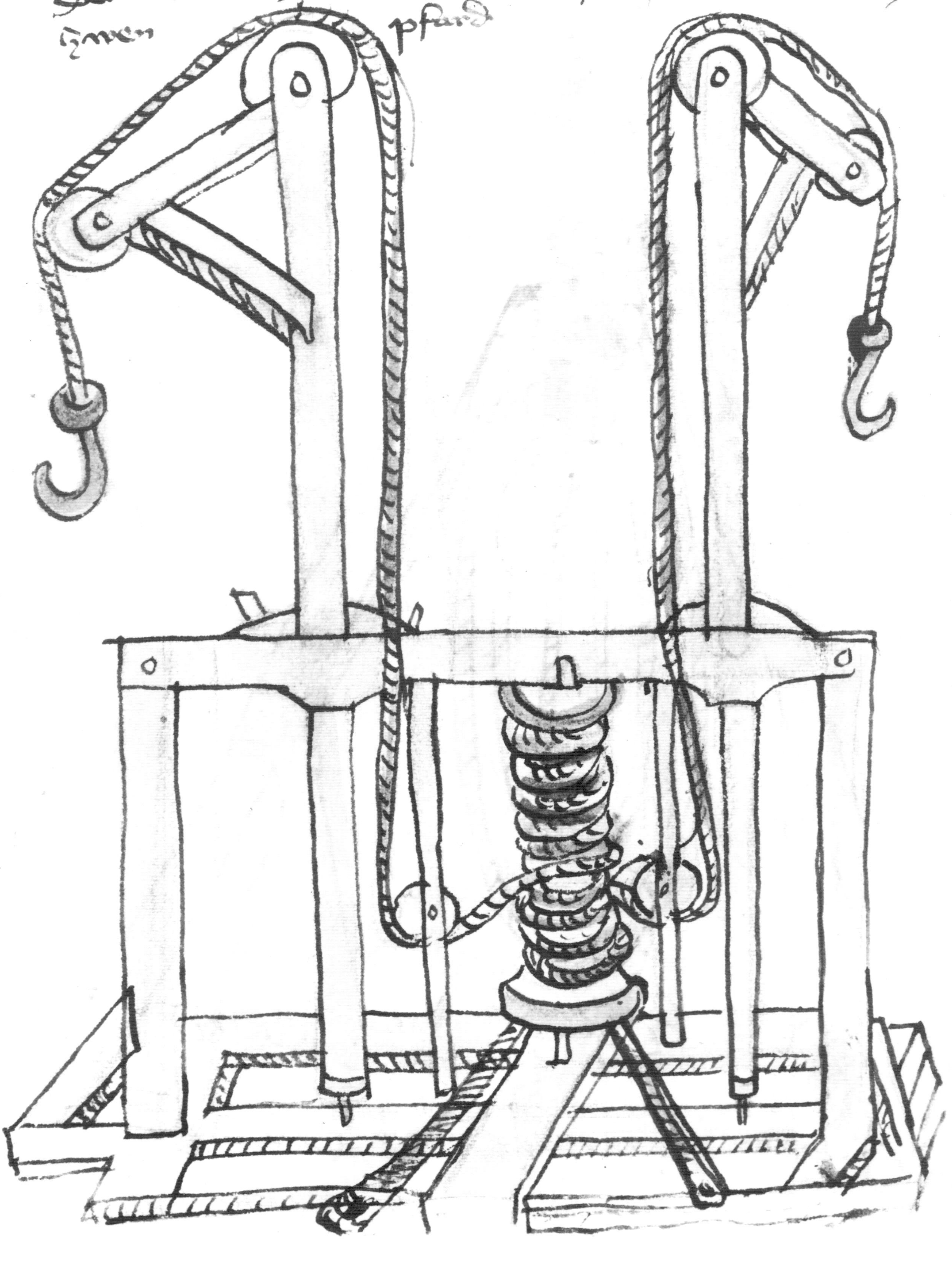

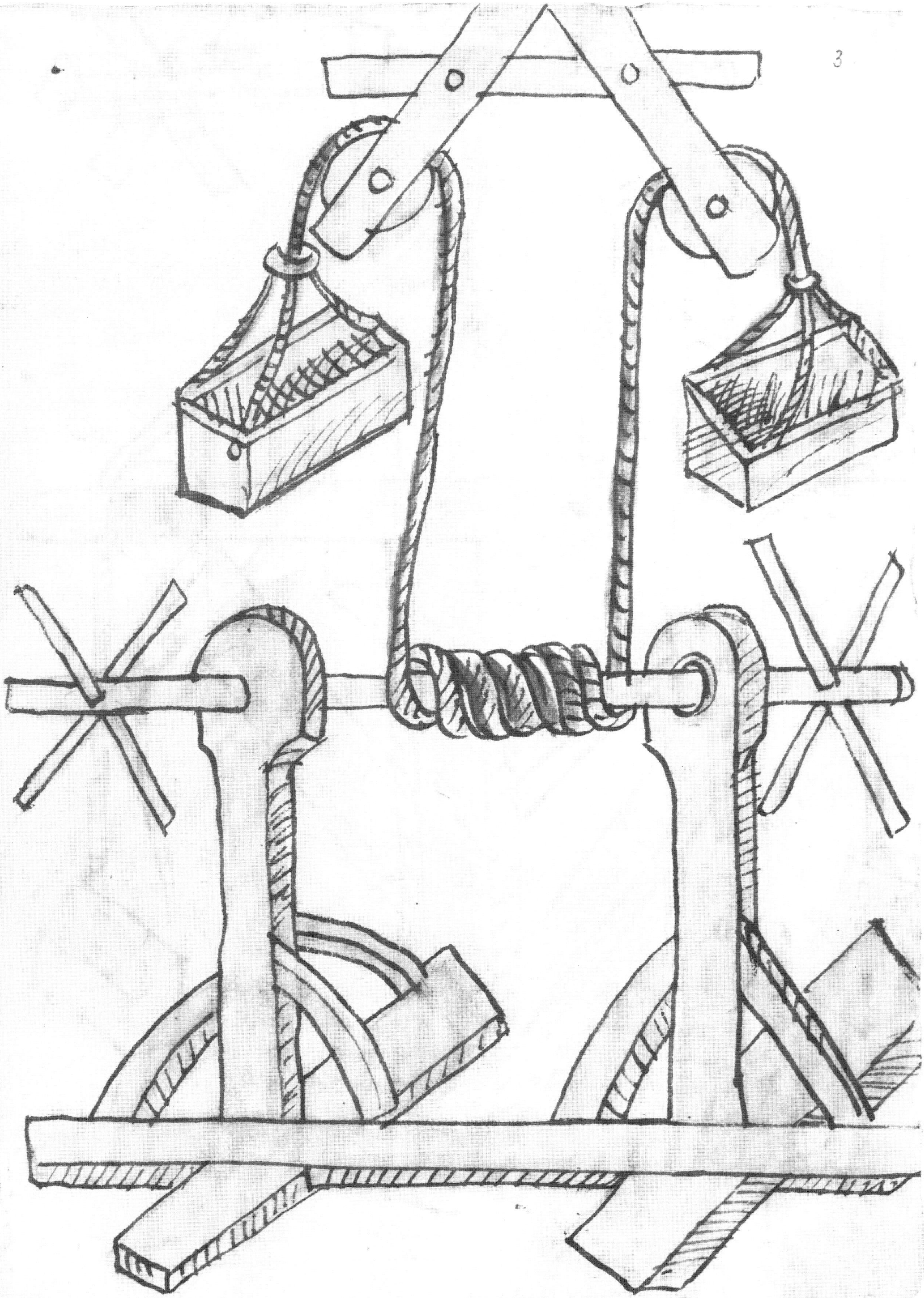

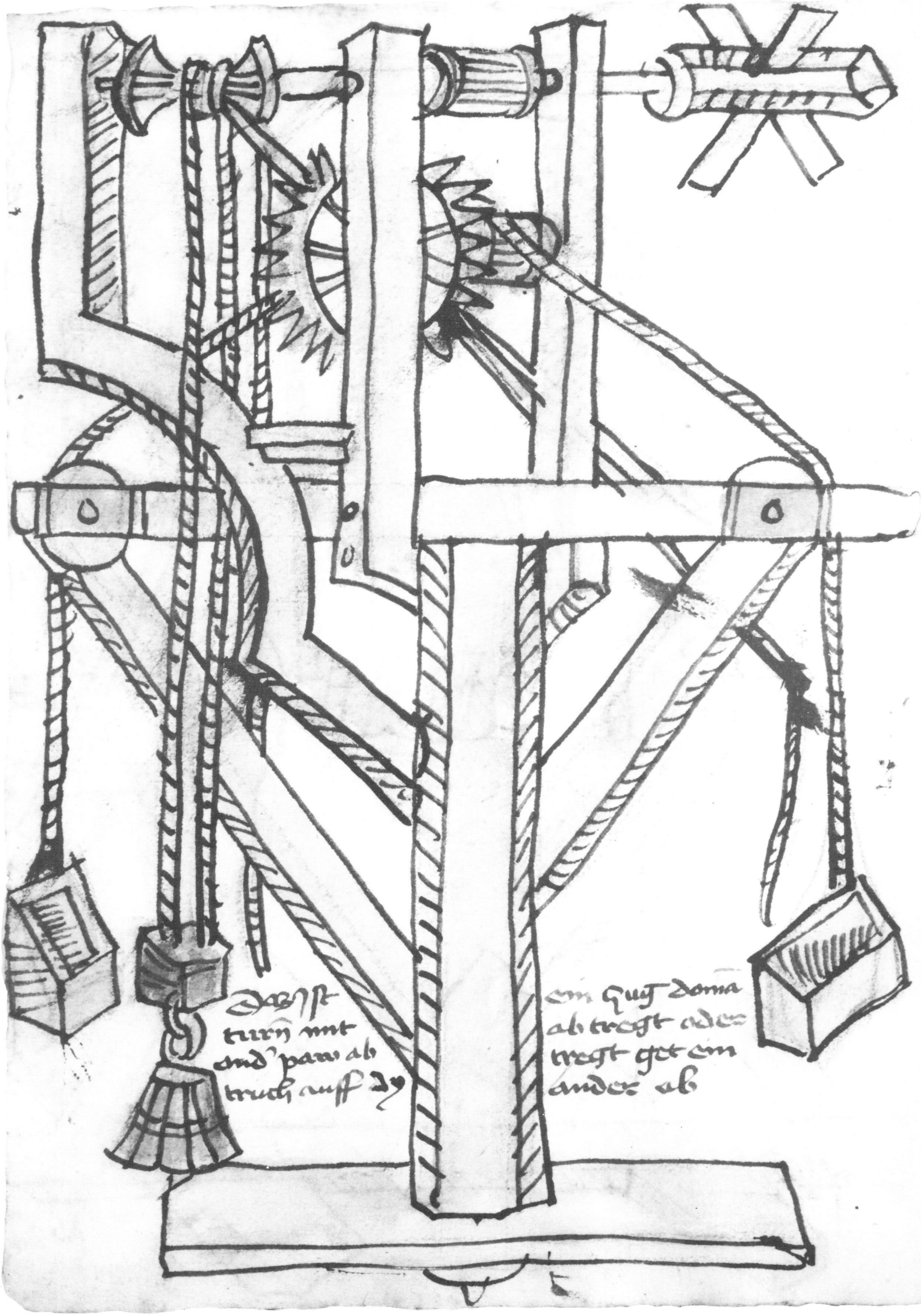

Das ist ein zug doma
turn mit ab tregt oder
vnd' paw ab tregt get ein
truch auff dy ander ab

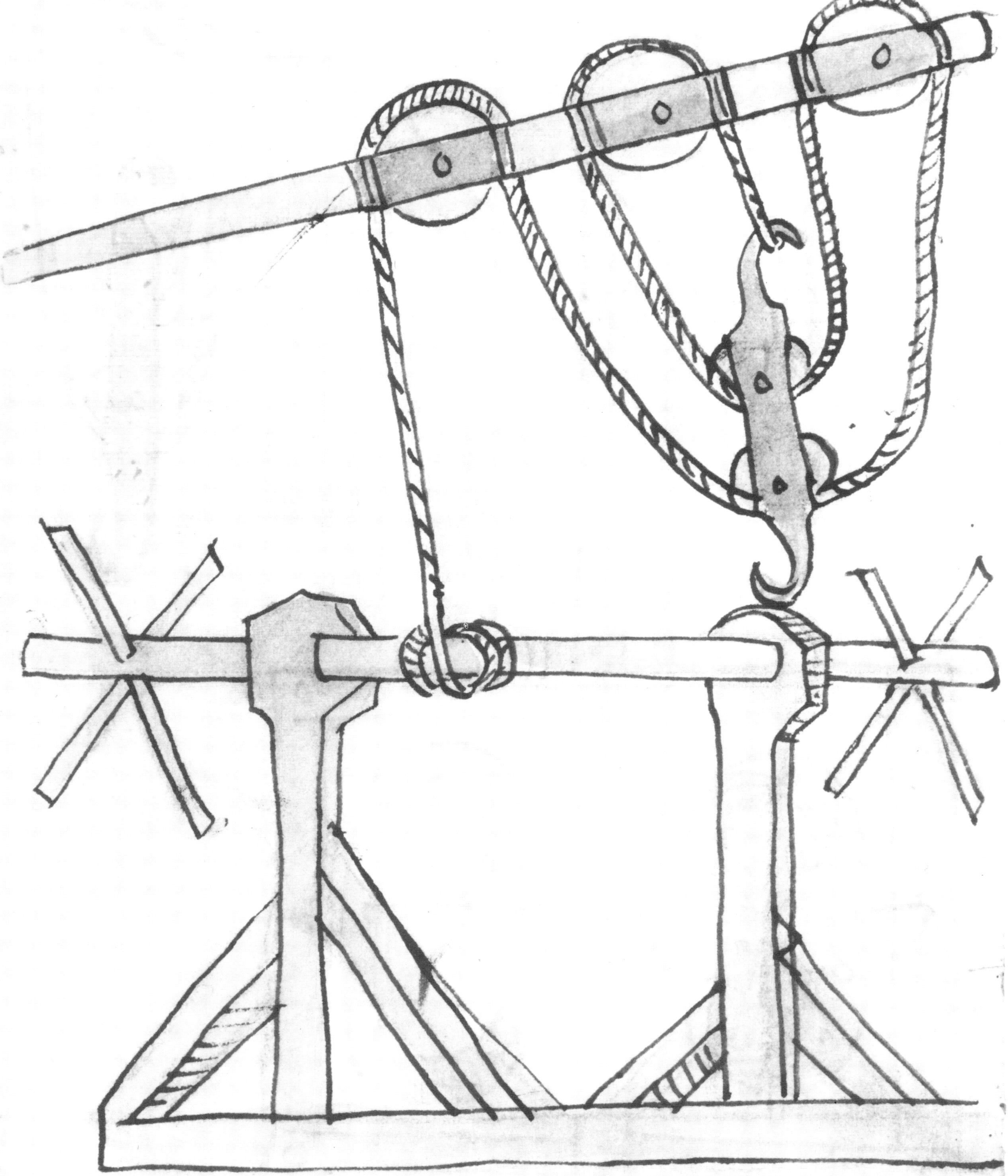

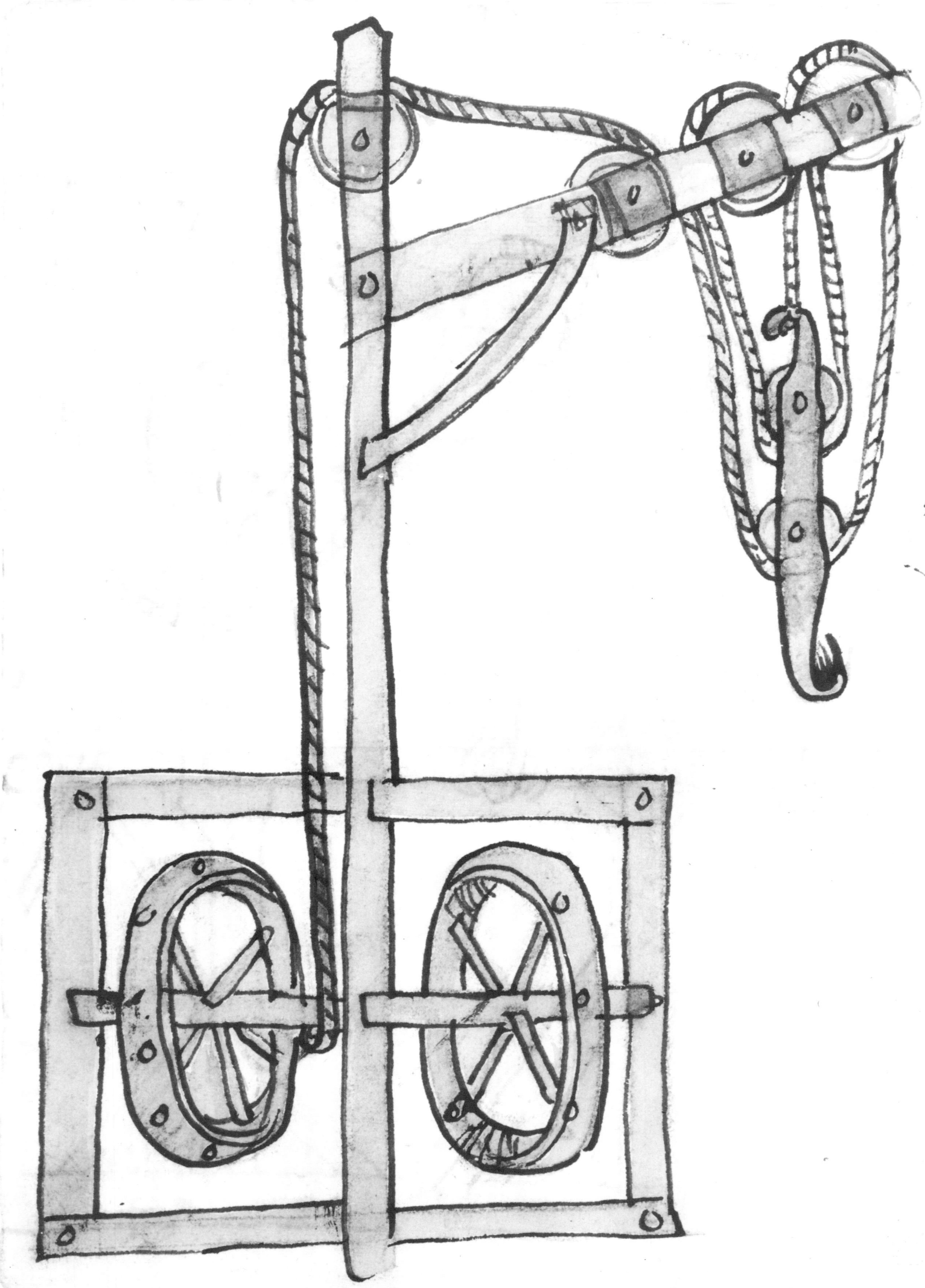

Item daz hauffent dy dimler dy man in trucke grabn wurfft
vnd dy halczer da an schlecht vnd ein prugk
macht daz man vber den grabn get
dar auß

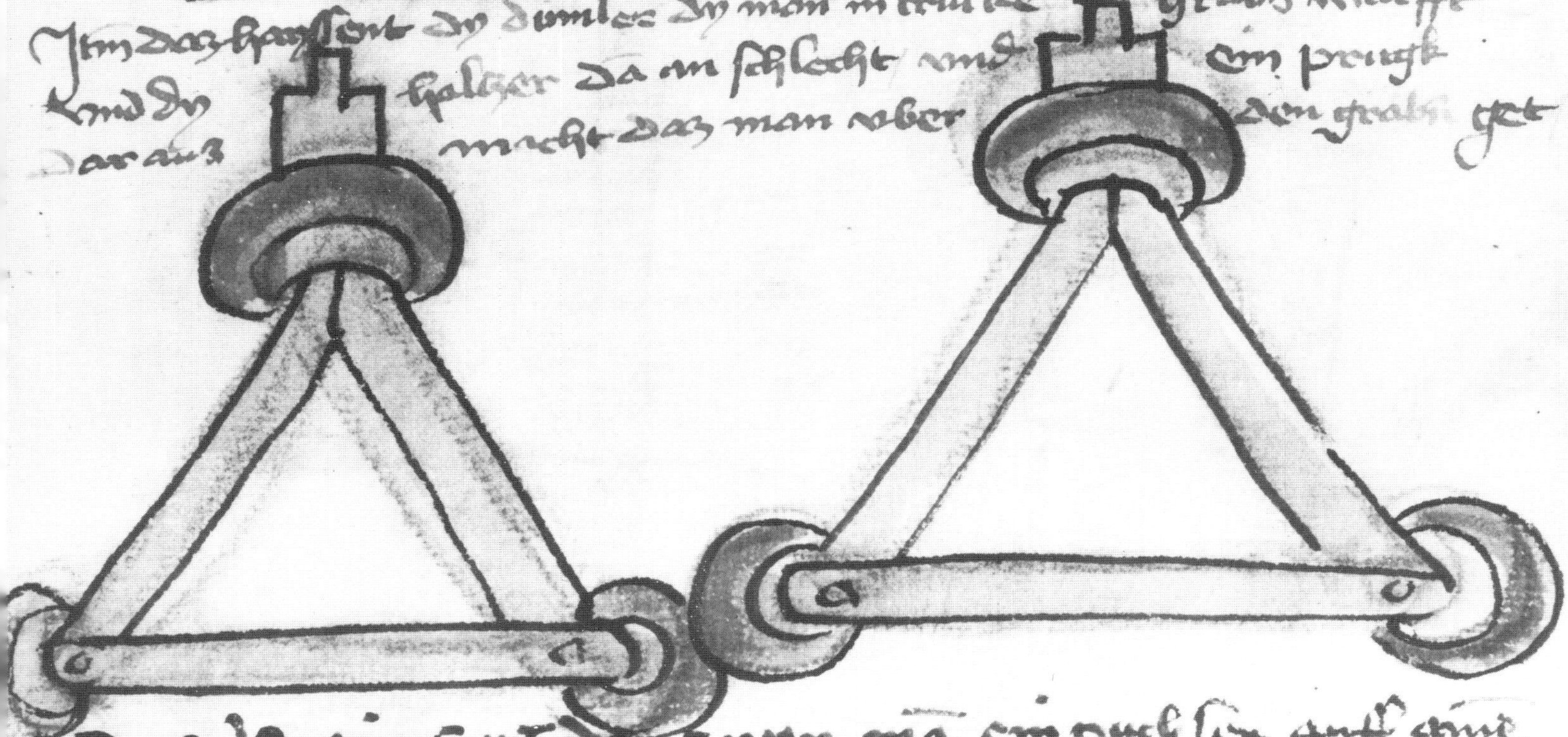

Daz ist ein zeug dy zwen mā ein puchssen auff eine
wagen heben dy sex zentn swar ist /

Item mit der kunst zwingt ma ein vall pruck auff dz tor
mit dreye schlossen verschlossen ist da schlacht man
die kracken oben in die pruck ein eysen anders
vnd da pint zwen starken pfar nole
eynander ein stuck die pruck mit einem
auff daz ist pewart ein slechts tost ein d
gulden

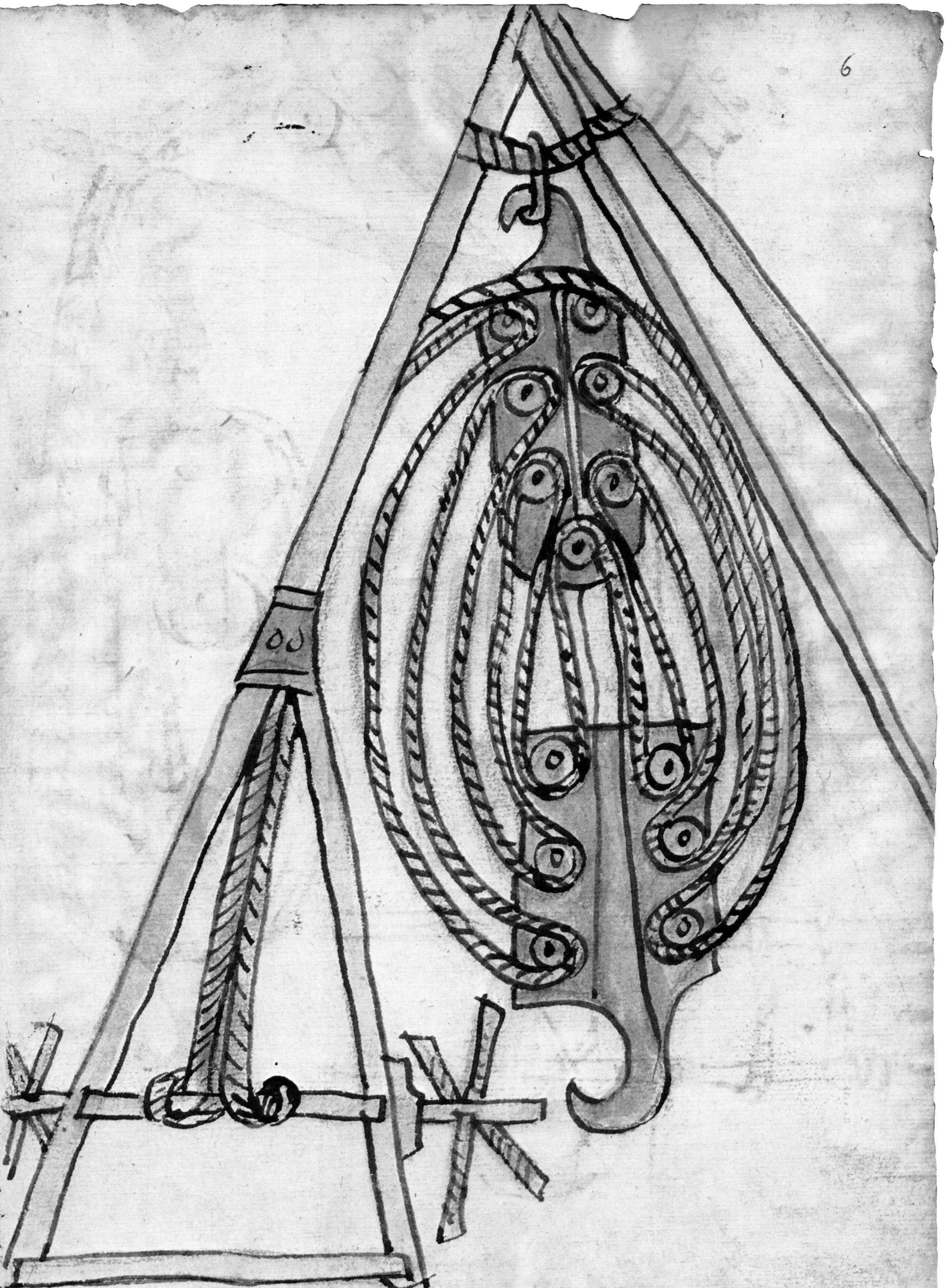

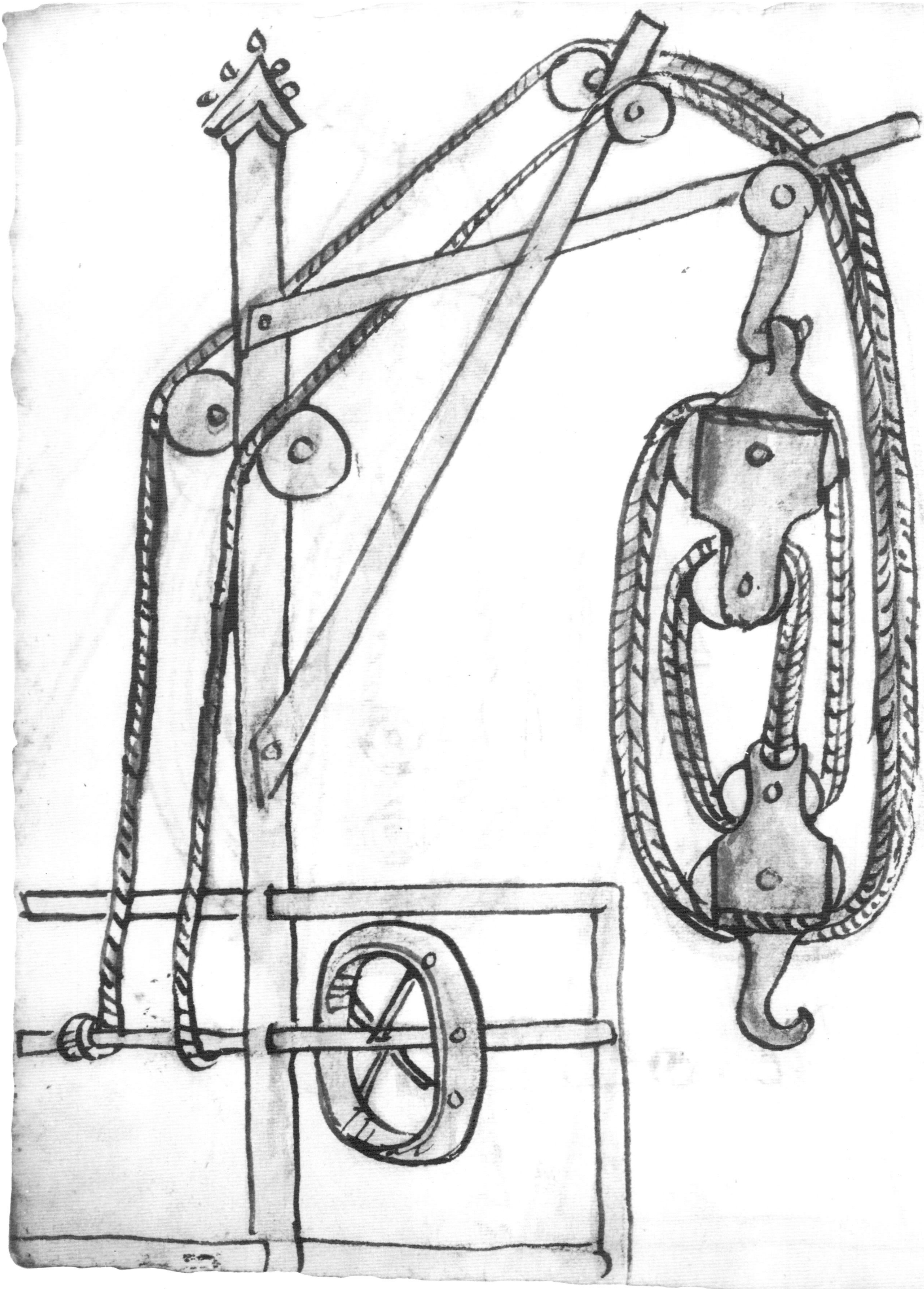

Item daz ist ein heber damit mā ein hebt 7
einer gaden hoch der da sicht waz man
in ein stat tutt dye da leyt auf ein ebn
da sicht mā waz mā darinne tut daz ist
pewart vñ gerecht

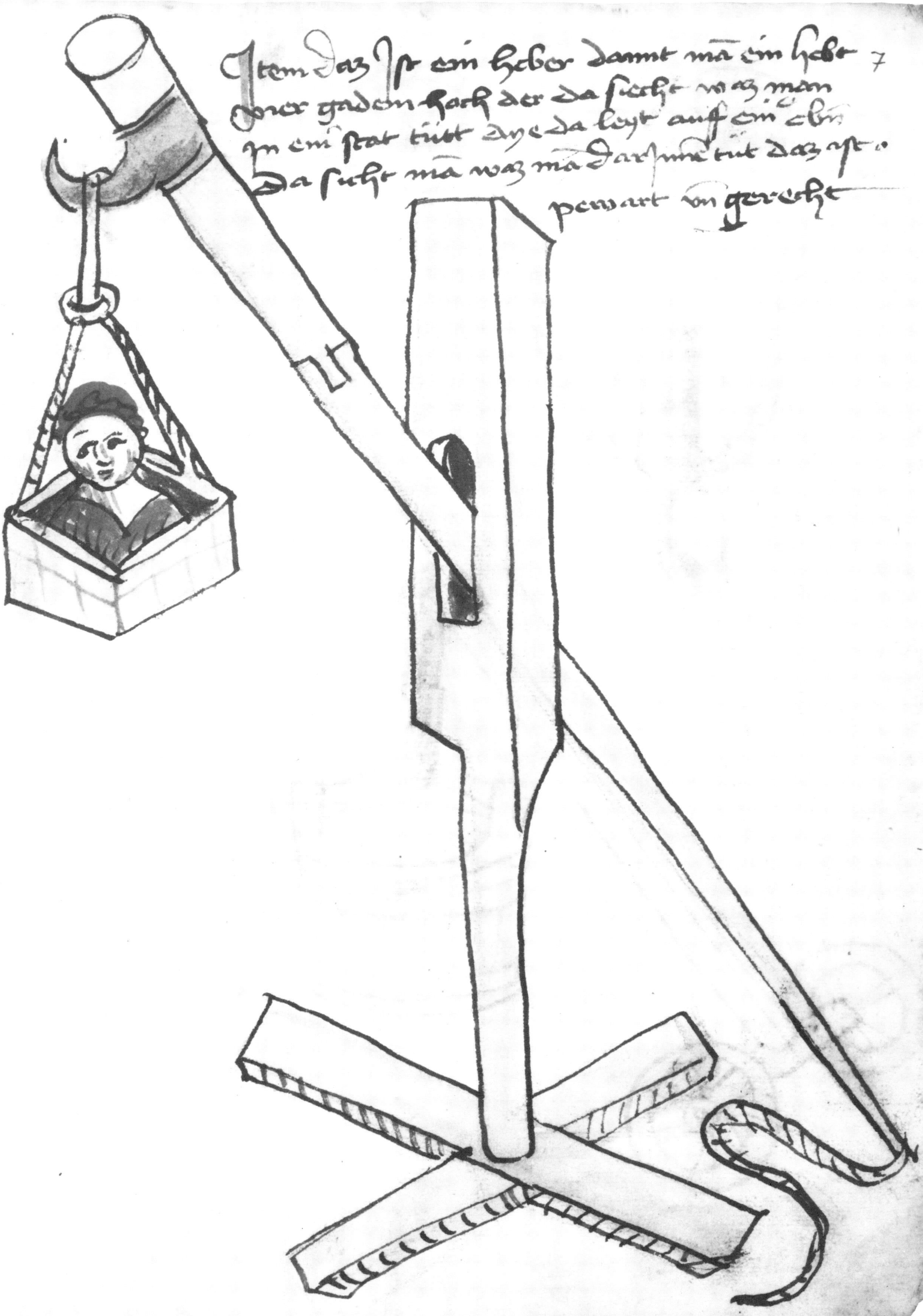

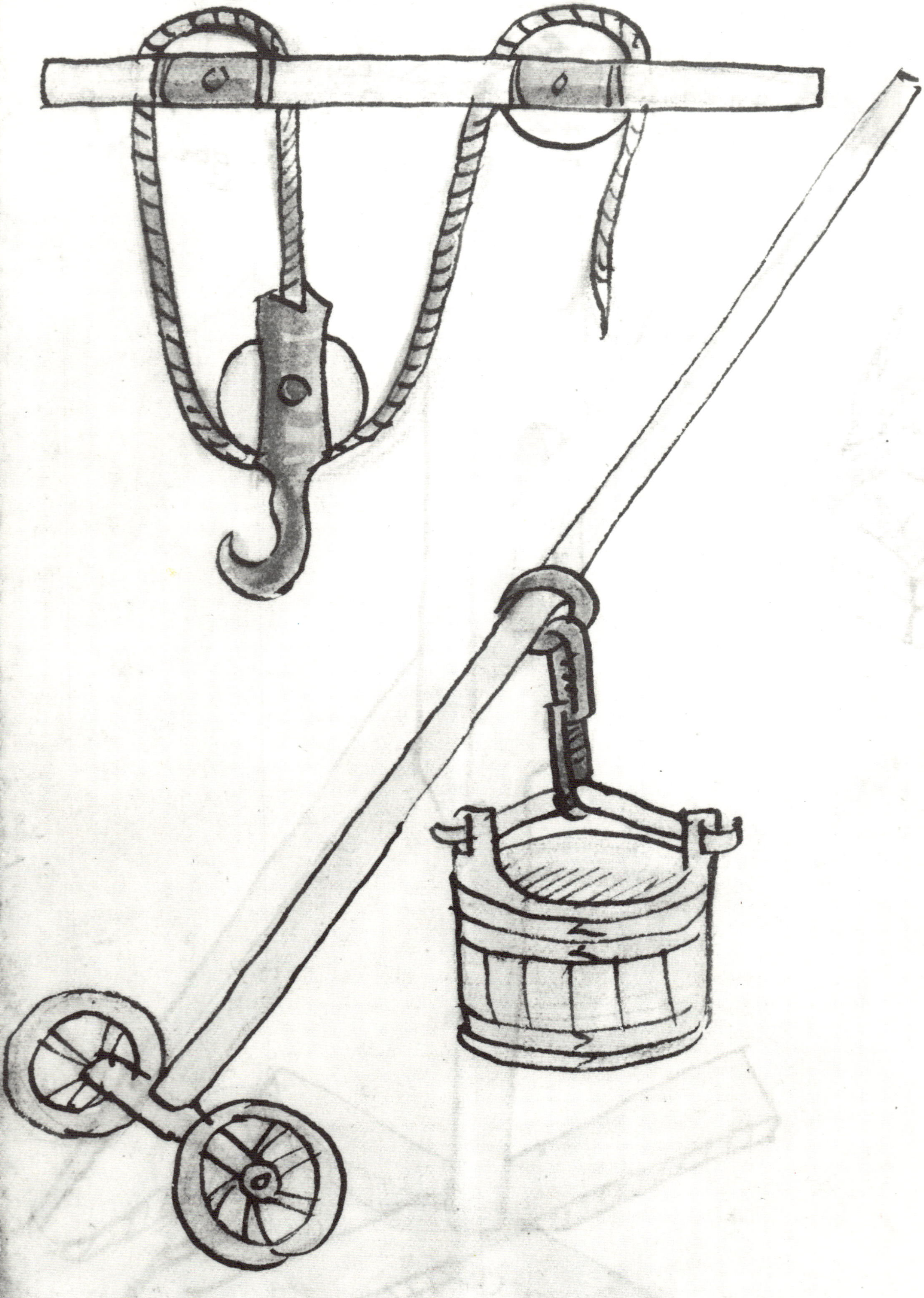

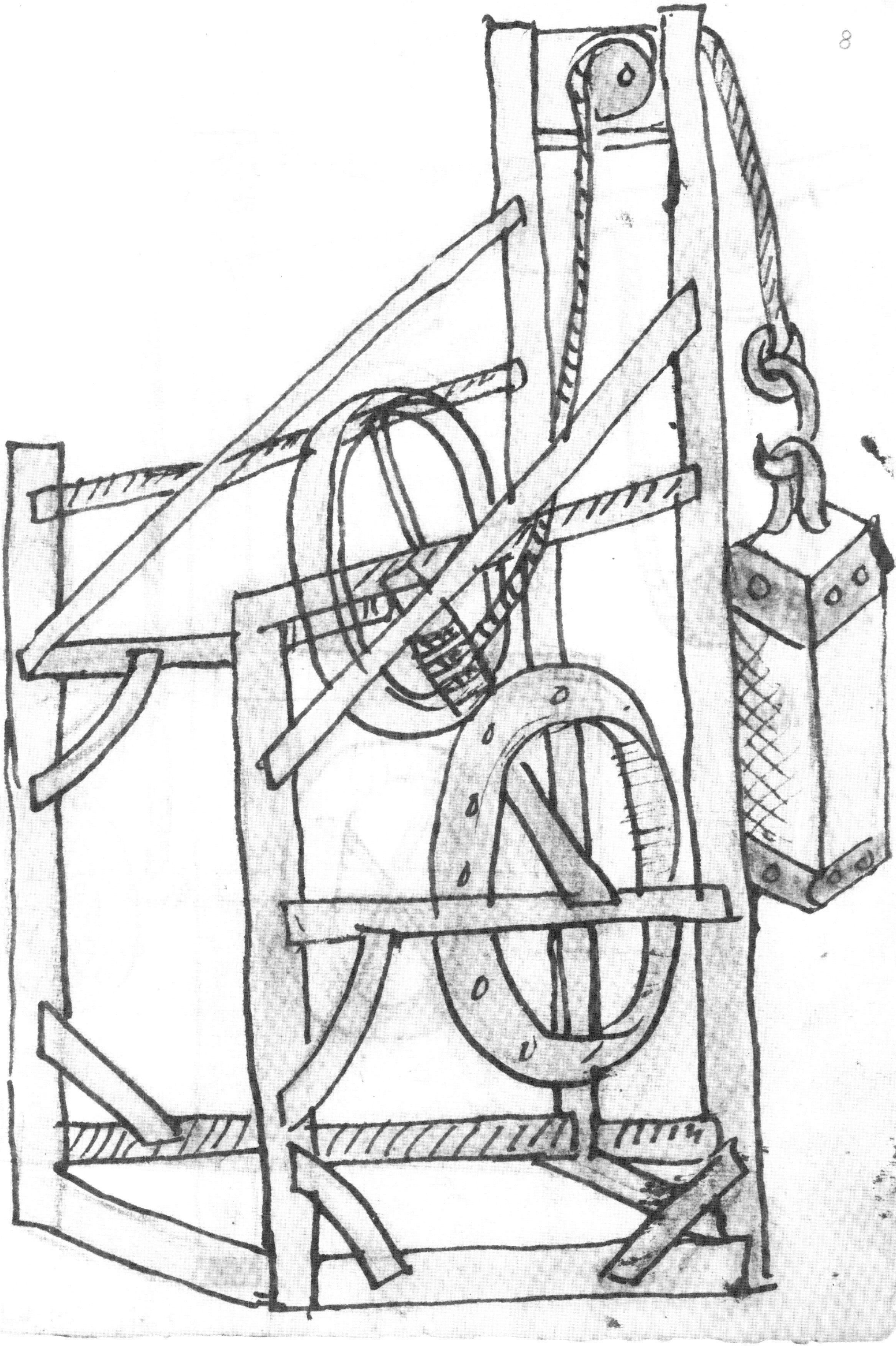

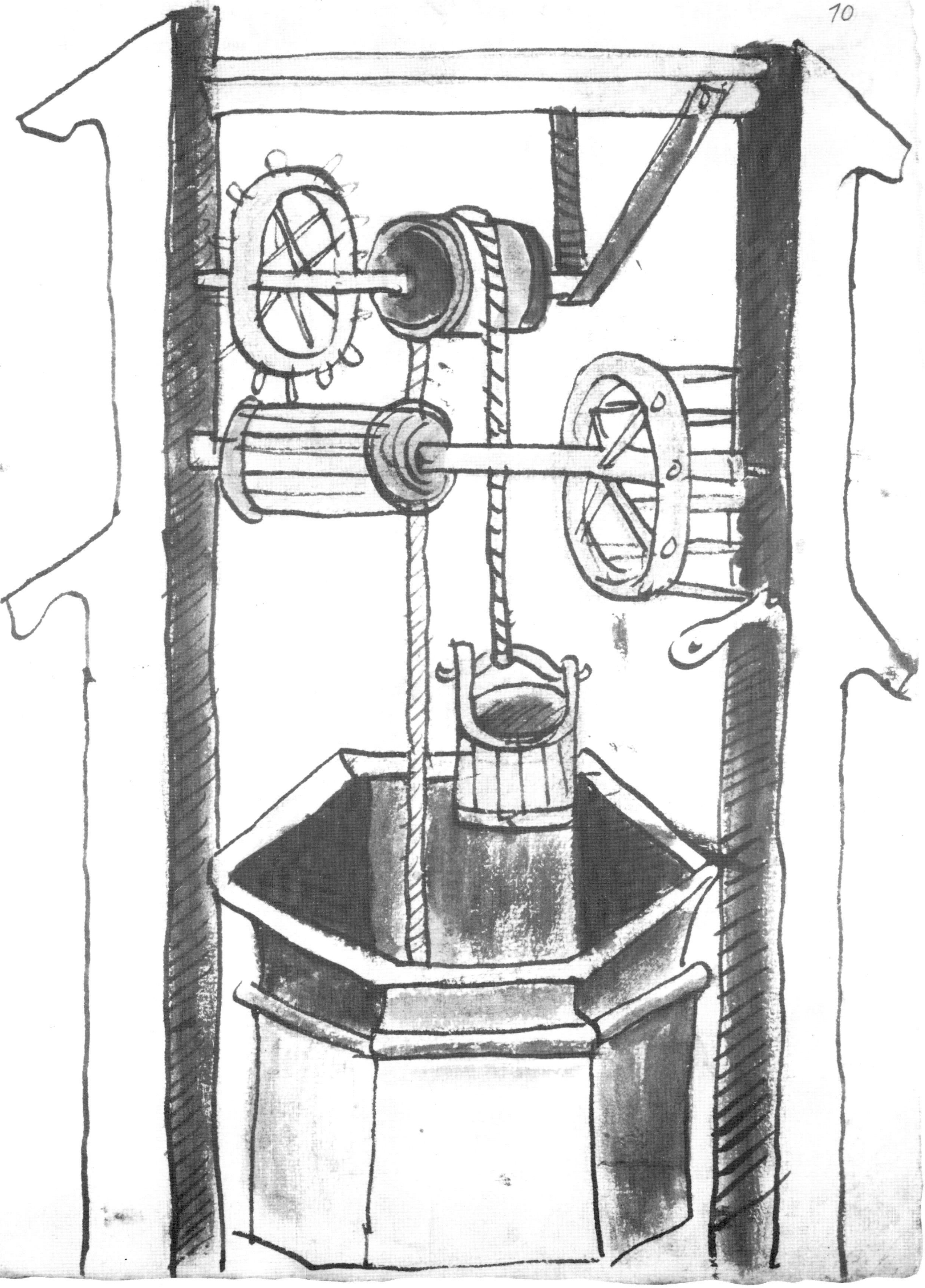

Item daz ist ain stampff damit mā pulver stost vnd dye stampff gent all drey in ain loch ainer auff der ander ab etc

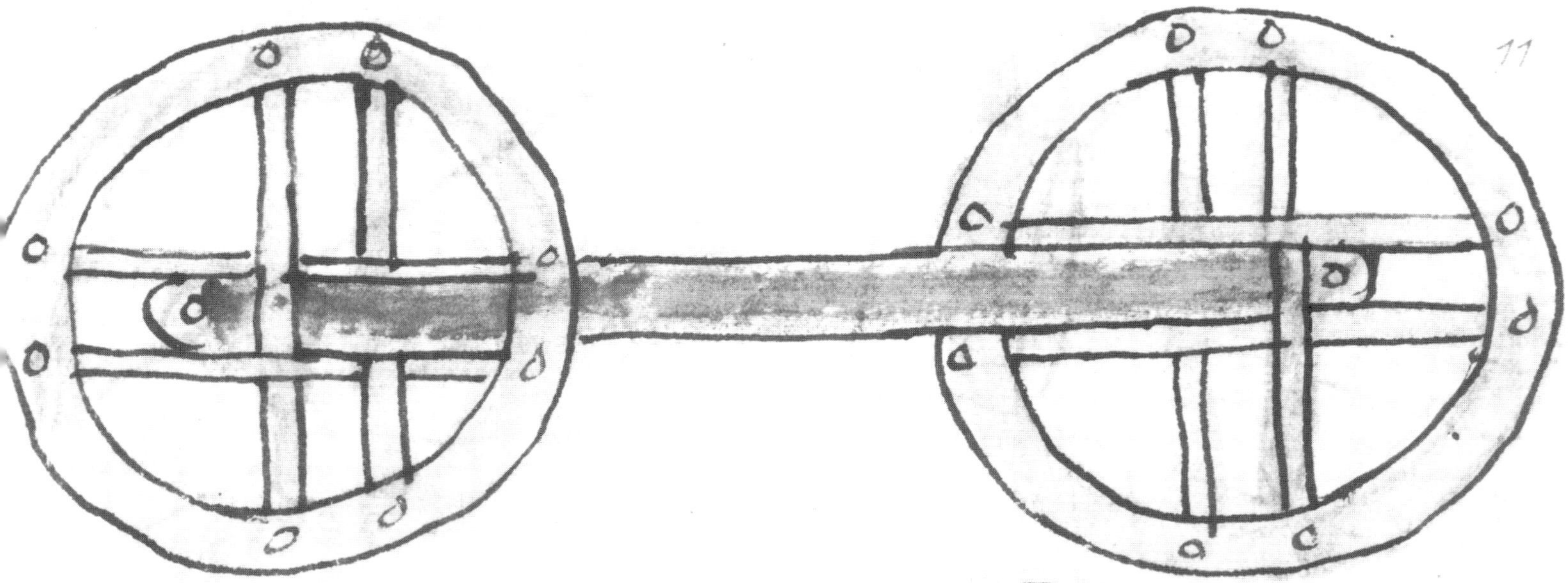

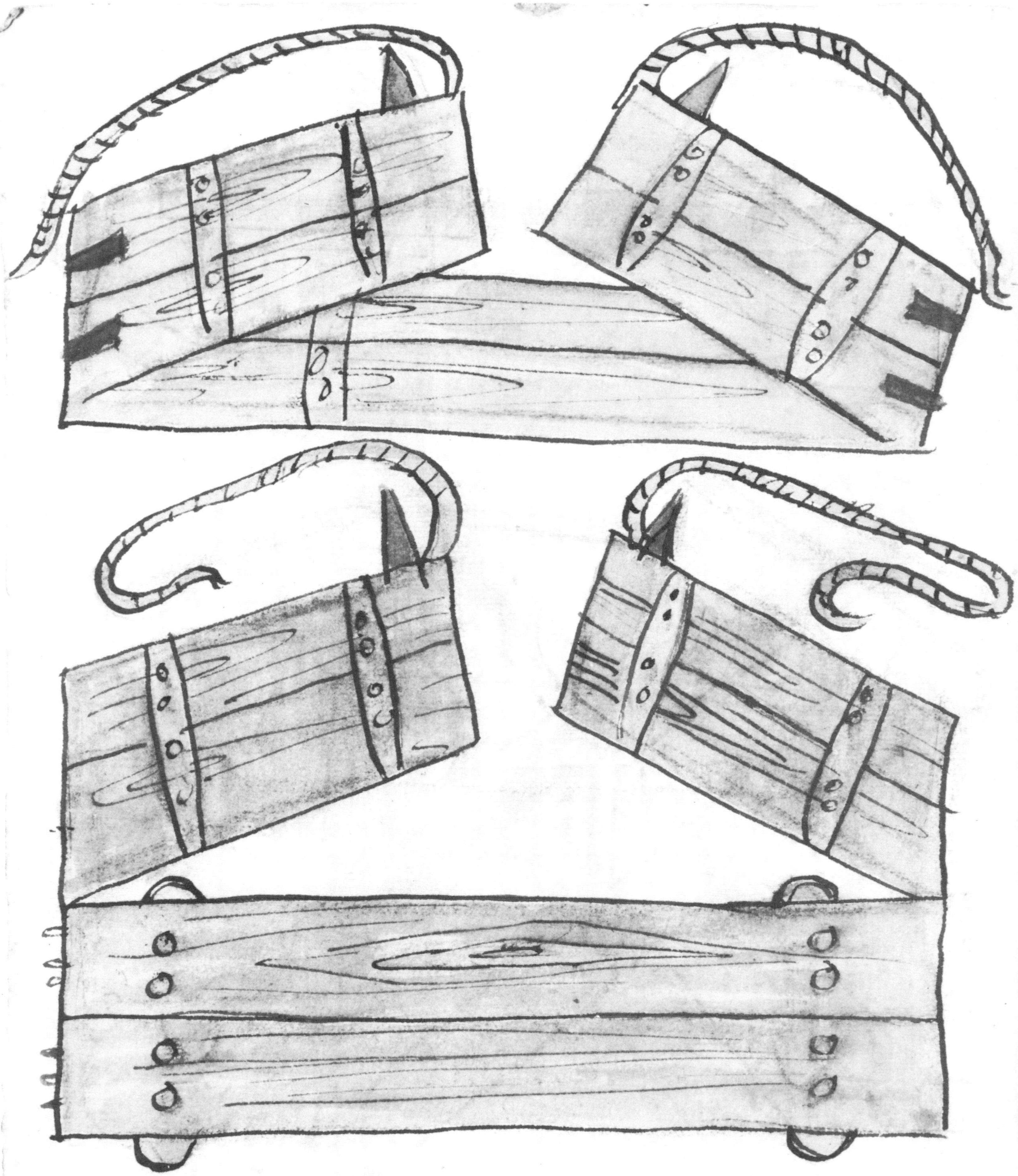

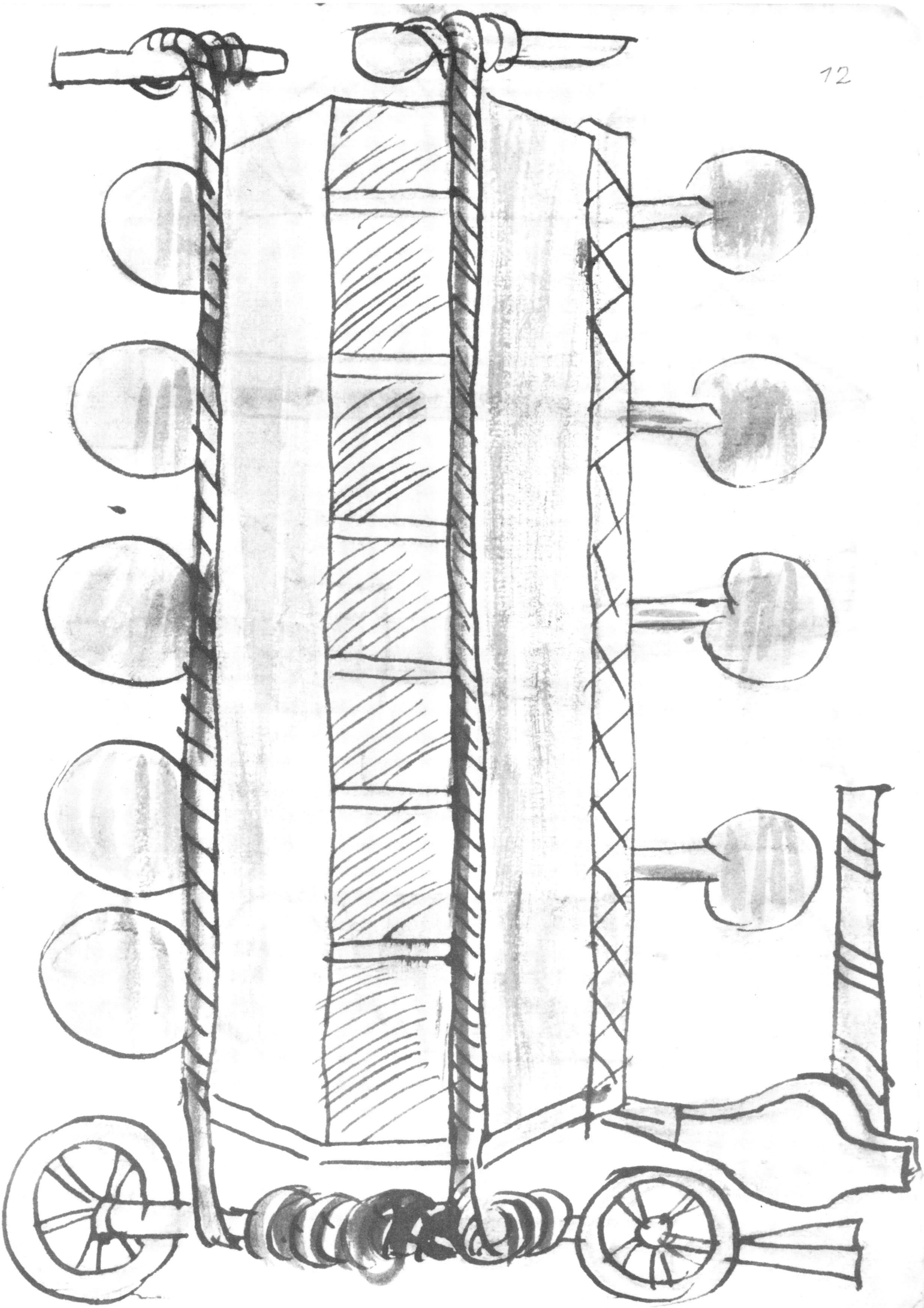

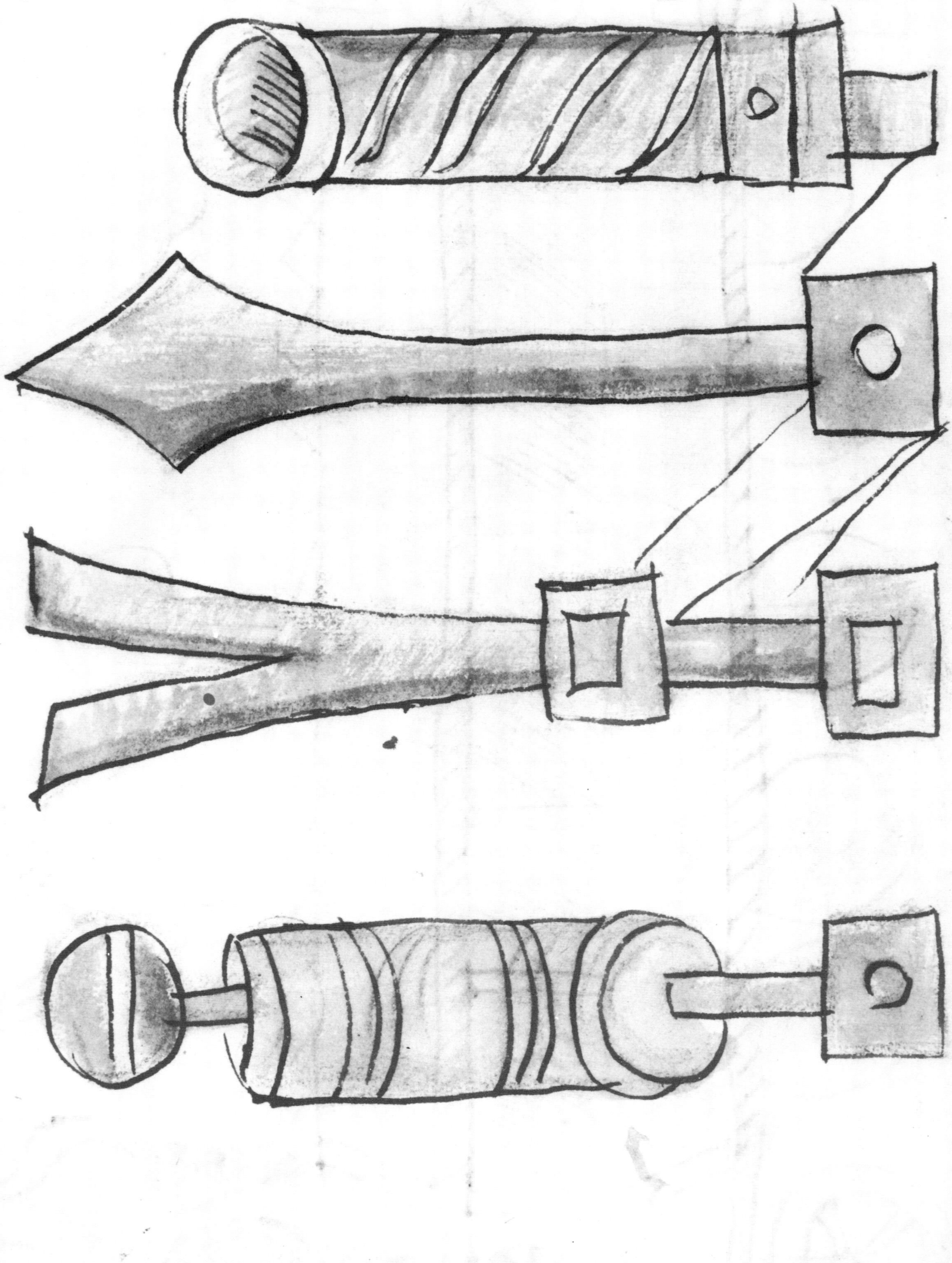

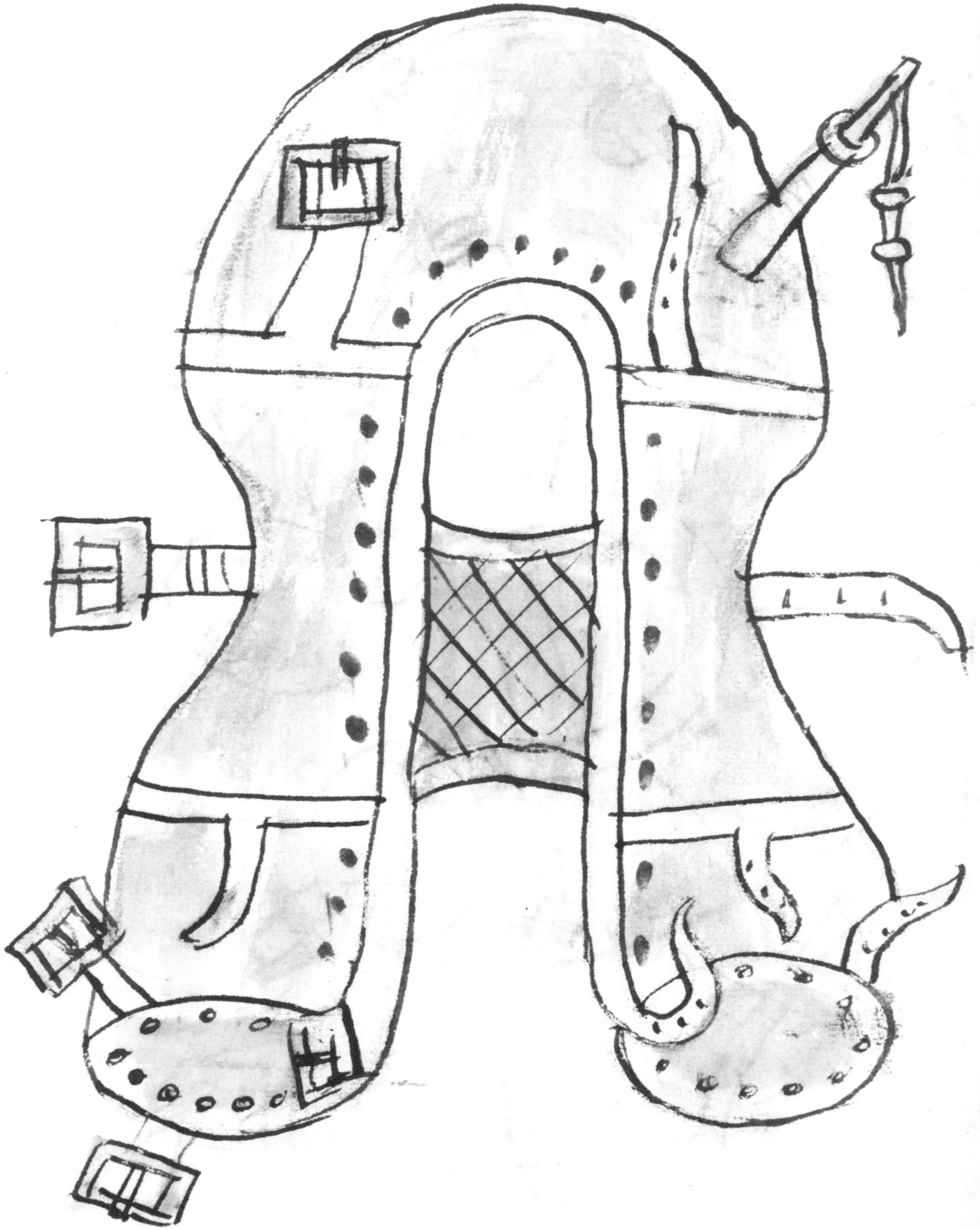

Daz ist ein gestell damit man mit rorn prot das habn
dy von nuerberg eins gemacht damit port man alltag 15
xv rorn daz yeglichs achtzehen schuch lanck ist von den
rorn macht man prunen mit tzef

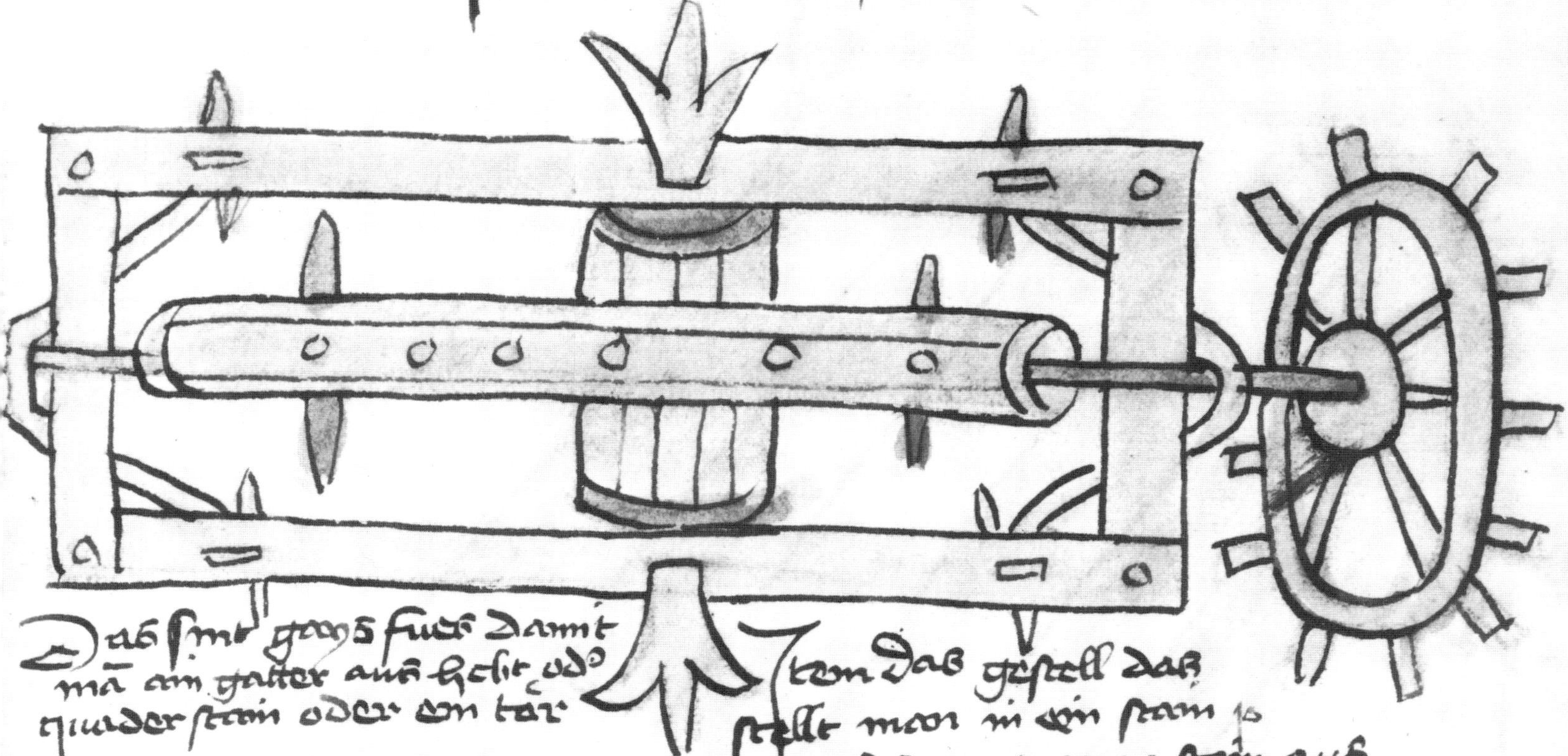

Das sint gayß fueß damit
mā ain gatter auff hebt oder
quaderstain oder ein tor

Item das gestell das
stellt man in ein stain so
pruch damit man stain auß
wigt so stosset man ain steng
an das lenger eysen da wigt
ain man schnellr mer wann vier
wider mit Nu ist es gar ein
nutzew arbayt

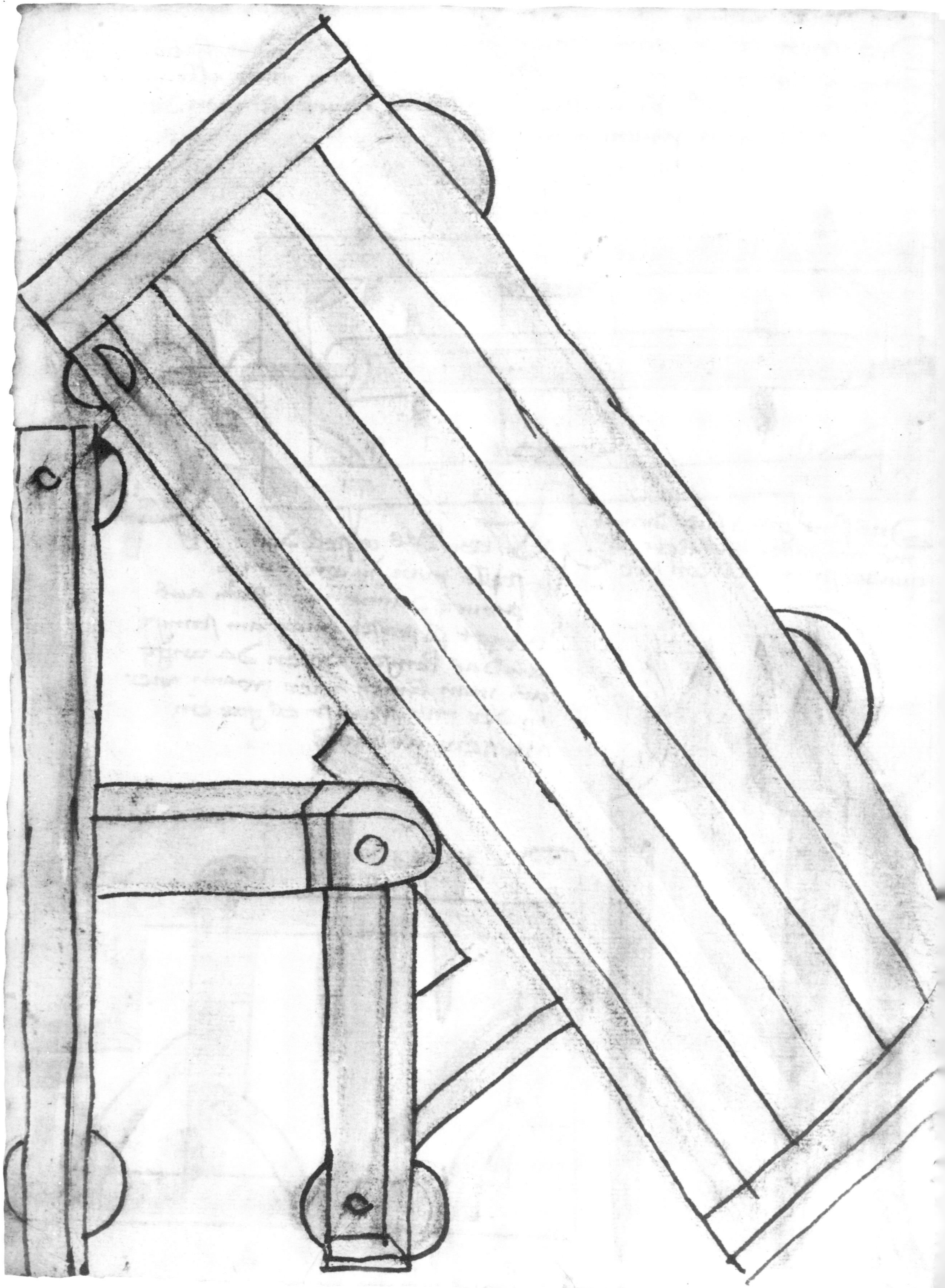

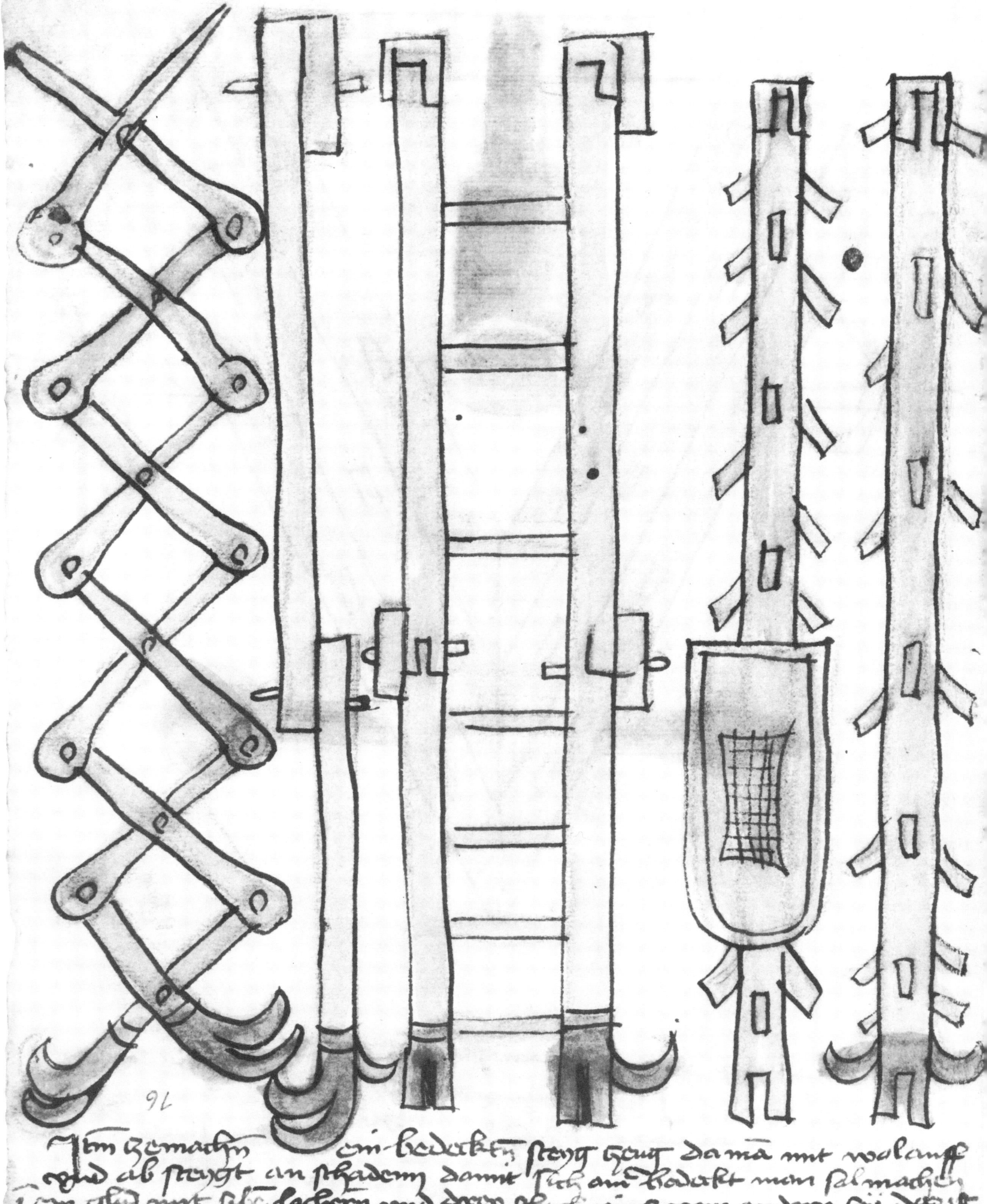

Item zemachen ein gedackten steyg zeug da ma[n] mit wol auff
vnd ab steygt an schaden damit sich ain gedackt man sol machen
i ein glid mit sibe[n] lochern vnd dreyn schuch aines vom andern sey des ist
xxi schuch hoch vnd dy selben palcken sullen all so sein mit dreyn glideren
Da steygt ein man lxiii schuch hoch etc

1421

Item den schirm hat her arkung vor satz gehabt da gem
hundert man wol darunder sicher der haspel ist in wen
dung vn wan ma zw der stat kumt so zeucht ma de
schirm auff vn schiust und lat in den wider zw gen
wint den den haspel und hintt sich so get der schirm
wider von stat und die klaint stien dar hintt an schad

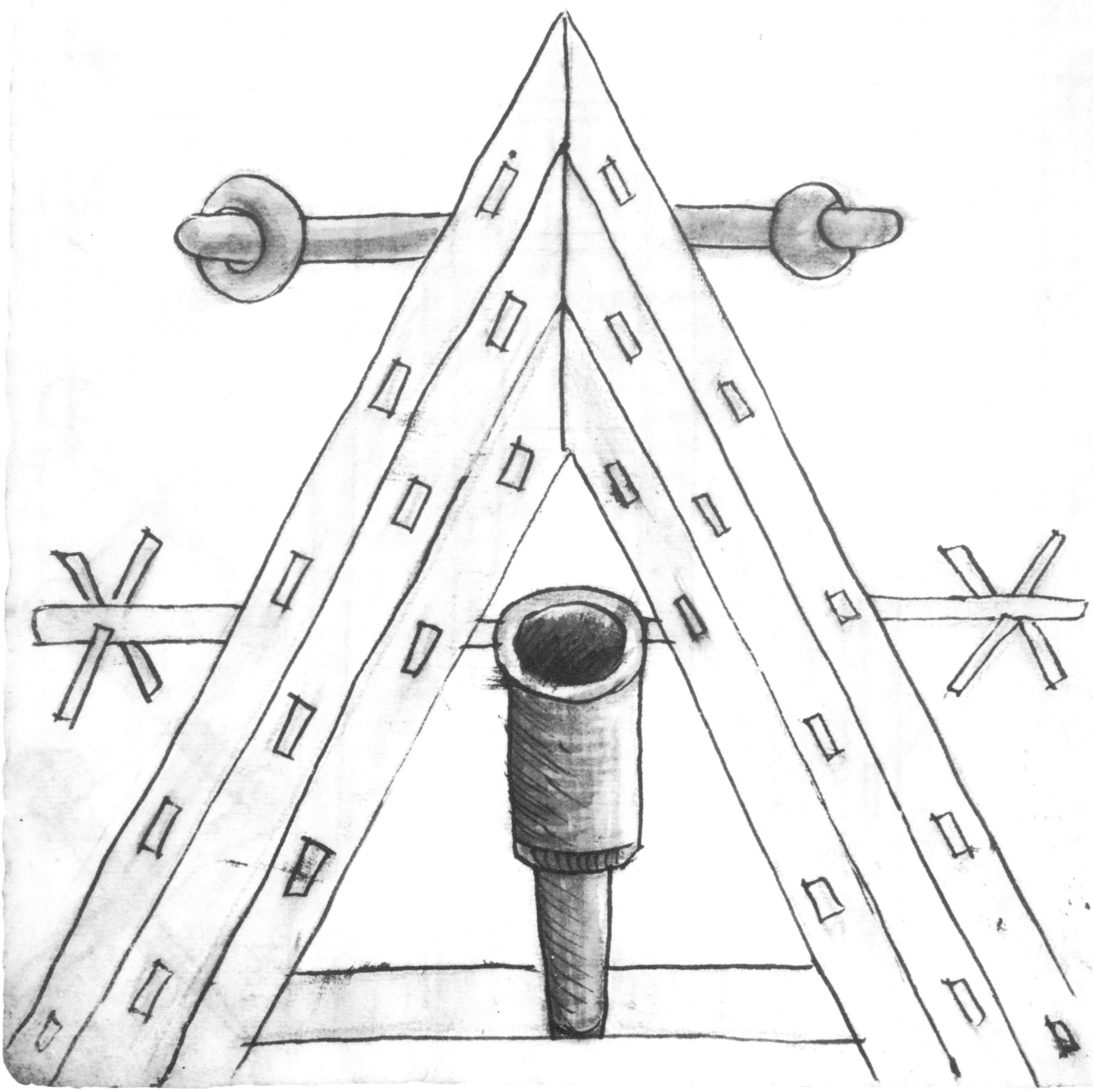

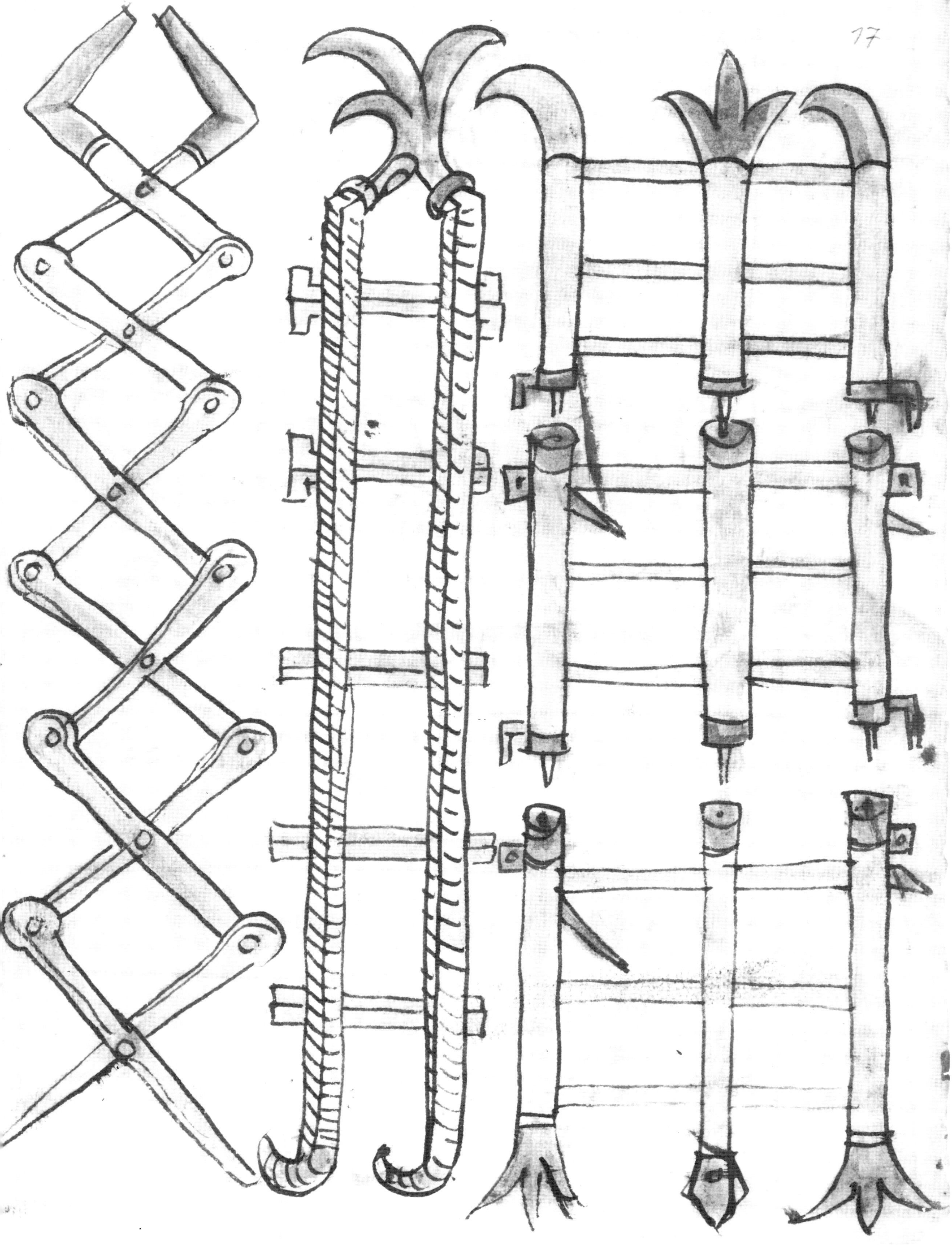

Item daz ist ein schiff daz get auff stillen wassern und hat .x. vettig da gelegen
.x. man zu zwein zeiten ein zwie faren und daz mag wol .xx. wappen tragen
und dy vier man dy daz schiff ziehen und dy vettigen gen in daz wass und in zwen
dinge hat yedlich vettich ein wendl den man umb treybt inwendig in dem
schiff so mag man faren auff dem wass als und zw und daz schiff sol vndeckt sein
daz man dy leyt nicht geschiessen mug und sol fornen ein schachen spitz haben
und an yedlich seyten ein neben spitz ein puchsen daz hayst ein
streyt schiff damain dye von katalon all andern schiffen oblige

Item daz sein stampff damit man korn stampft wan ma nit
mul gehaben mag den sin treybt ein man umb an de rad
Eins get auff der ander nyder etc

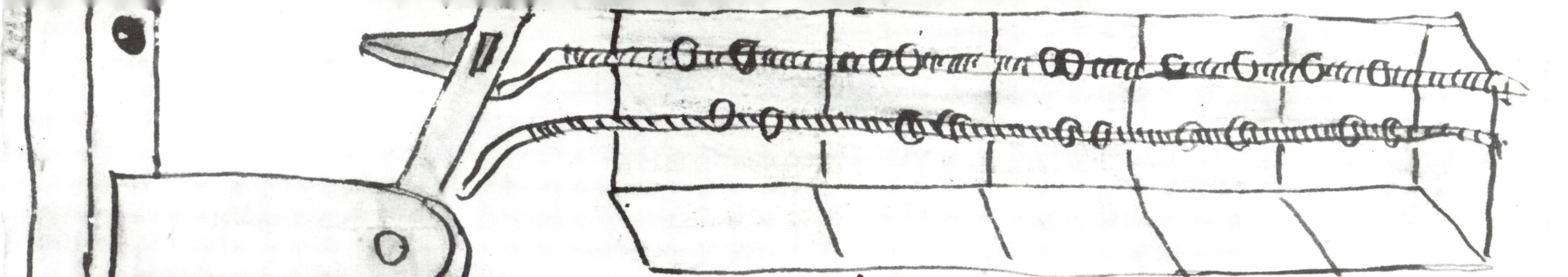

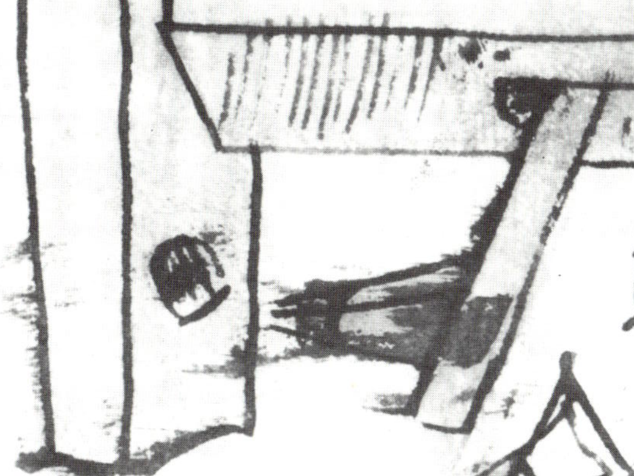

Item wer den mul machen well der sol neme zwai stuk teuchsel eysen und
sol machen den furm des eysen als sy da gemalt ist / und sol oben sein
haupt und unten stechlein sein und sullen da da eysen inen gett sol auch
stechlein sein und die mulstain dreyer schuch prait und ma sol sy zwicken
mit den zwaye zwygein zwinen man so mel des sy vertig

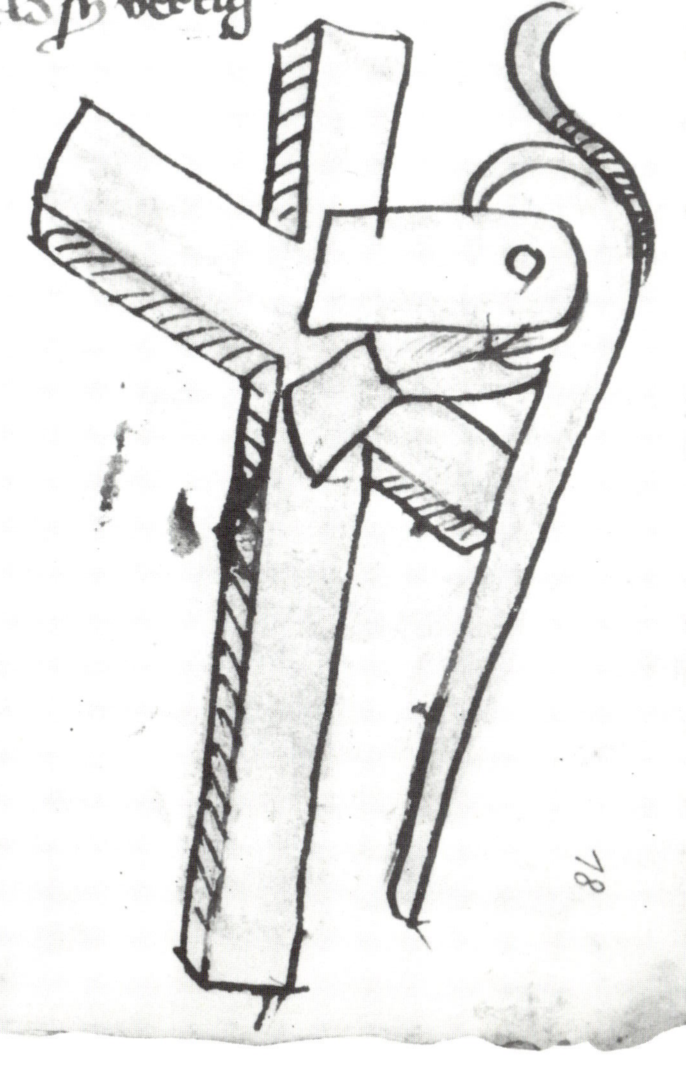

Item das ist ain wassz mull da das Rad nach d° prunt
auff dem wassz leyt vn sein mull parum stet ob sich
vnd ist auch ain gerechte mul vn sy darff chain
chomp Rad ain pabst wo kam der hat sy erdacht

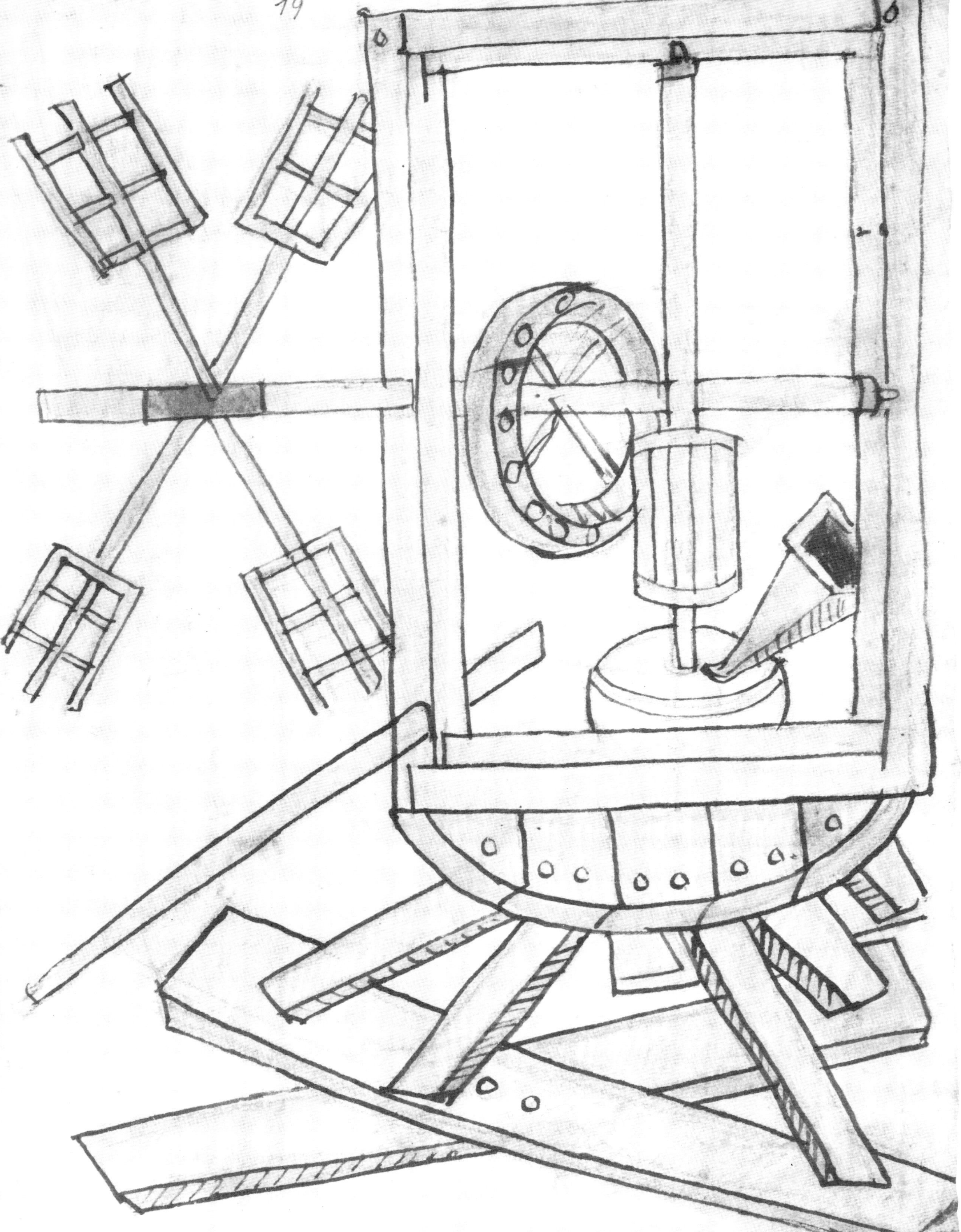

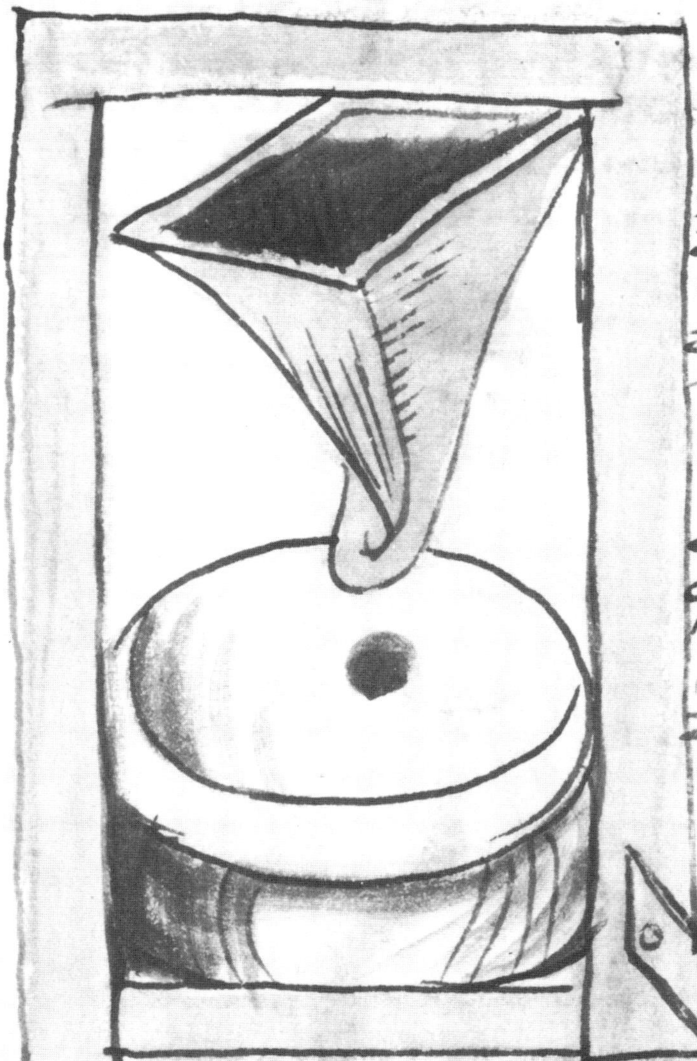

Item zemachen ein mul ma sol
machē ein mul gestell x schuch
lanck und acht schuch weyt un der
gestell sal acht schuch weyt sein und
ein flegel der sal man auff seine well
pawm gen mitten an dem well pawm
da sal ein kamp rad sten vier schuch
hoch und kompt mit sy kamē in
sein fur getrib mit sex tribeln in sein
stag in sein mul eysen gestalt als ein
end mul eysen in sein mul stain der
obrist dreyer schuch prayt un ein hant
prayt un aines halbin schuch dick so
wirt es ein recht vertigen mul dye
da melt als vil als ein halbe pach
mul

Item sein well pawm ist stossen
an ainem gevierten happfem
der sol

Jtm zemachen ein trett mul so sol ma mache ein mul gestell
xviij schuch vnd sein trett rad xvj schuch hoch vn sein kampf
rad vj schuch vnd sein iiij schuch preit wil ma daz es
als vil mall so sol ma haven man daran stelln daz ist
ein vertigew mul dar auf sich ein yedlicher muller wall
verrichtt

Item ze machen ein mul dy da melt als vil als ein wasser mul
So sal ma machen ein gestell x ii schuch prait vnd achte 21
weyt vnd ein schwinck rad xii schuch hoch vnd ein schuch dick
vnd den selben wellpawm da sal an sein ein chrumpes eysen
daz hoch ist daz sal vier schuch lanck sein vnd dar sal ma an
legen mit ein stangen vntz daz daz rad an sein lauff chumpt
so sal ma es nicht lassen stil stan anders ez wirt hart an ge
pringen von sein mul werck sal sein als ein wasser werch
enger gechempt sein wan sein wass mul so hat ma eine
gutten wass mul dy ein gutte
natturfft melt damit man
bewart ist

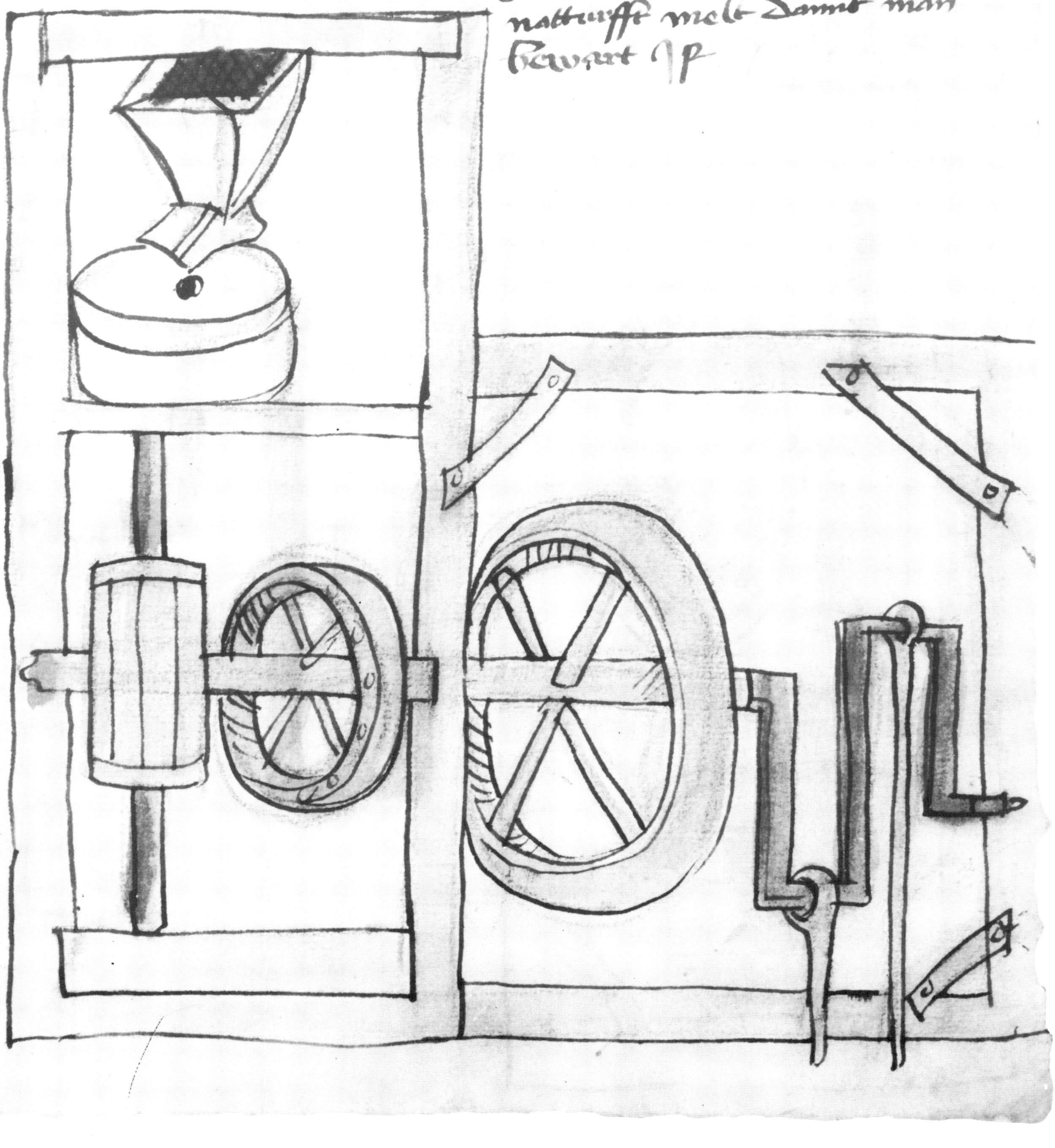

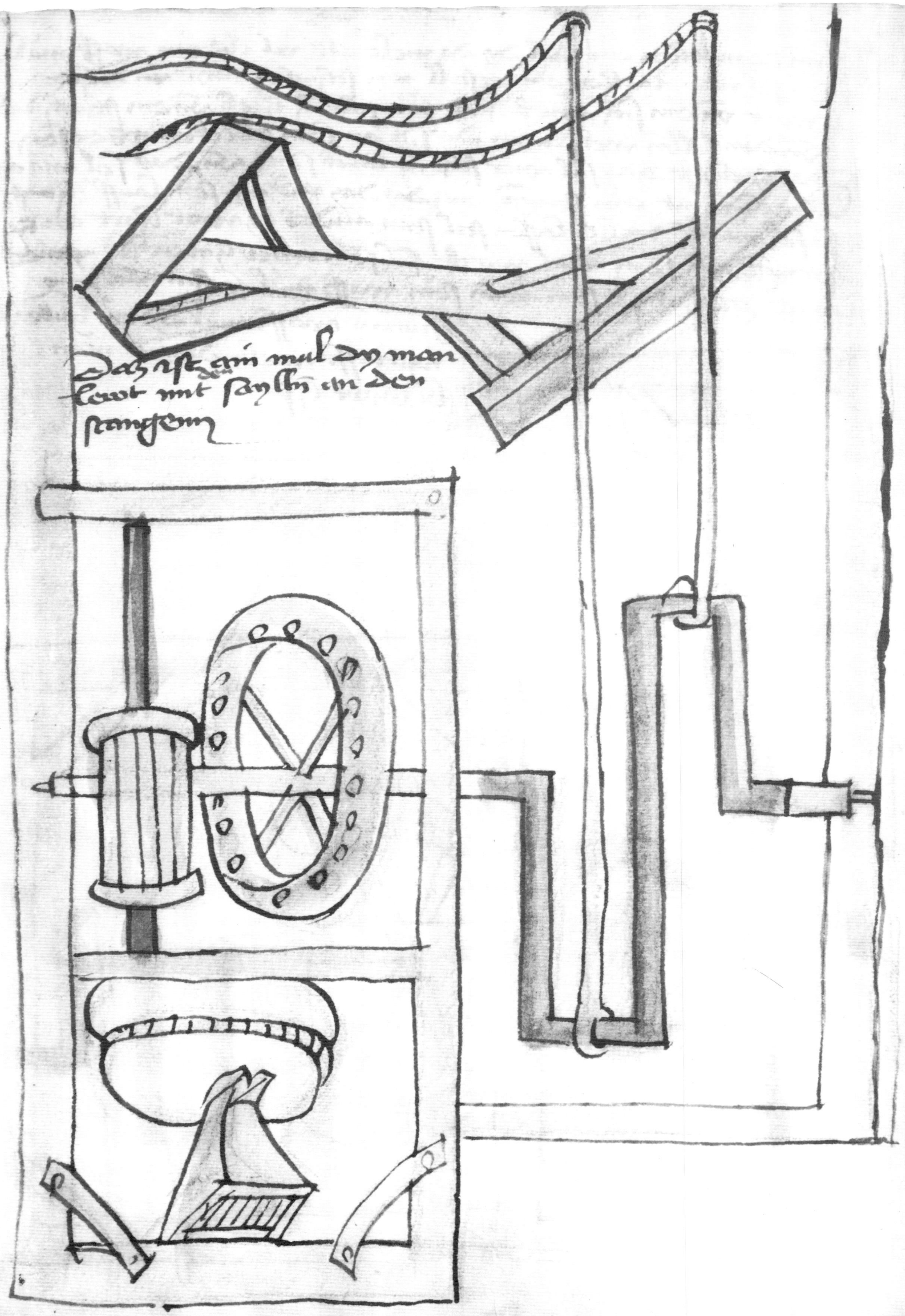
Daz ist ein mul dy man
lewt mit sayllen an den
stangen

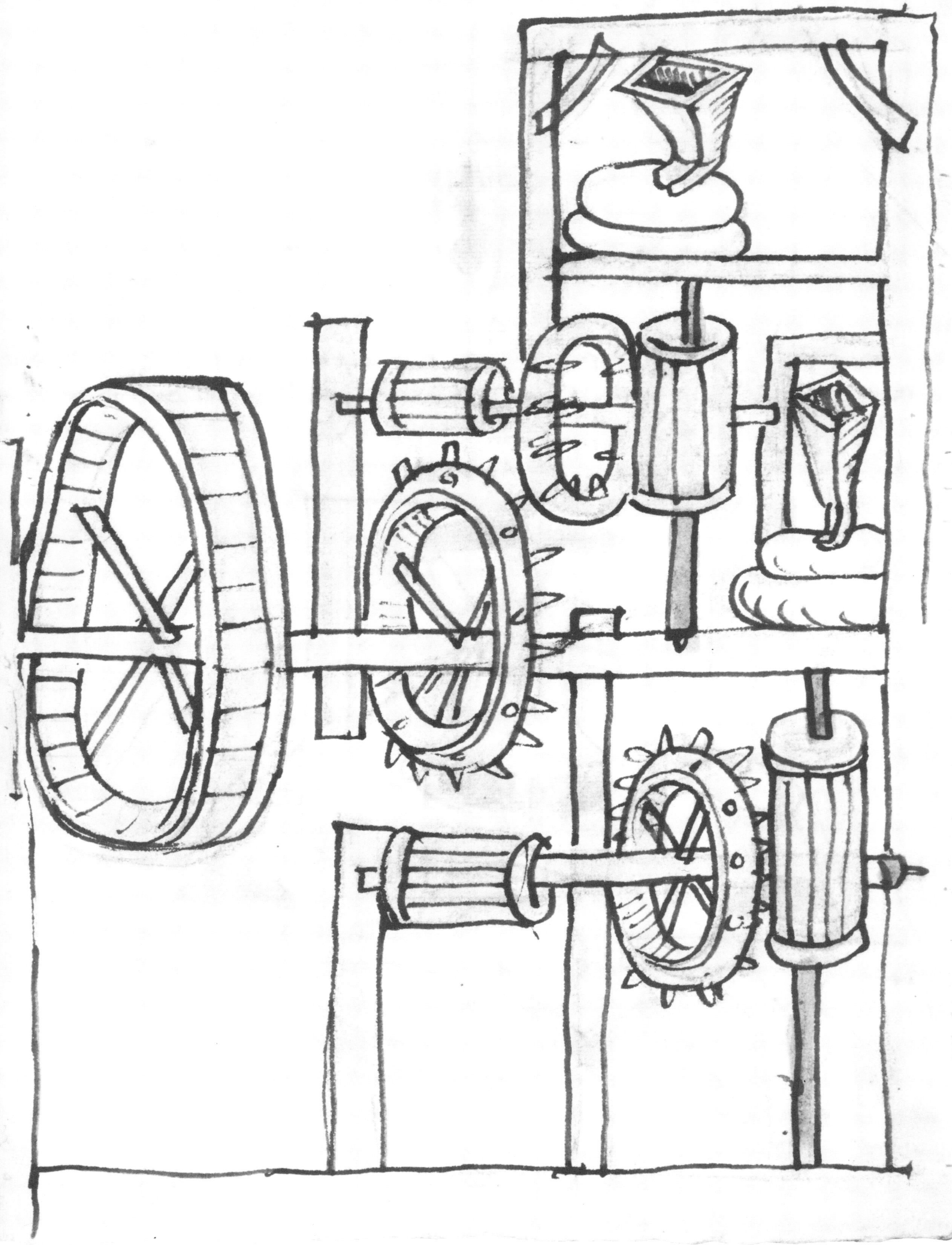

Item daz ist der hussen wagen paryß dar auf dy hussen vechten dy ist gut vn gerecht

It daz ist ein palier mull als dy hachn maist zu venedig hab darauff ma allerlay gestain paliert dy bedarff in schabn dy erst ist pleyen pleyen dy ander zymer dy tritt kupfern

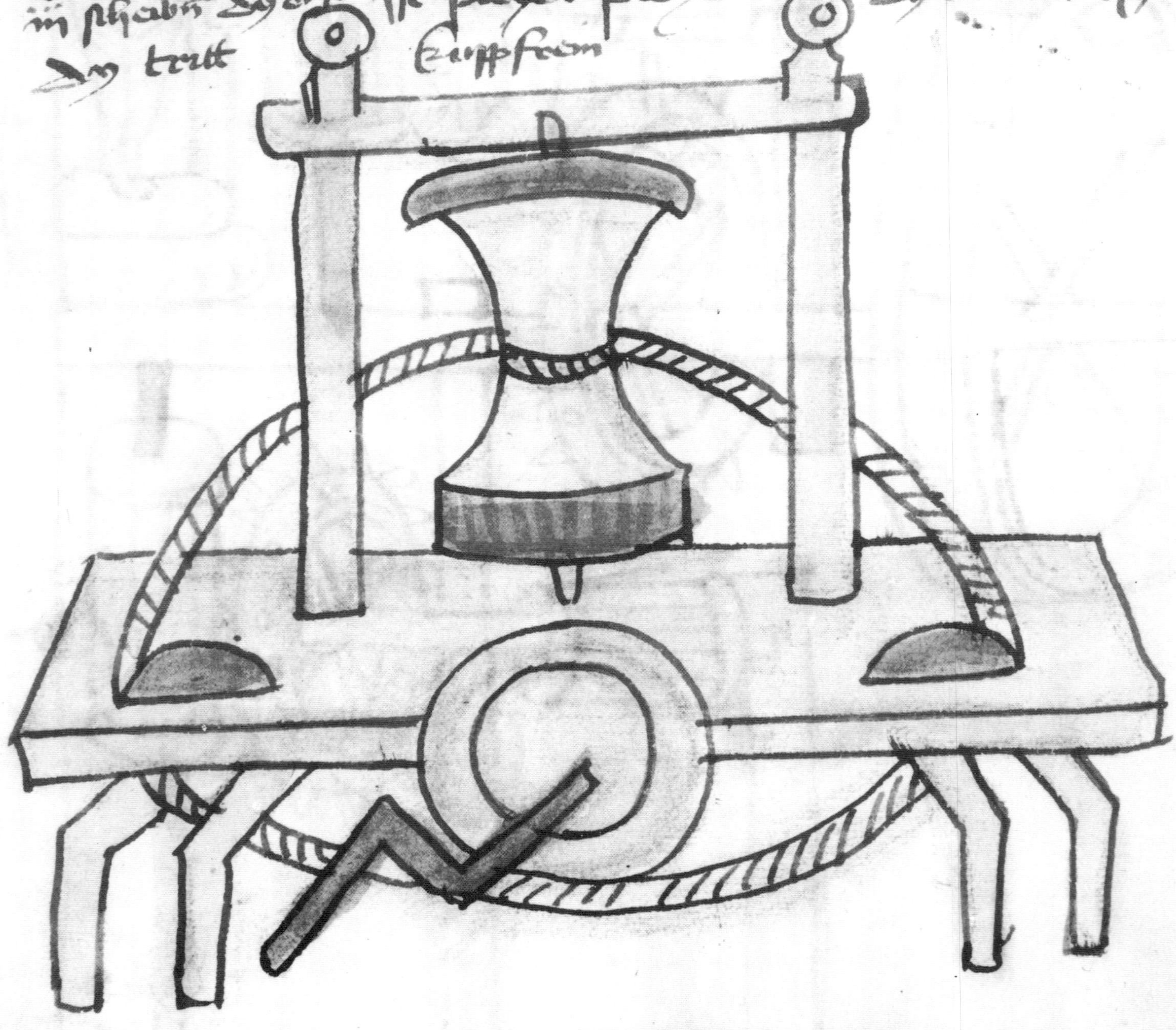

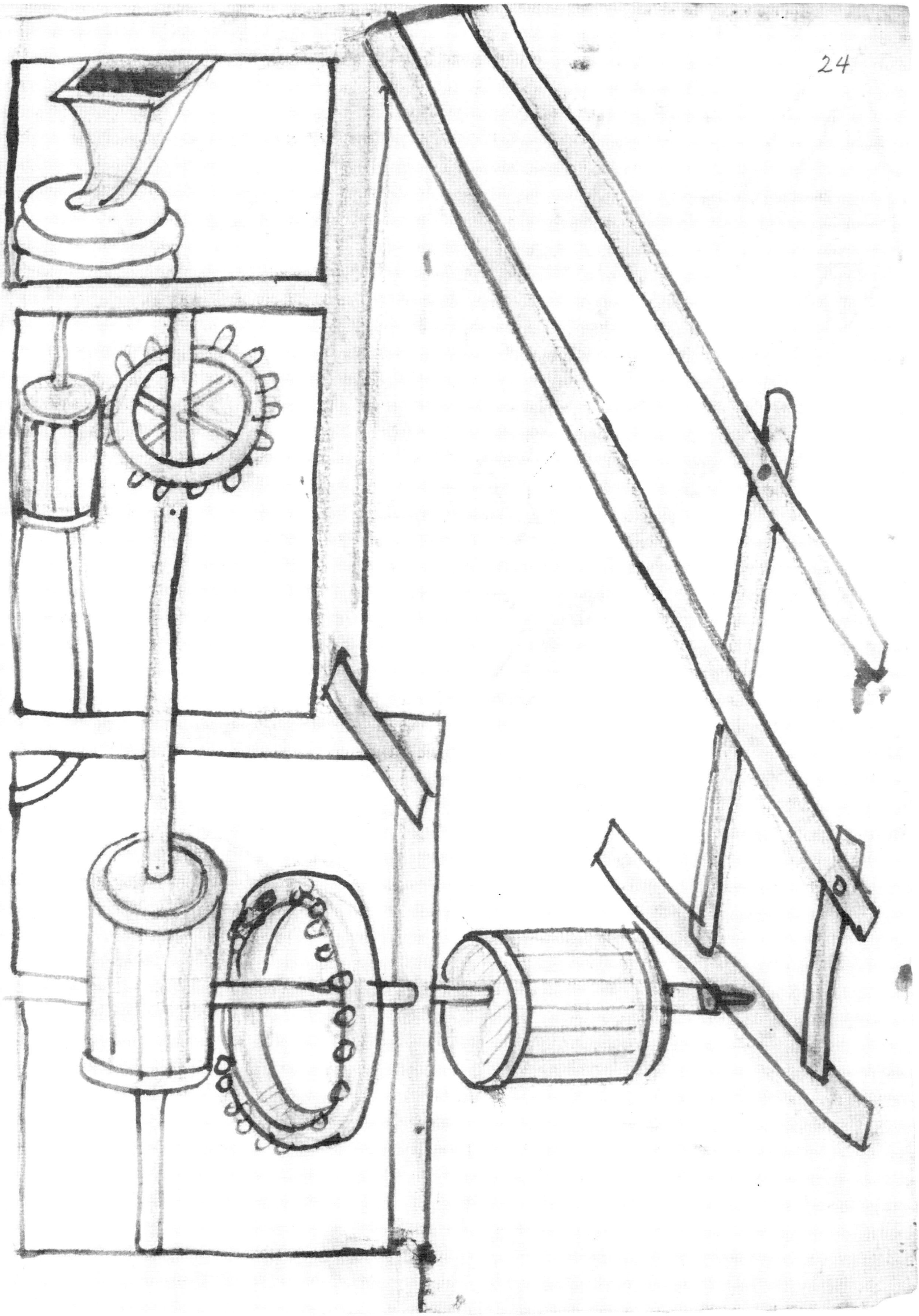

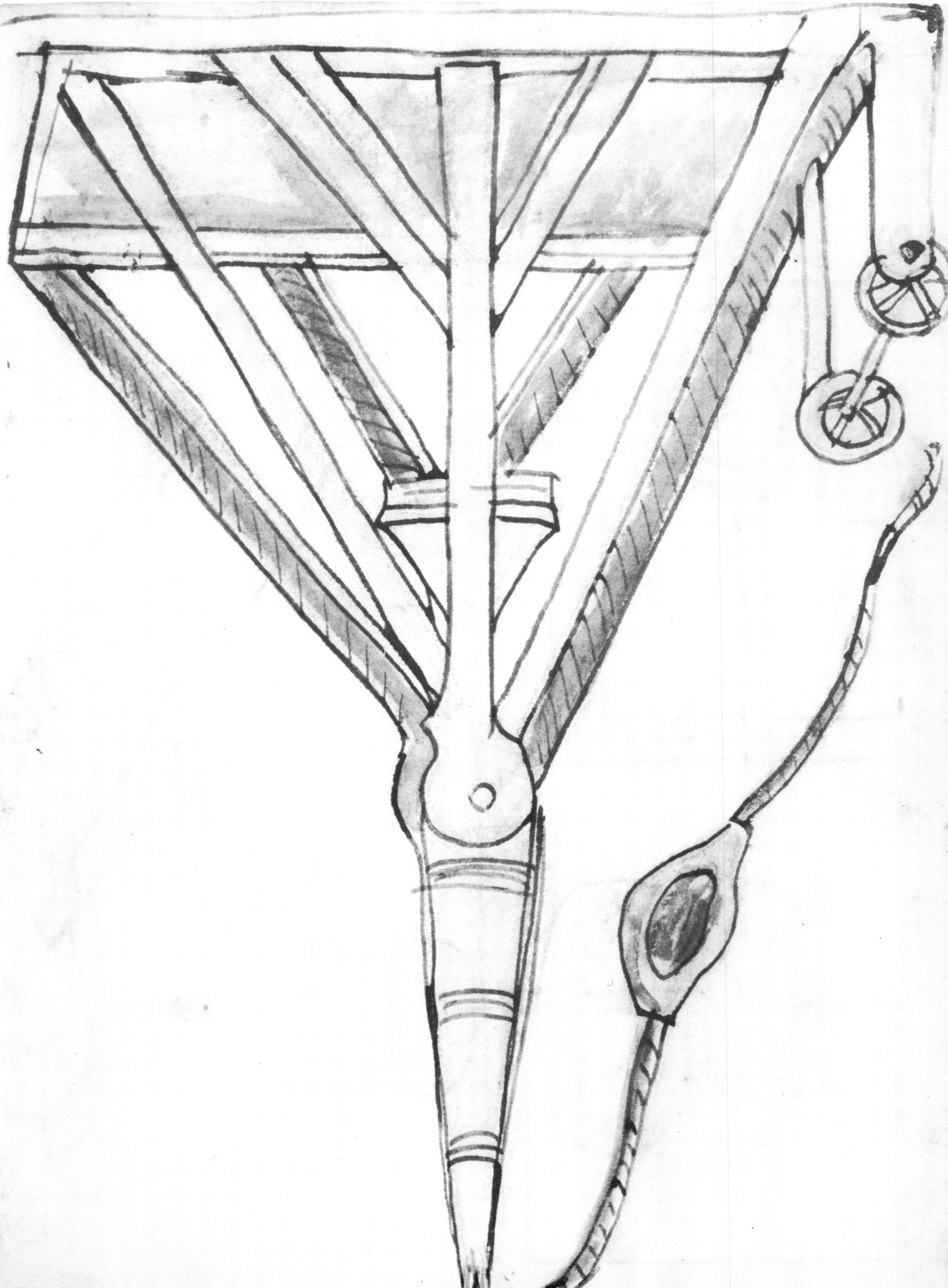

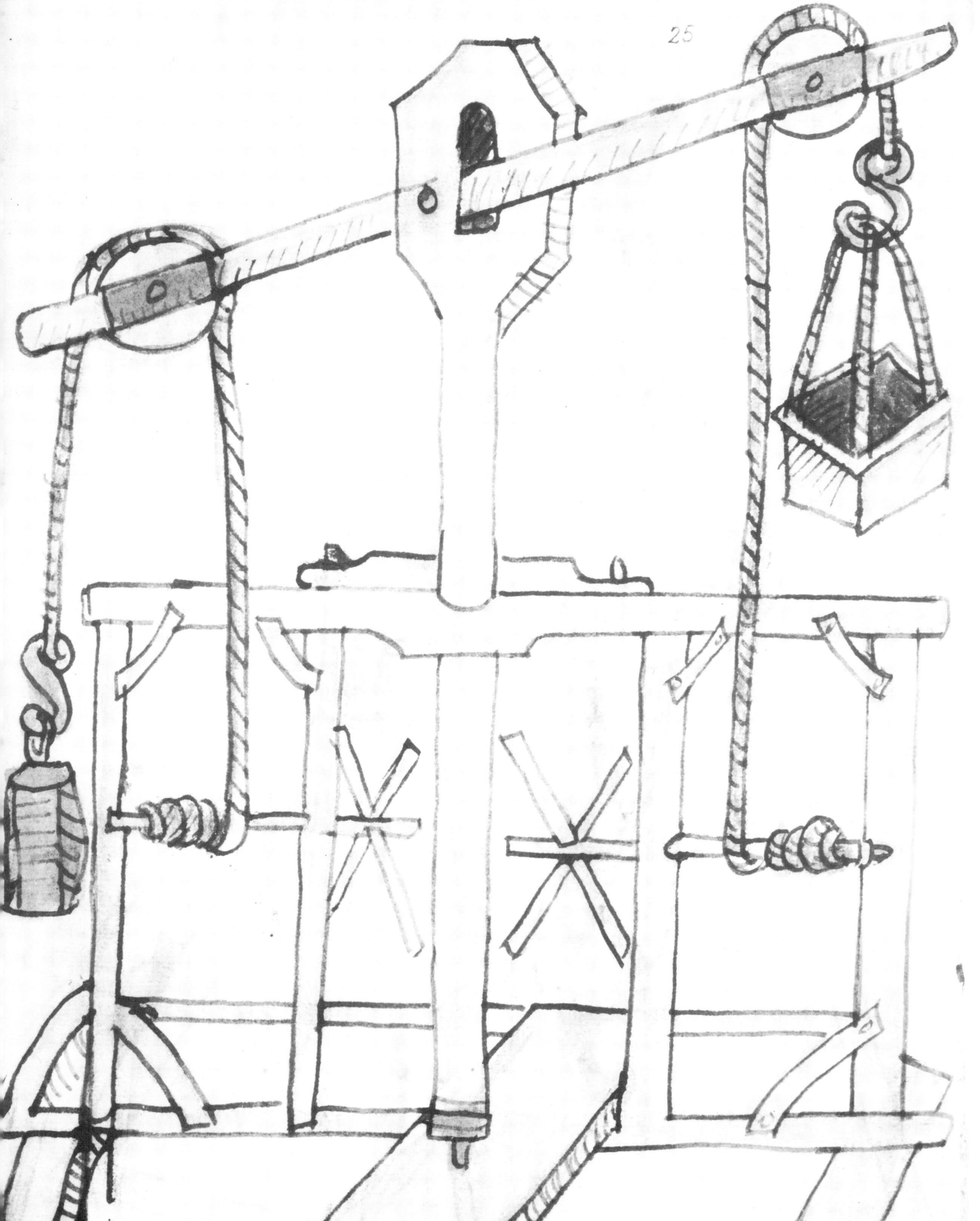

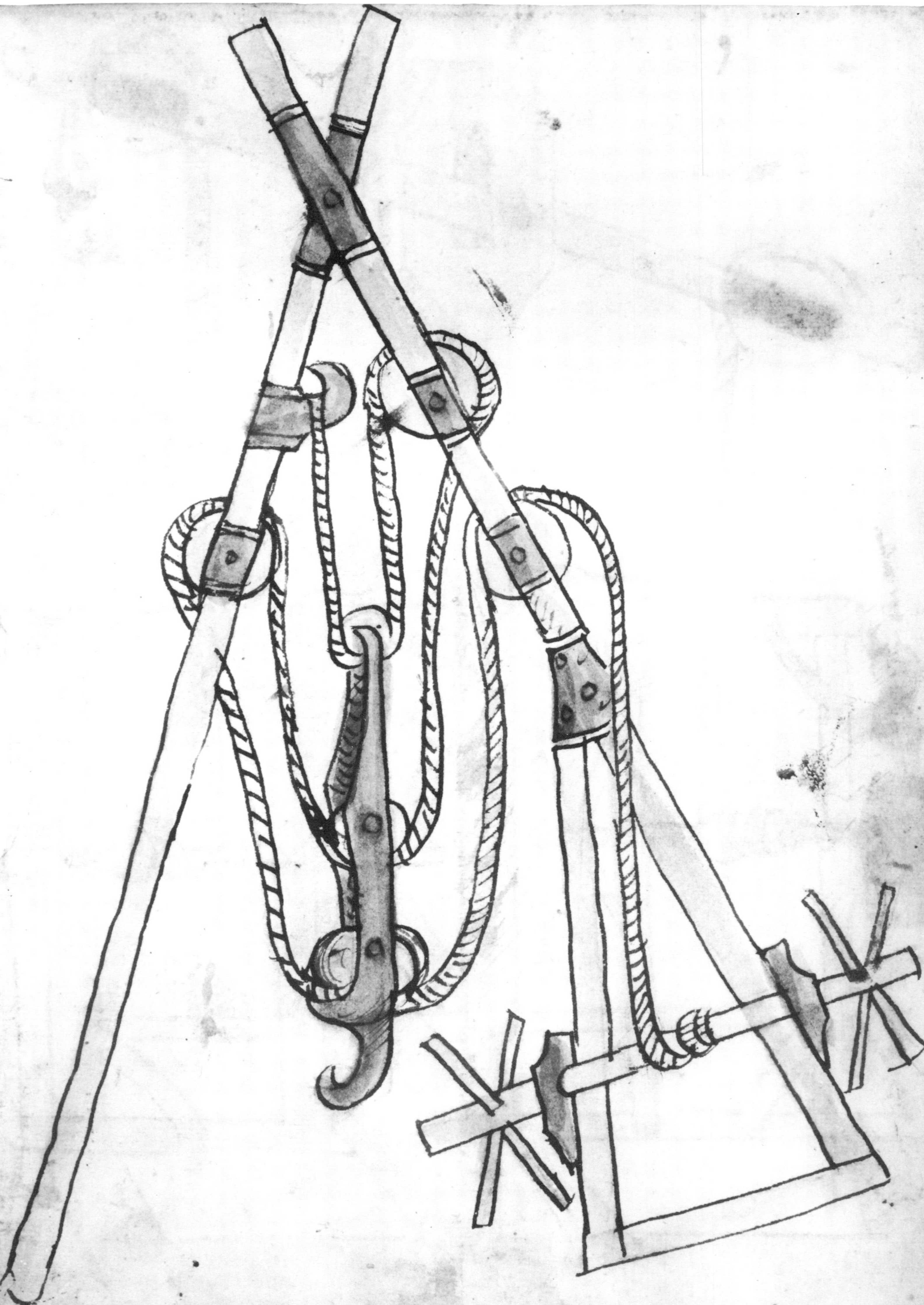

Item ze machen ein prugk da man vber wasser graben get
So sol man dy seyller dreyer schuch praeyt machen
vnd ein halben hoch od dick vnd dy palichen dy darauff gehorn
der sol yedlich vi schuch lanck sein vnd hol inn geheft
mitten auff dye seyller vnd dy hol vnd den palichen vnd dy
seyller mit gehorn auff daz wasser vnd an dem end von
der seyller stett sein hanthab vnd wann mans lengern
wil so sol ma dy glyder in ein ander stossen

Item ze machen dyssen wagen daz sol sein halb tayl sein
als ein schiff geclamert vnd glider auff gien als ein
disch vnd als lang vnd als praint als ein ander
wagenn vnd in wendig ein druchlein nach der lenge
des wagens da ligenn sein stangk leder in vnd sein fues
vnd wann es in das wasser chumpt so ist es ein schiff
vnd ein pruck vnd auff dem lannde so ist es ein wagen
das da durffen zway pfärt zyechen vnd das ist ein
gutt vertigew chunst etc

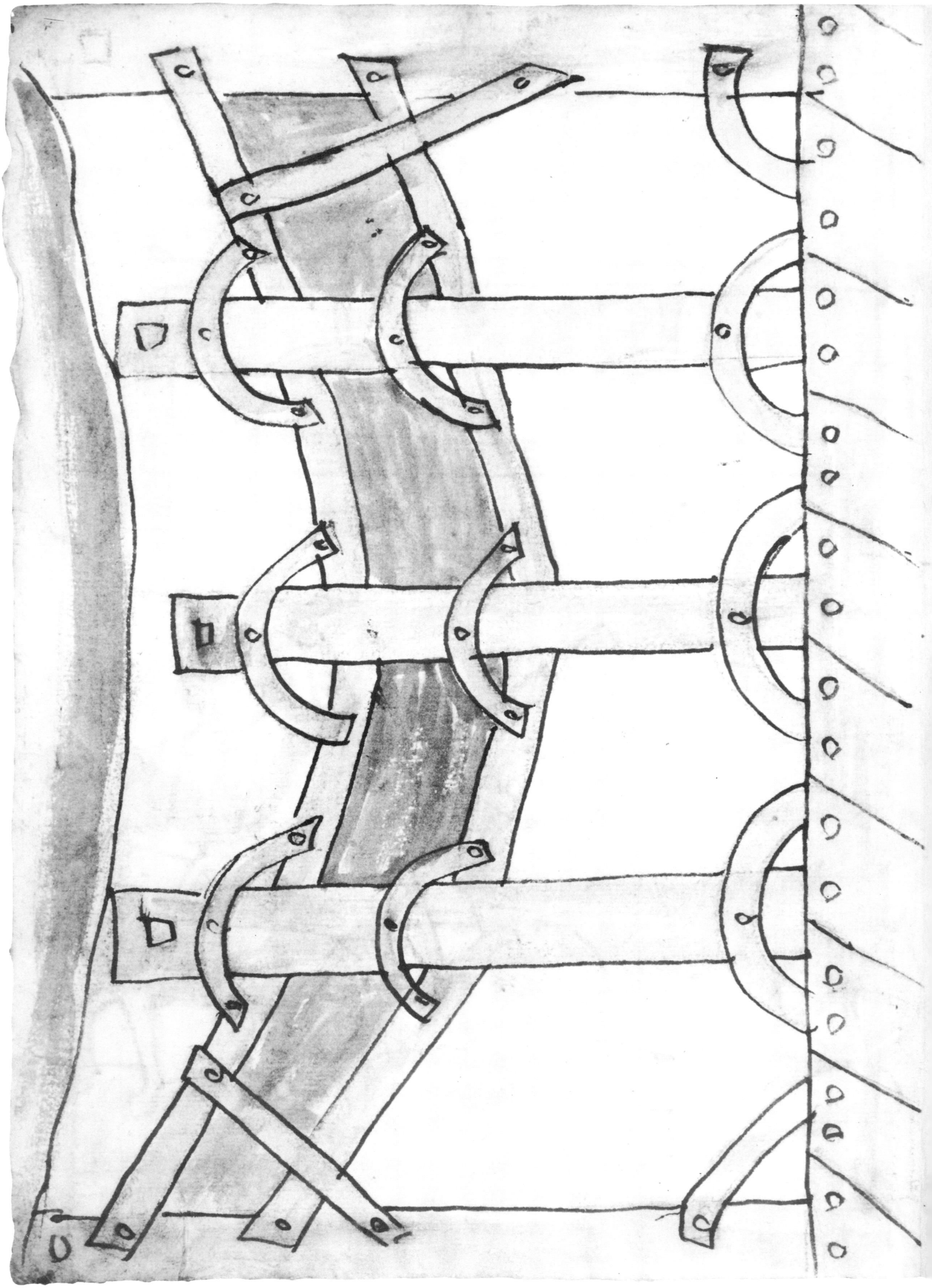

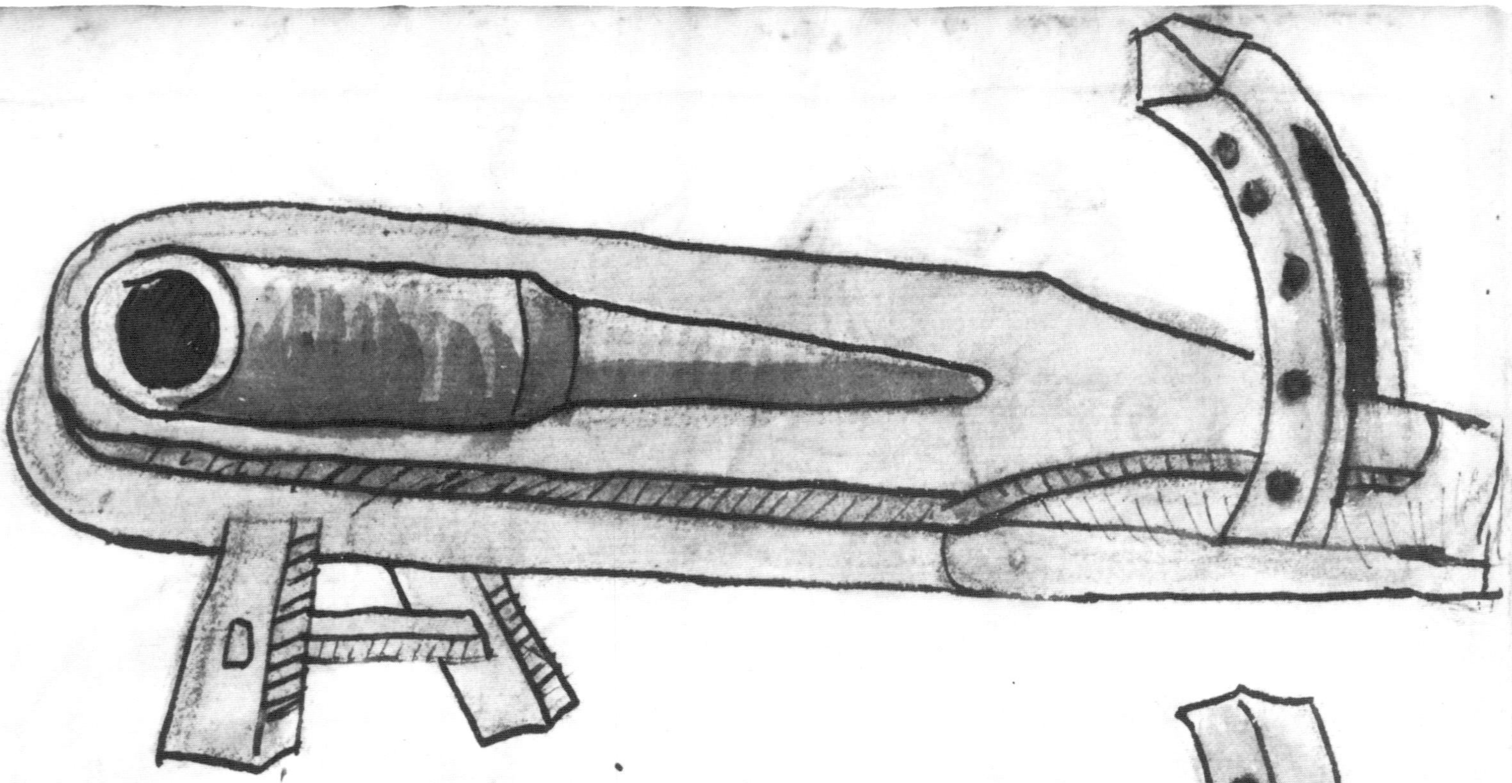

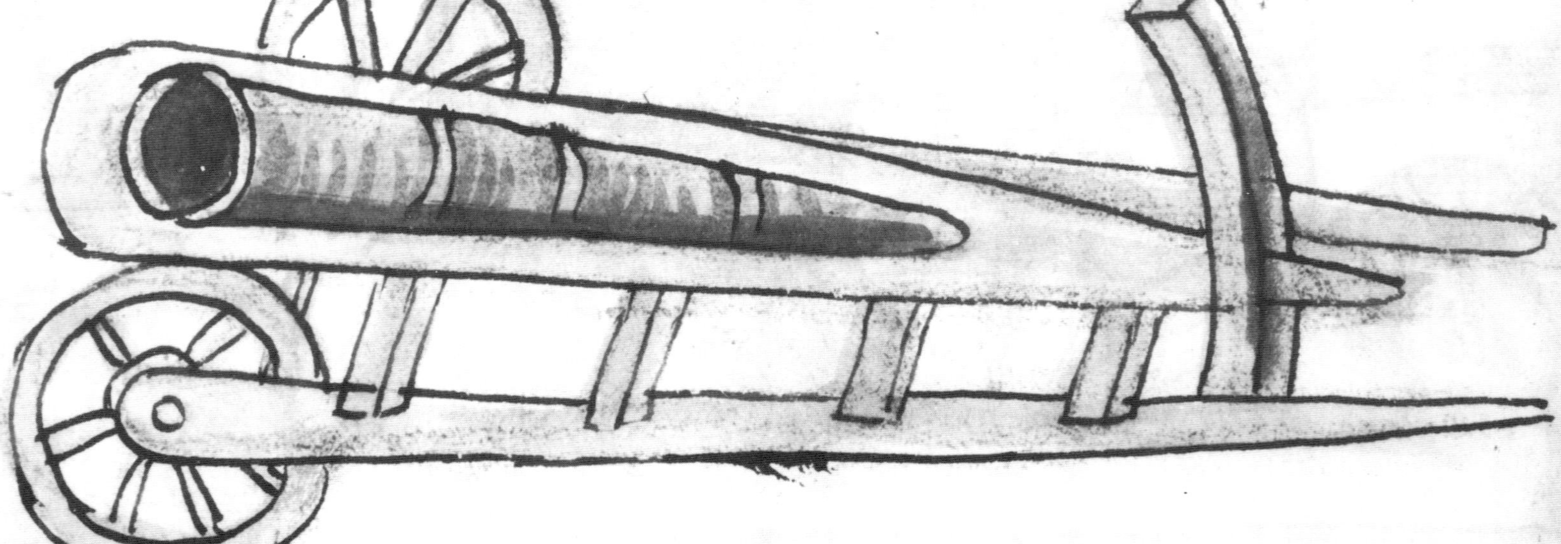

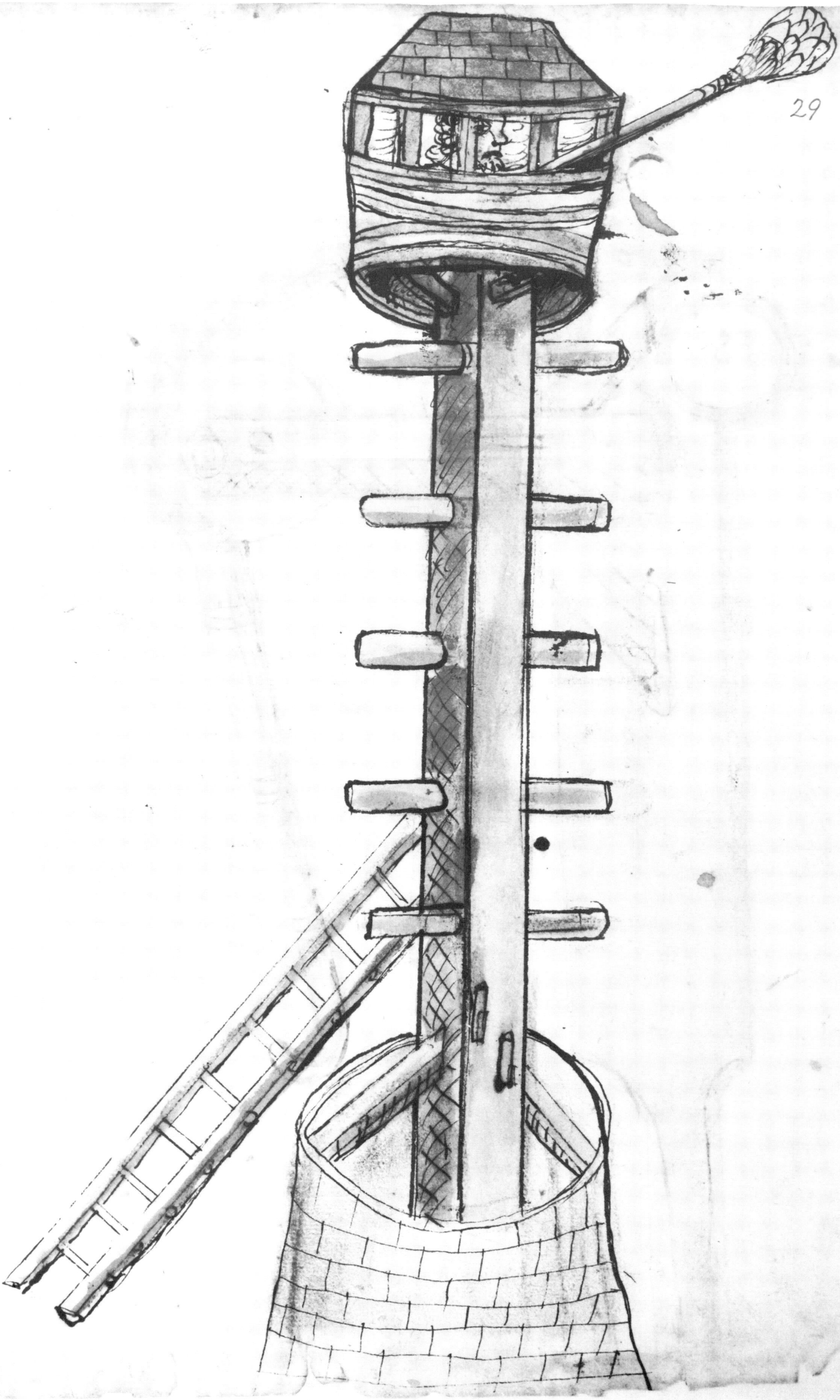

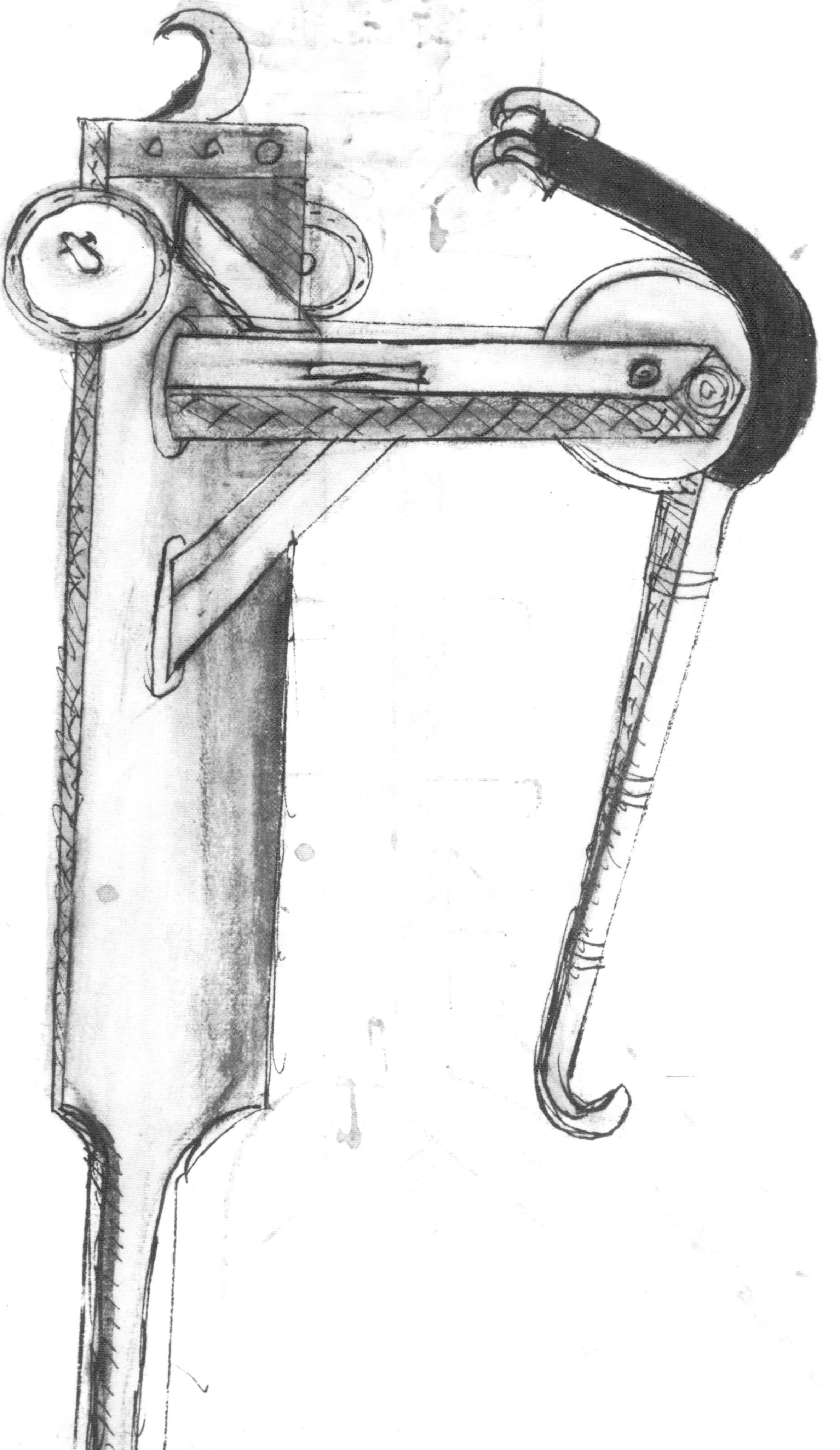

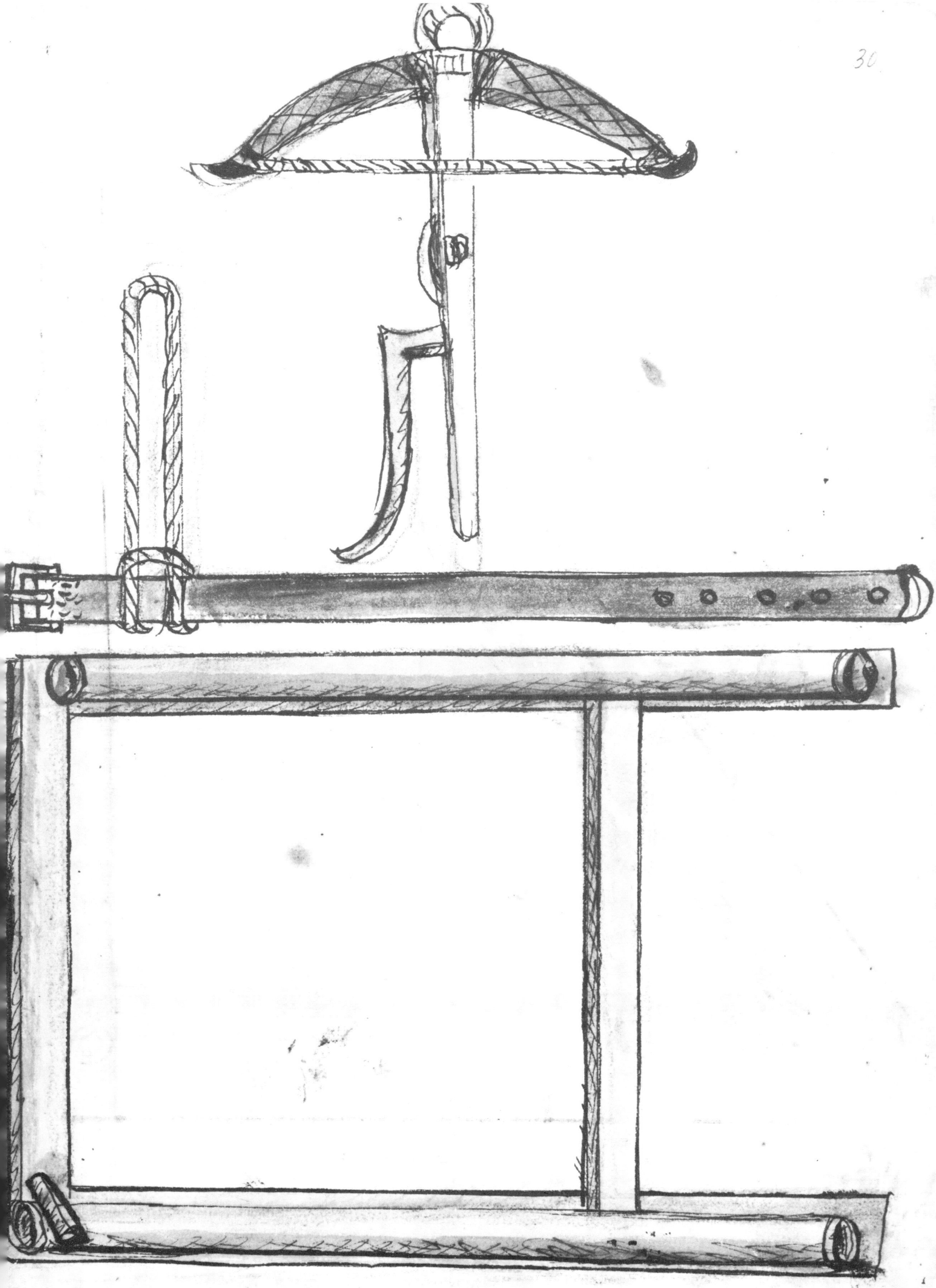

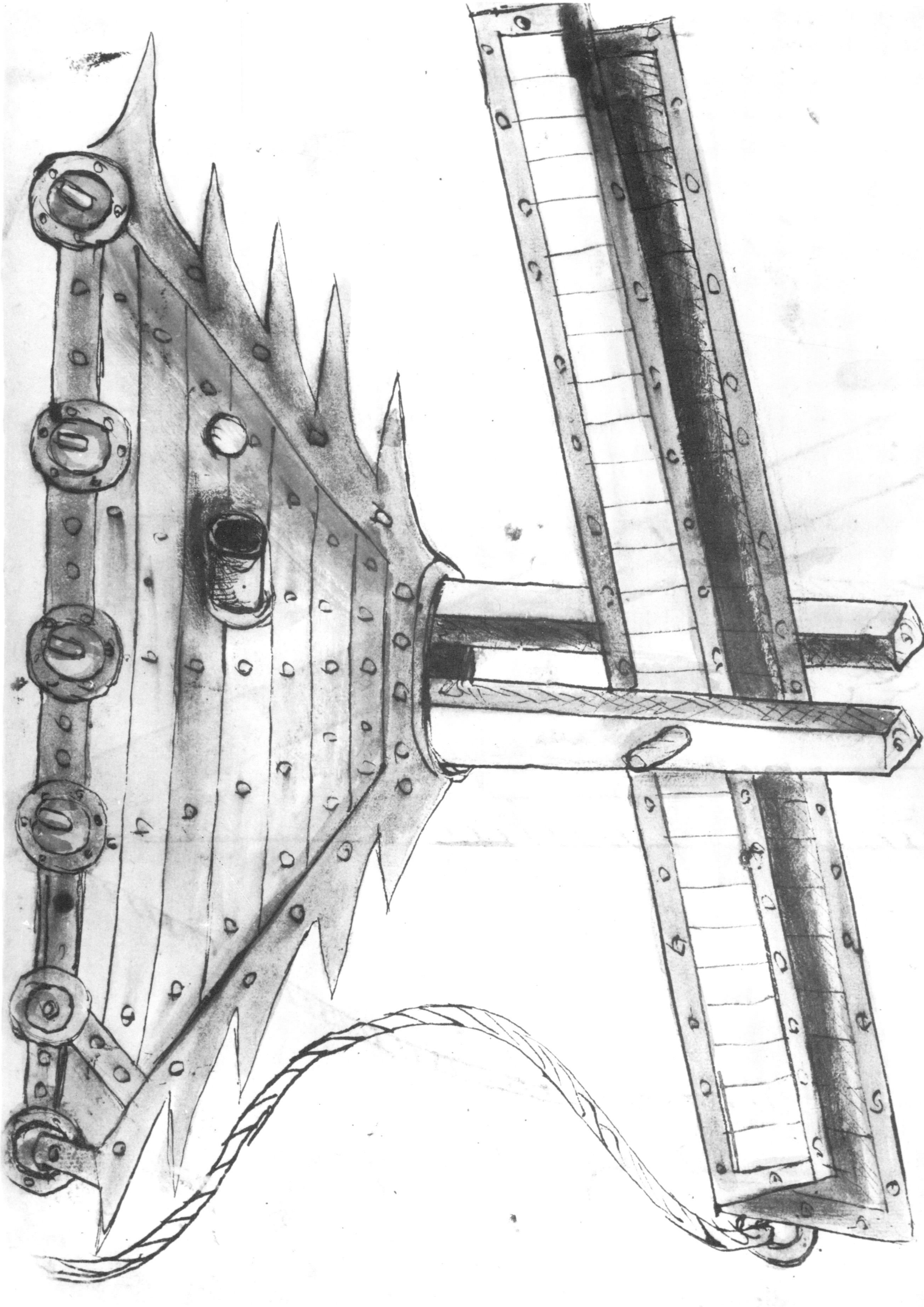

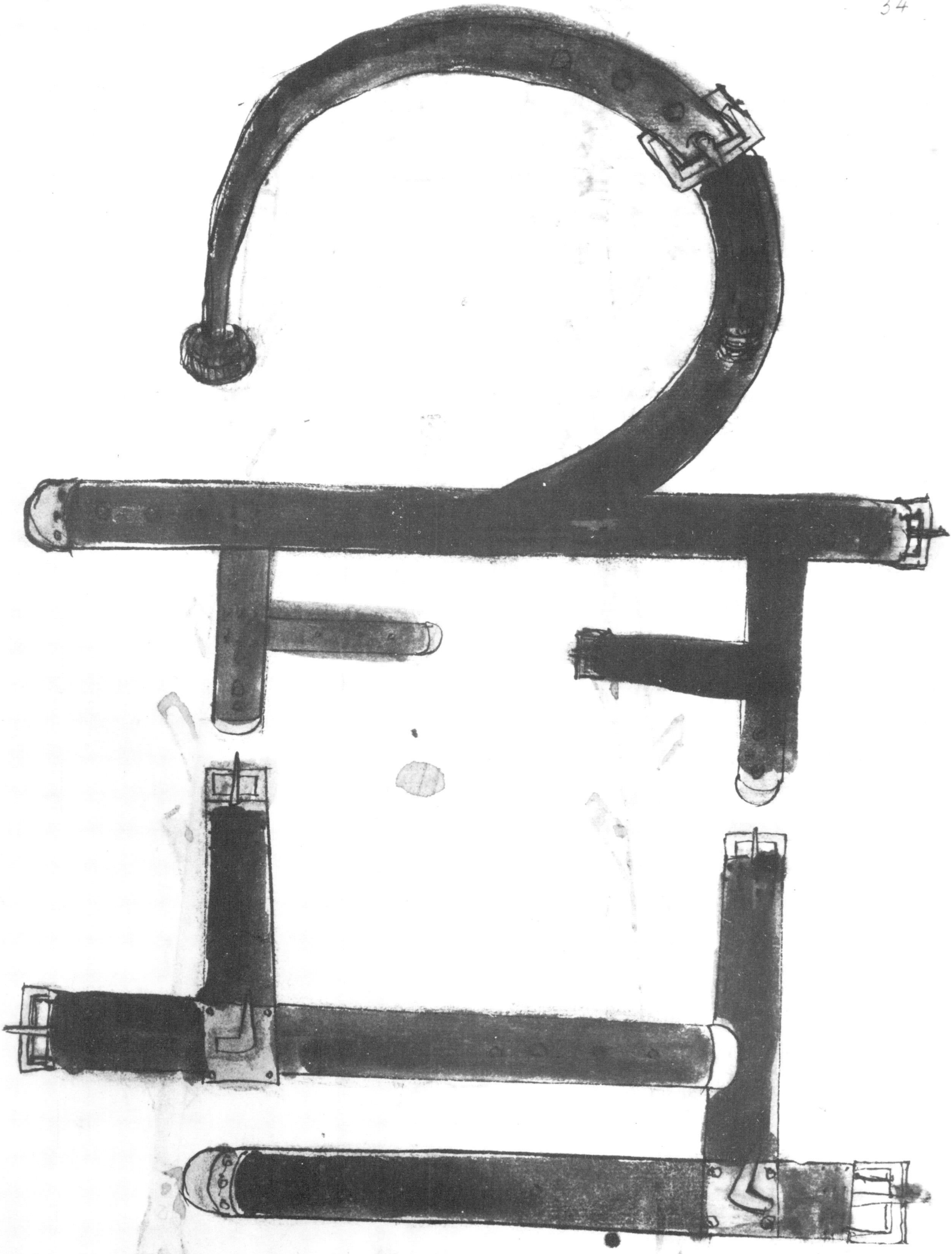

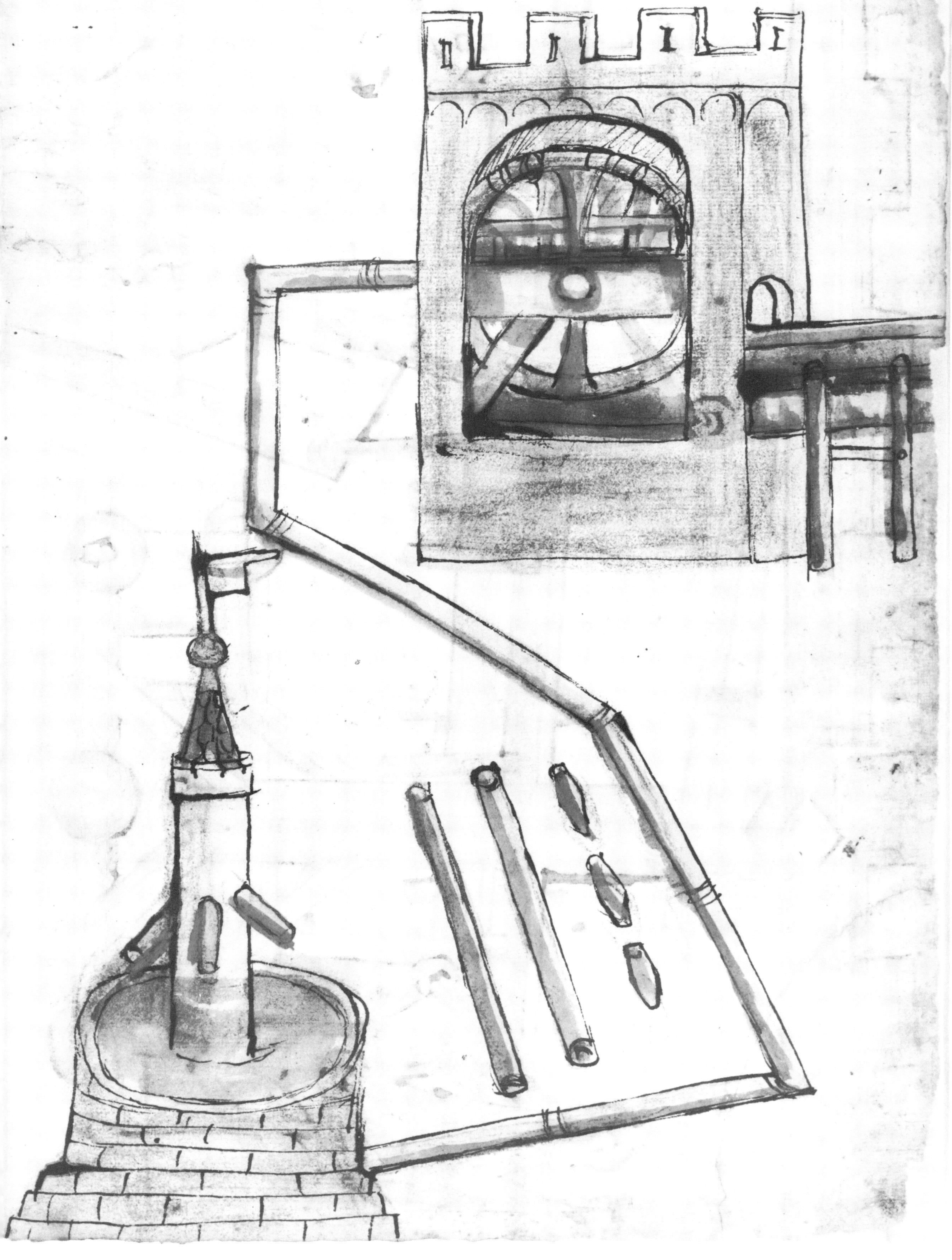

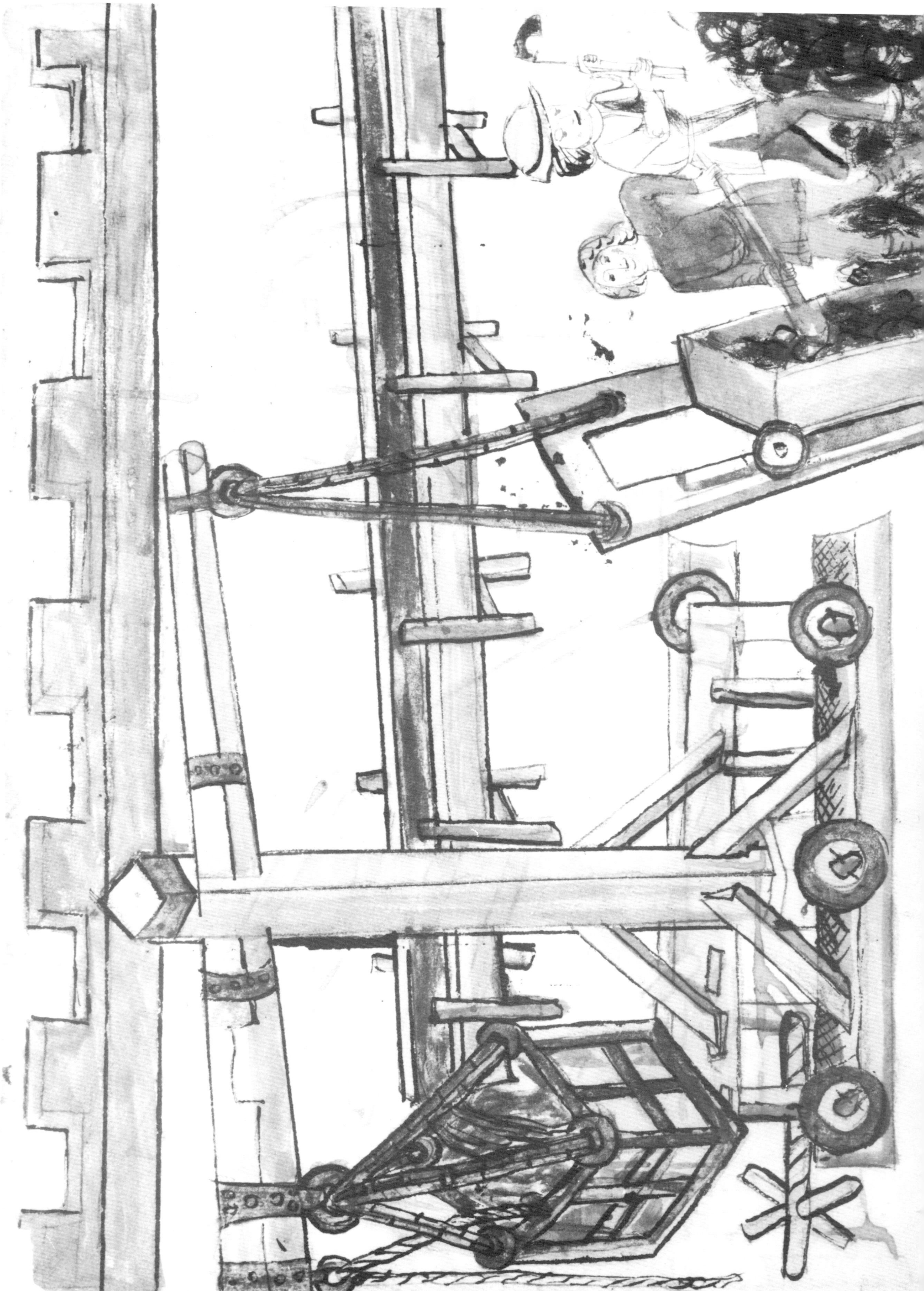

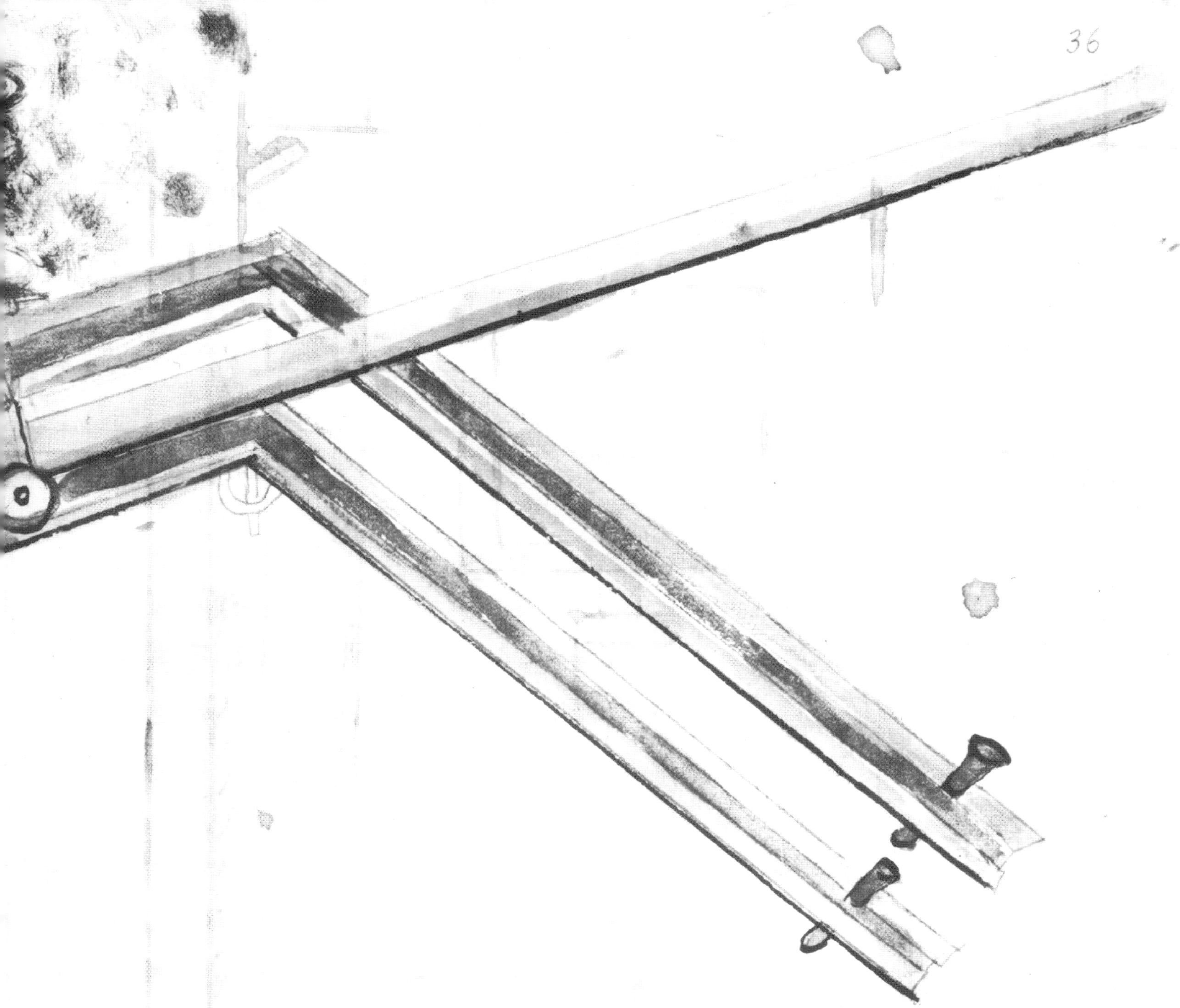

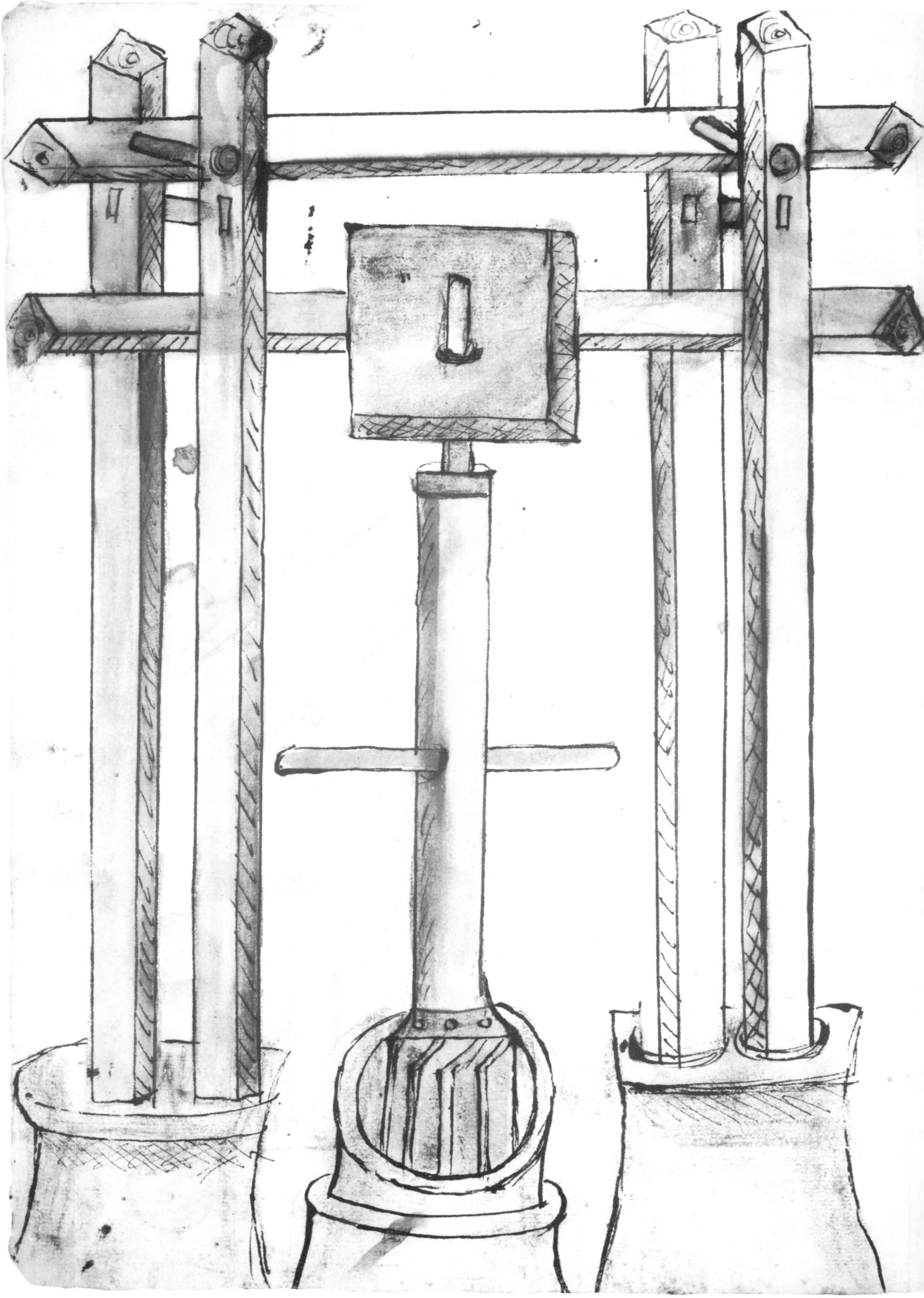

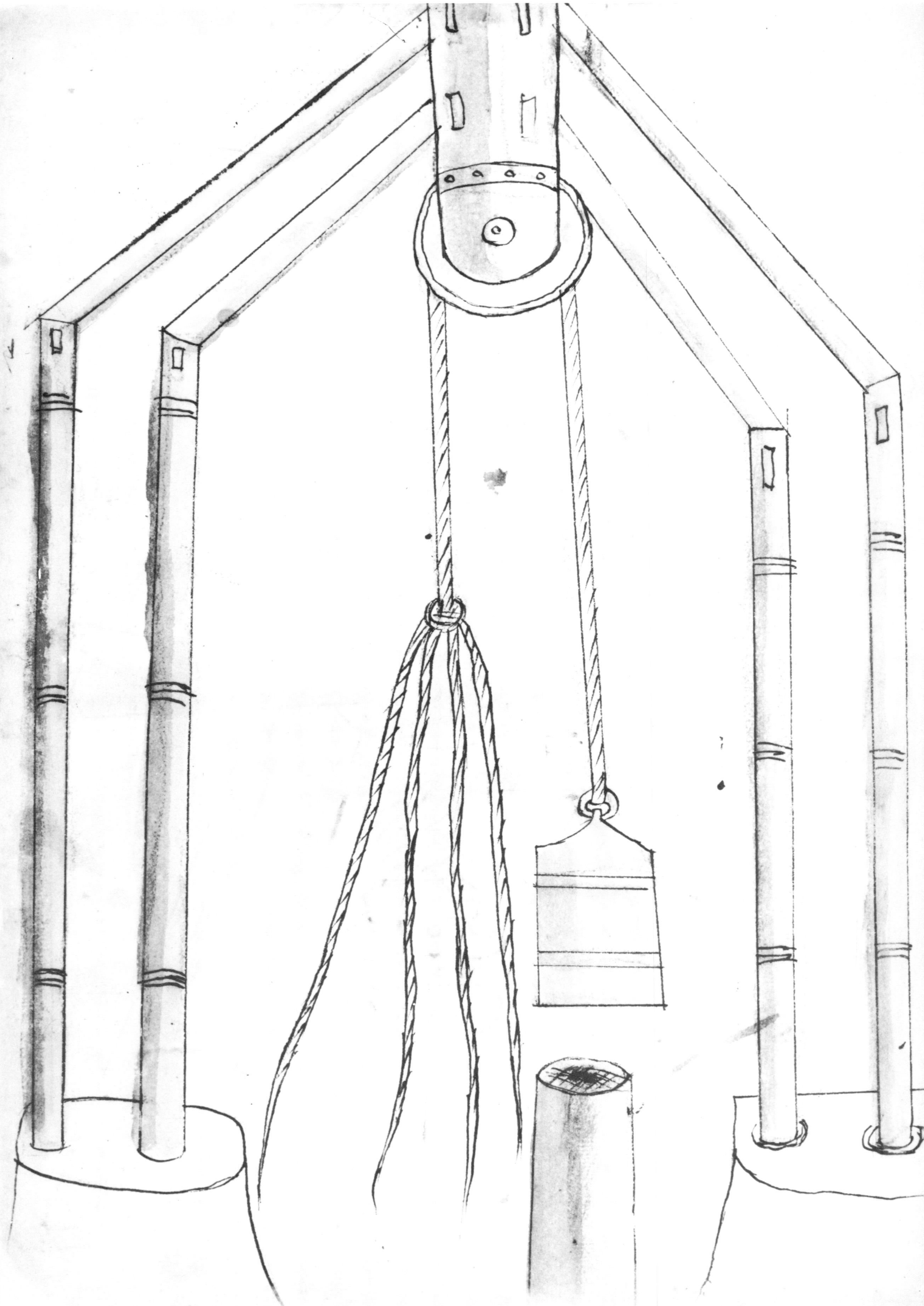

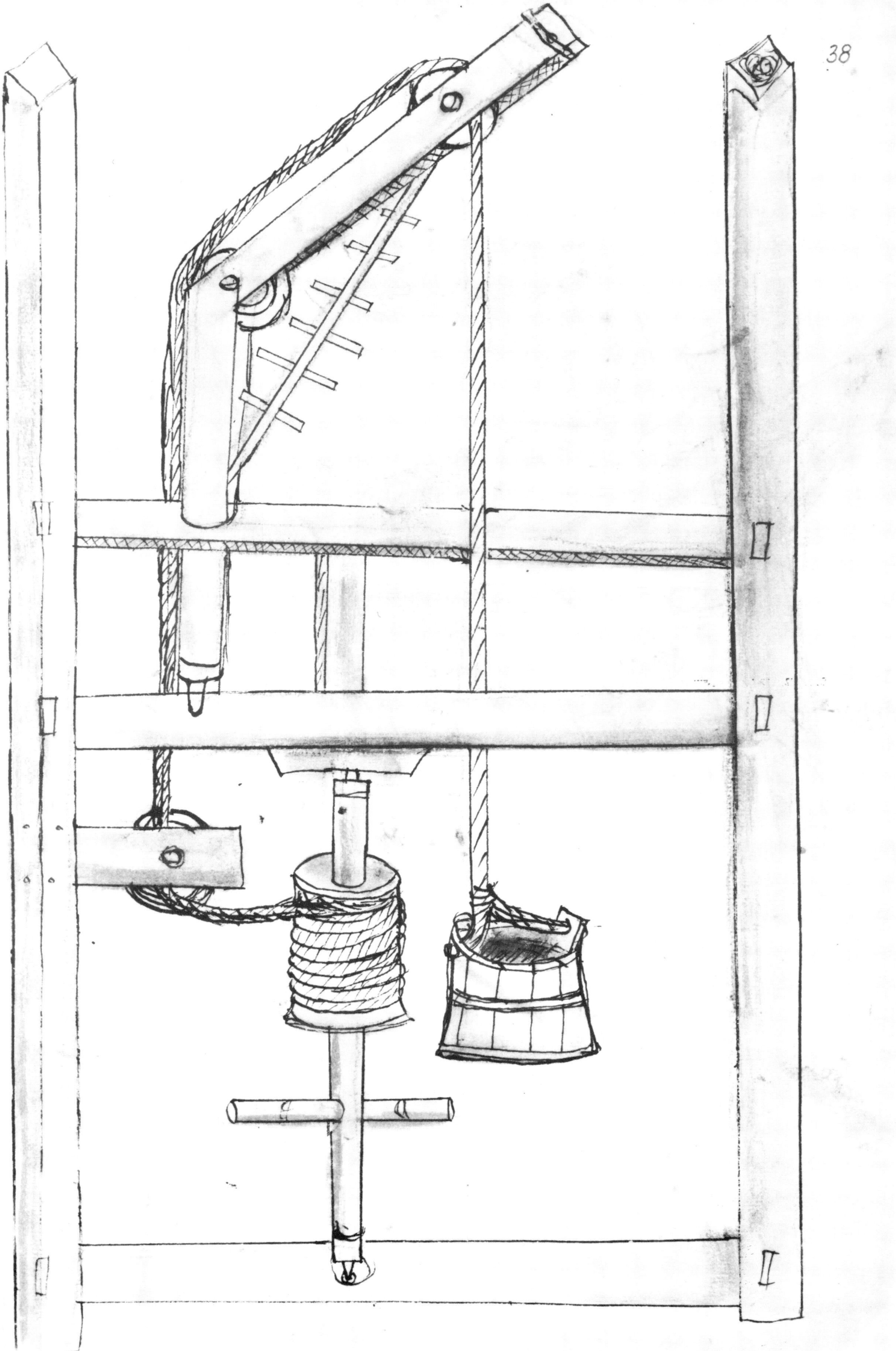

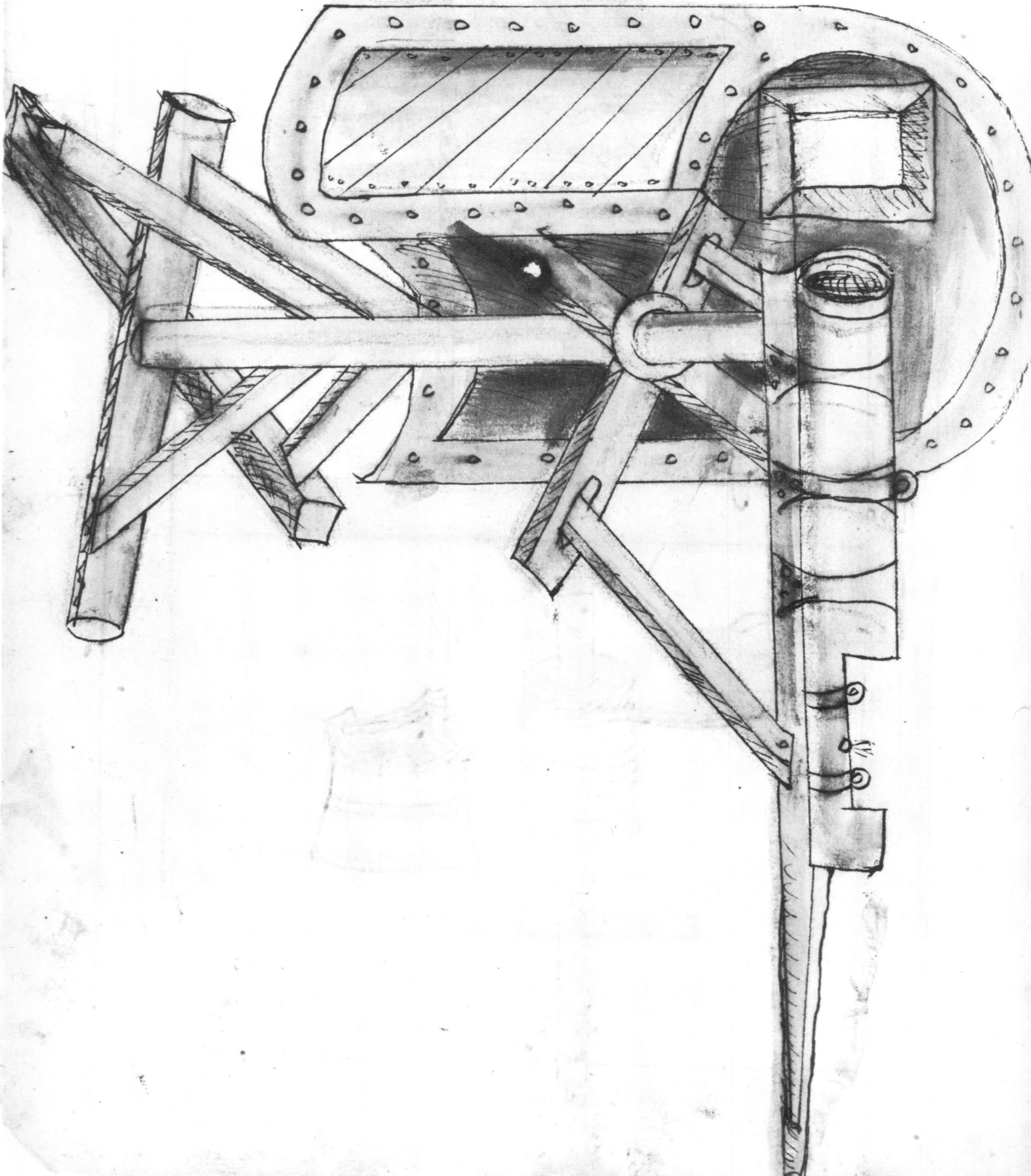

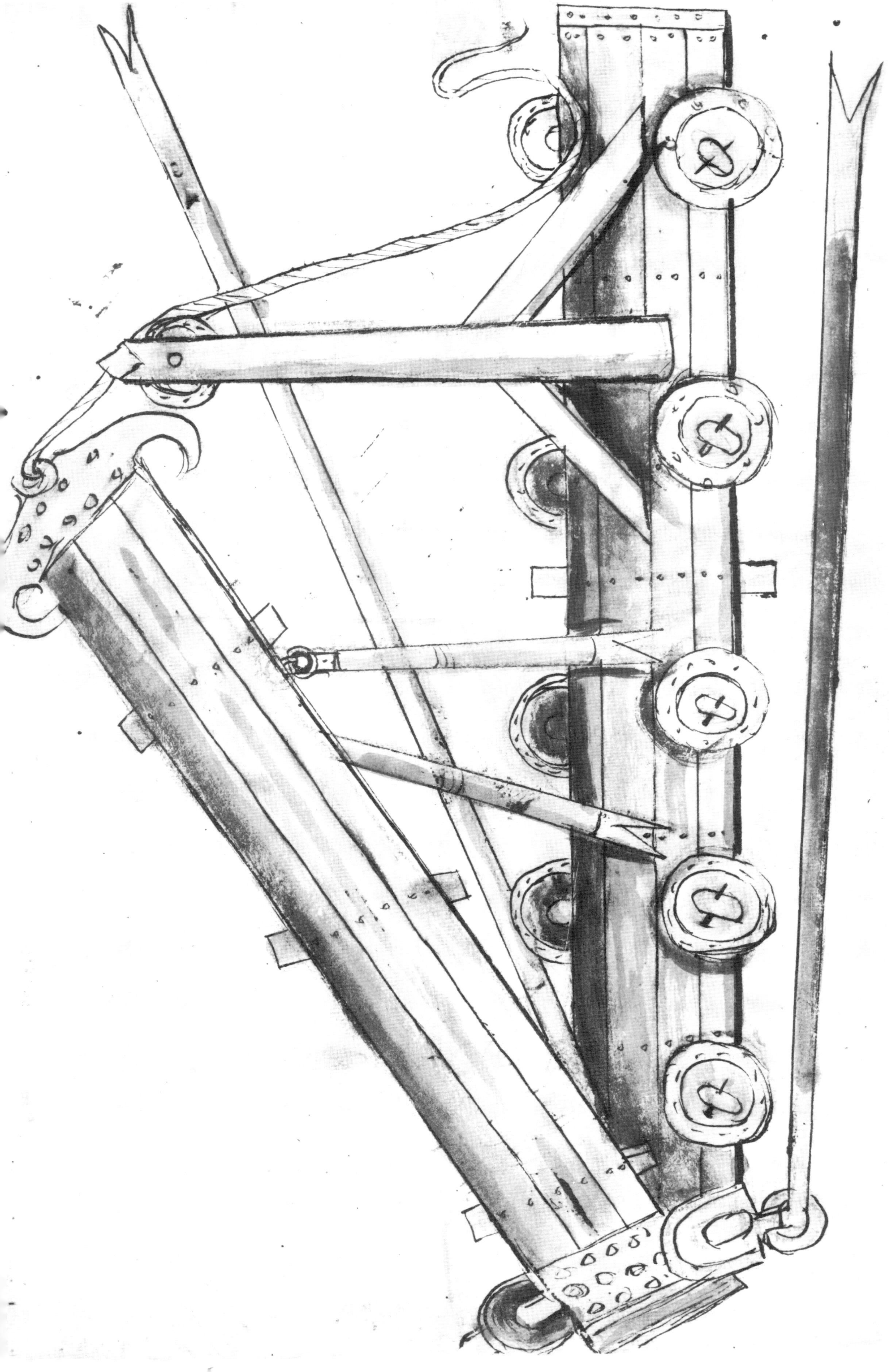

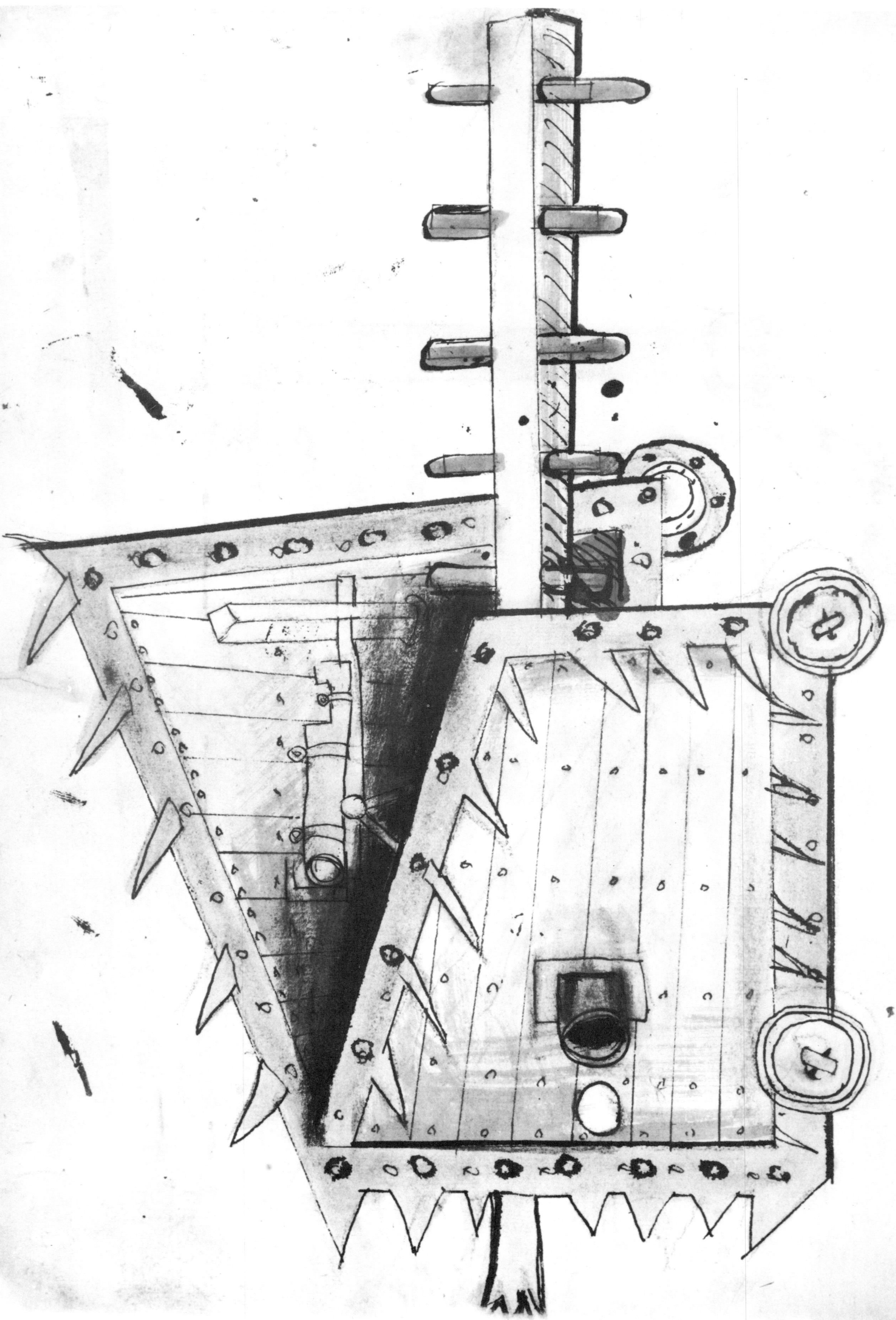

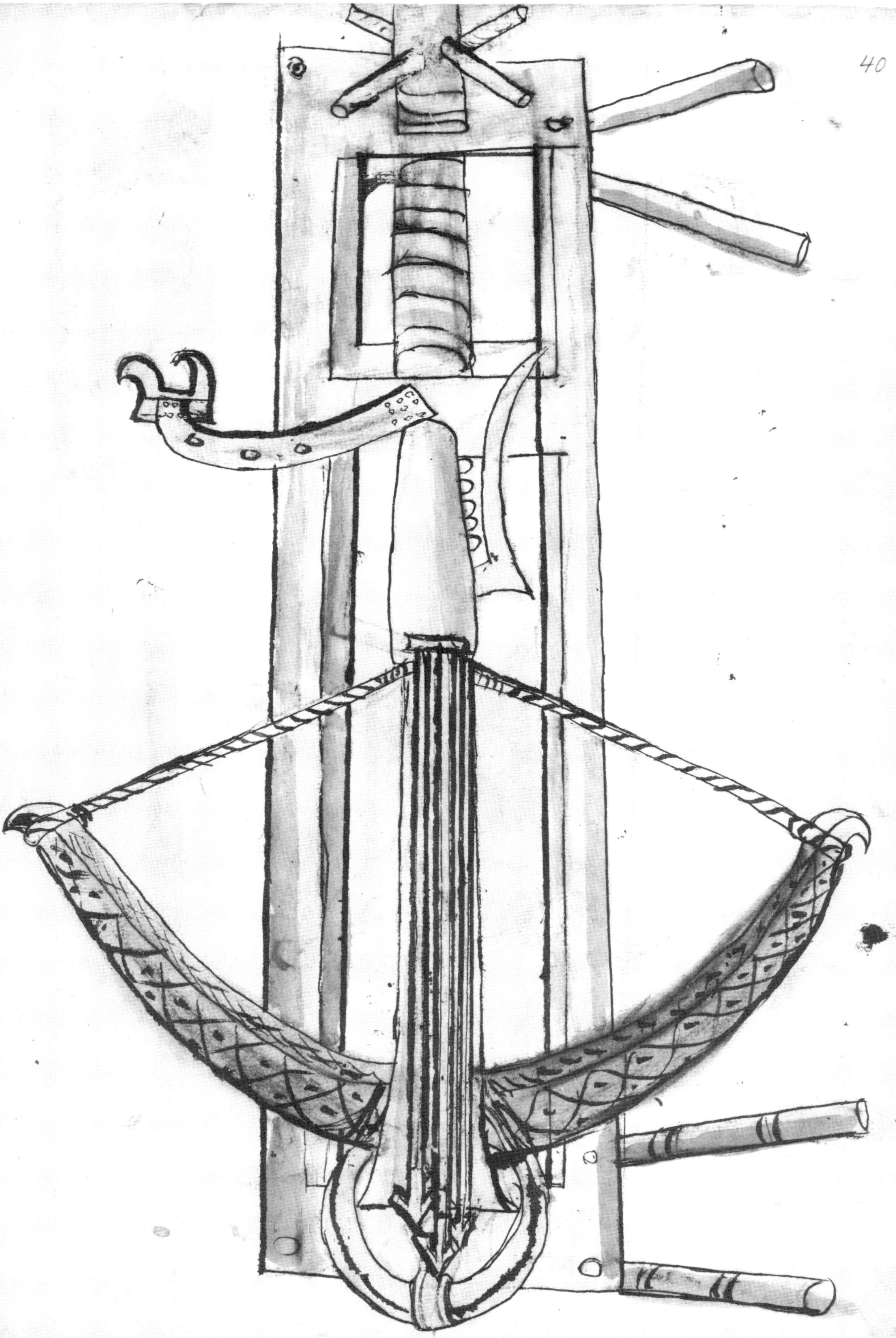

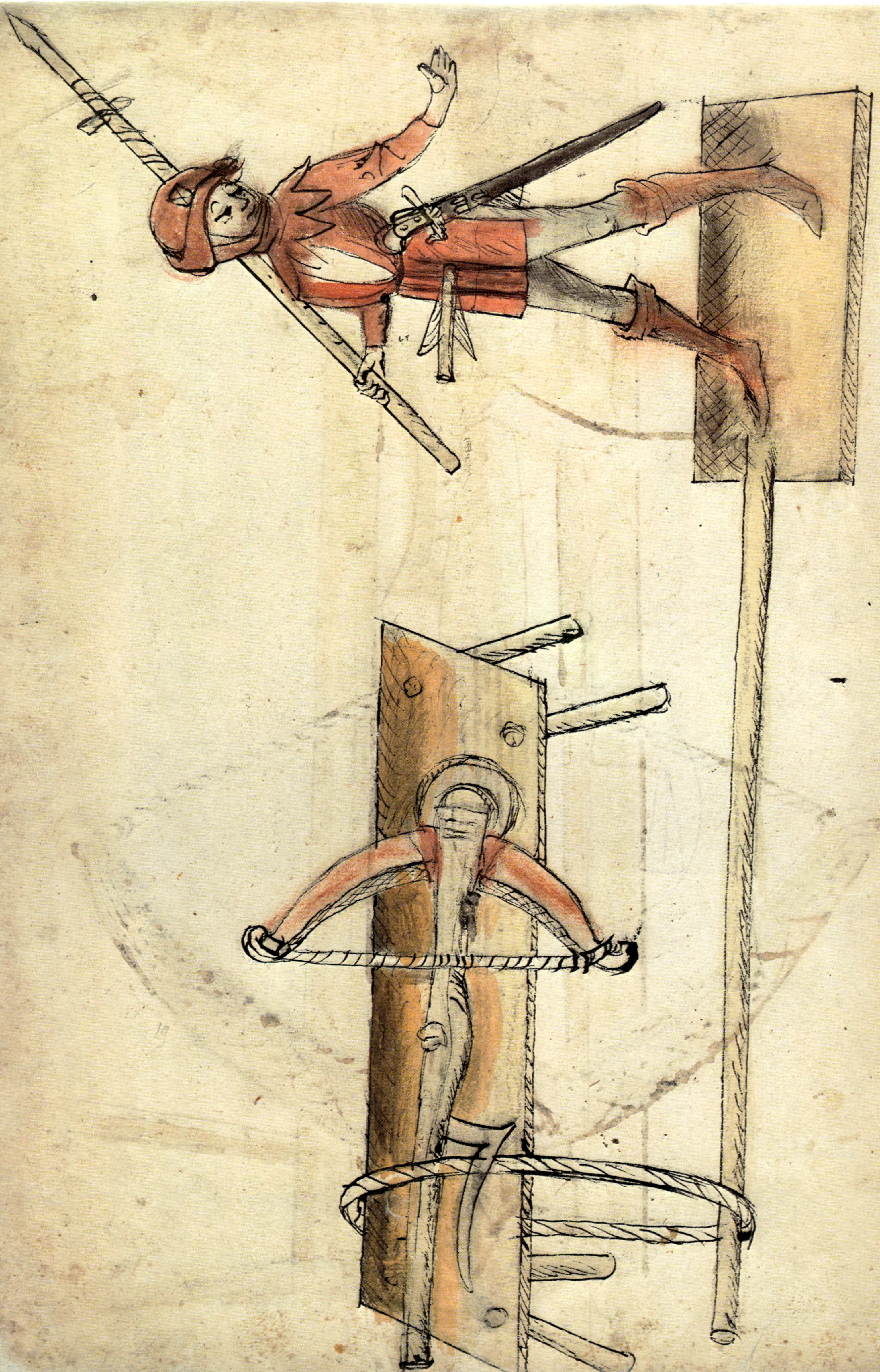

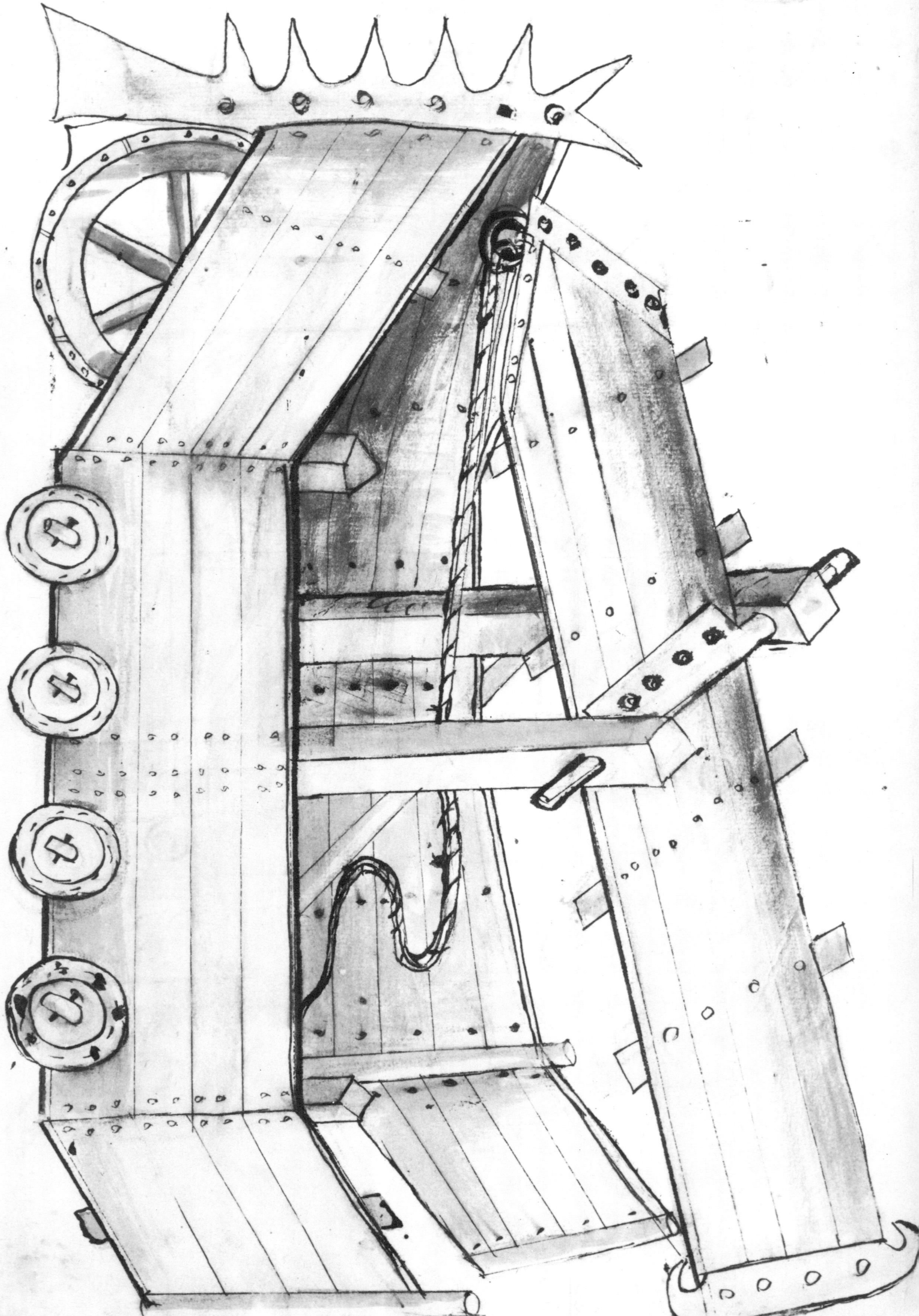

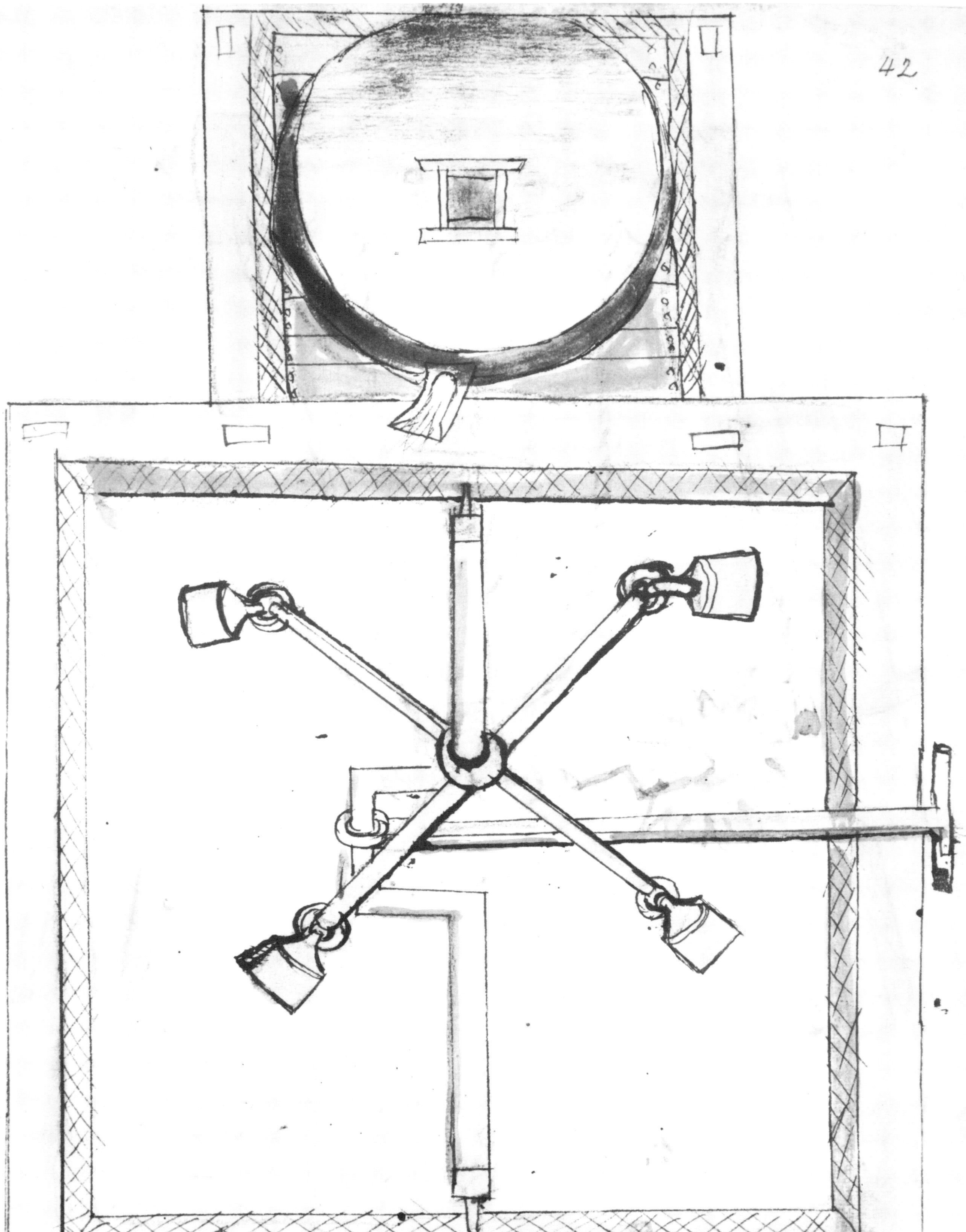

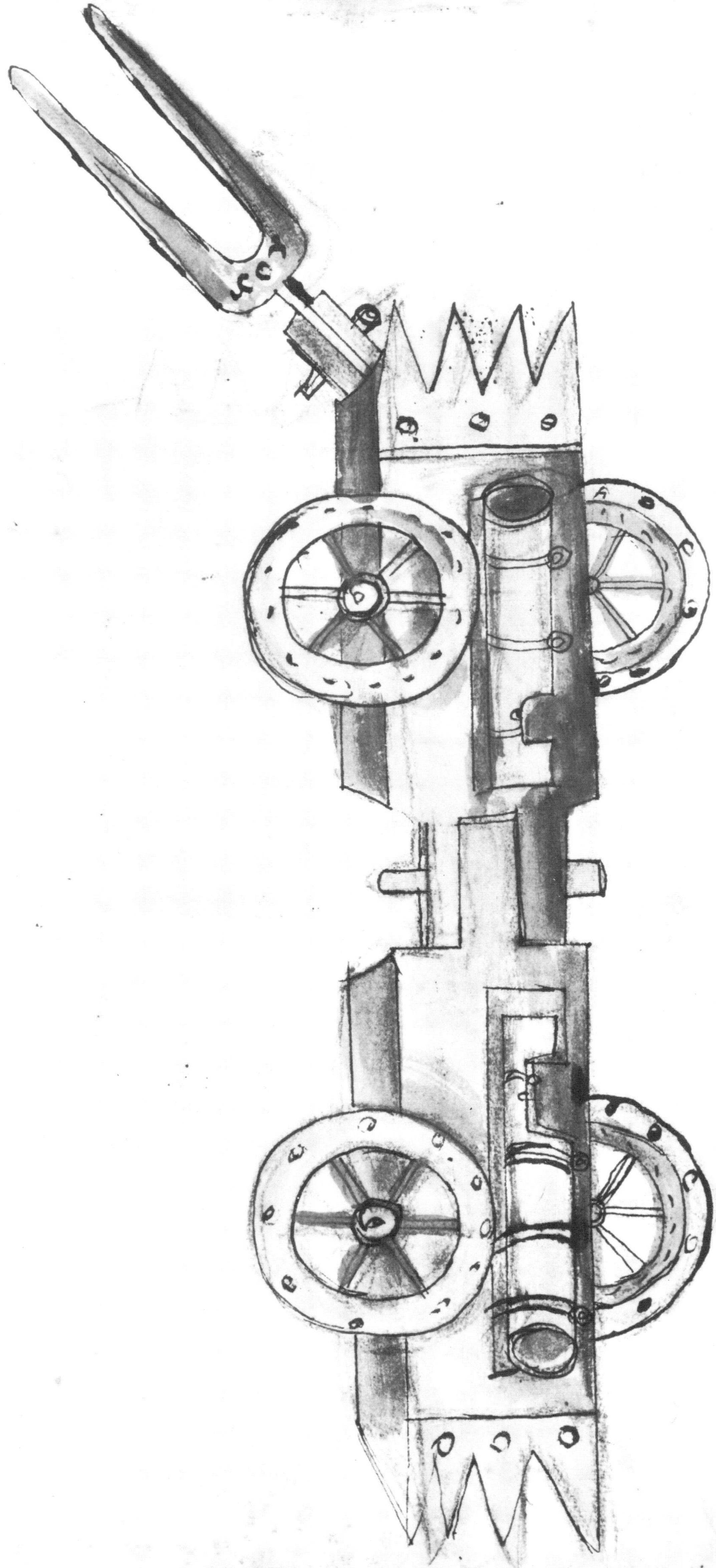

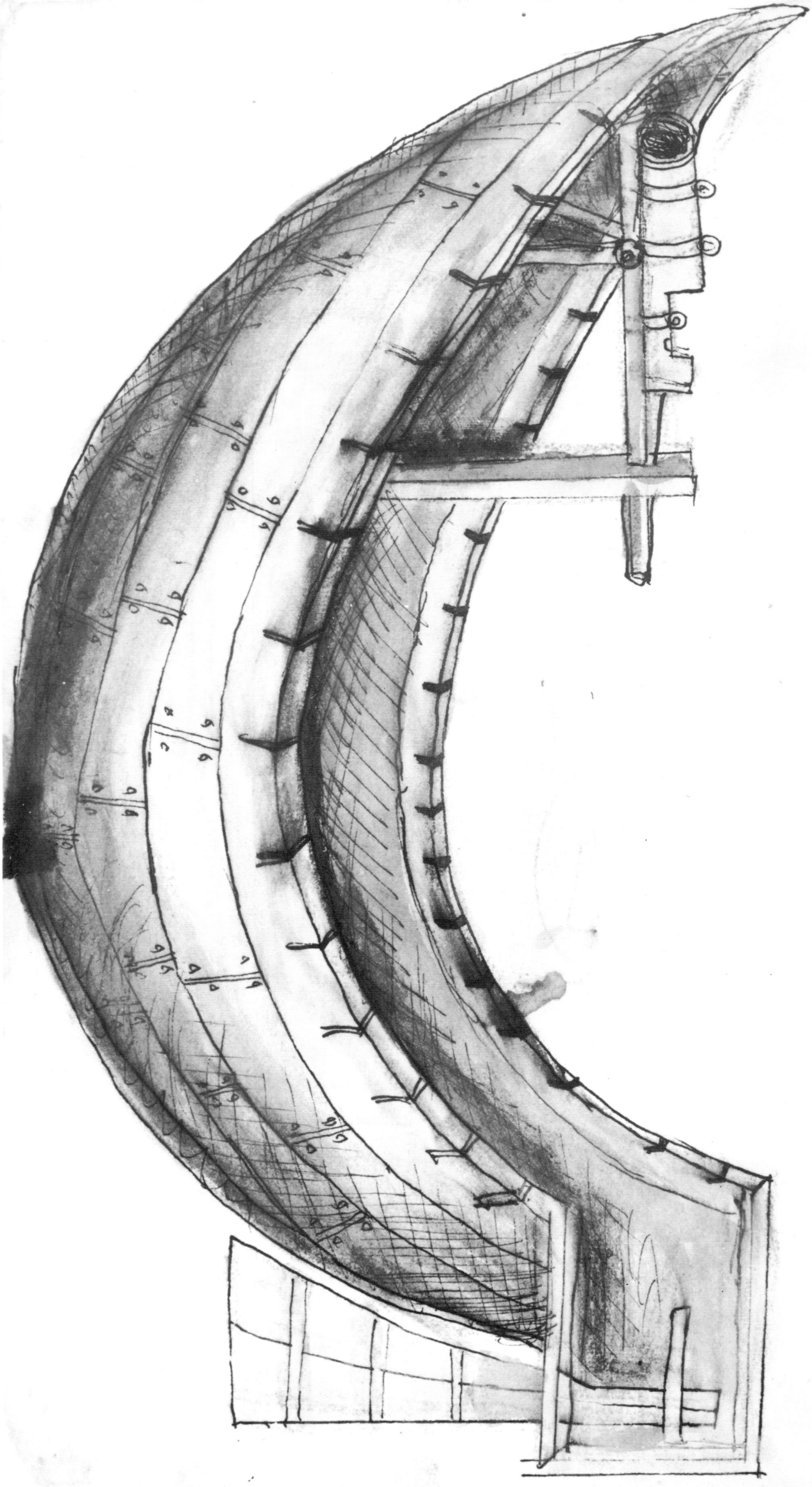

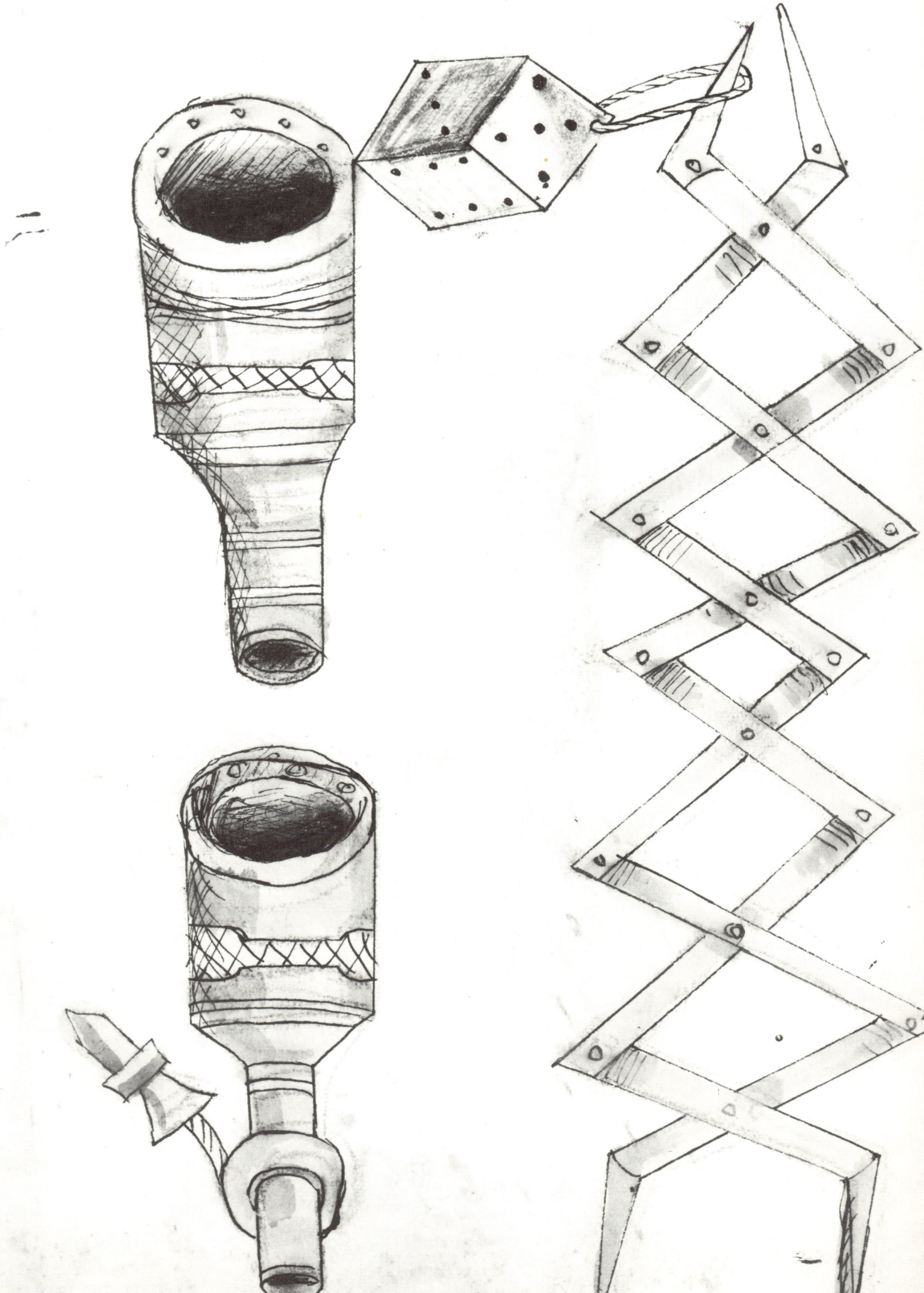

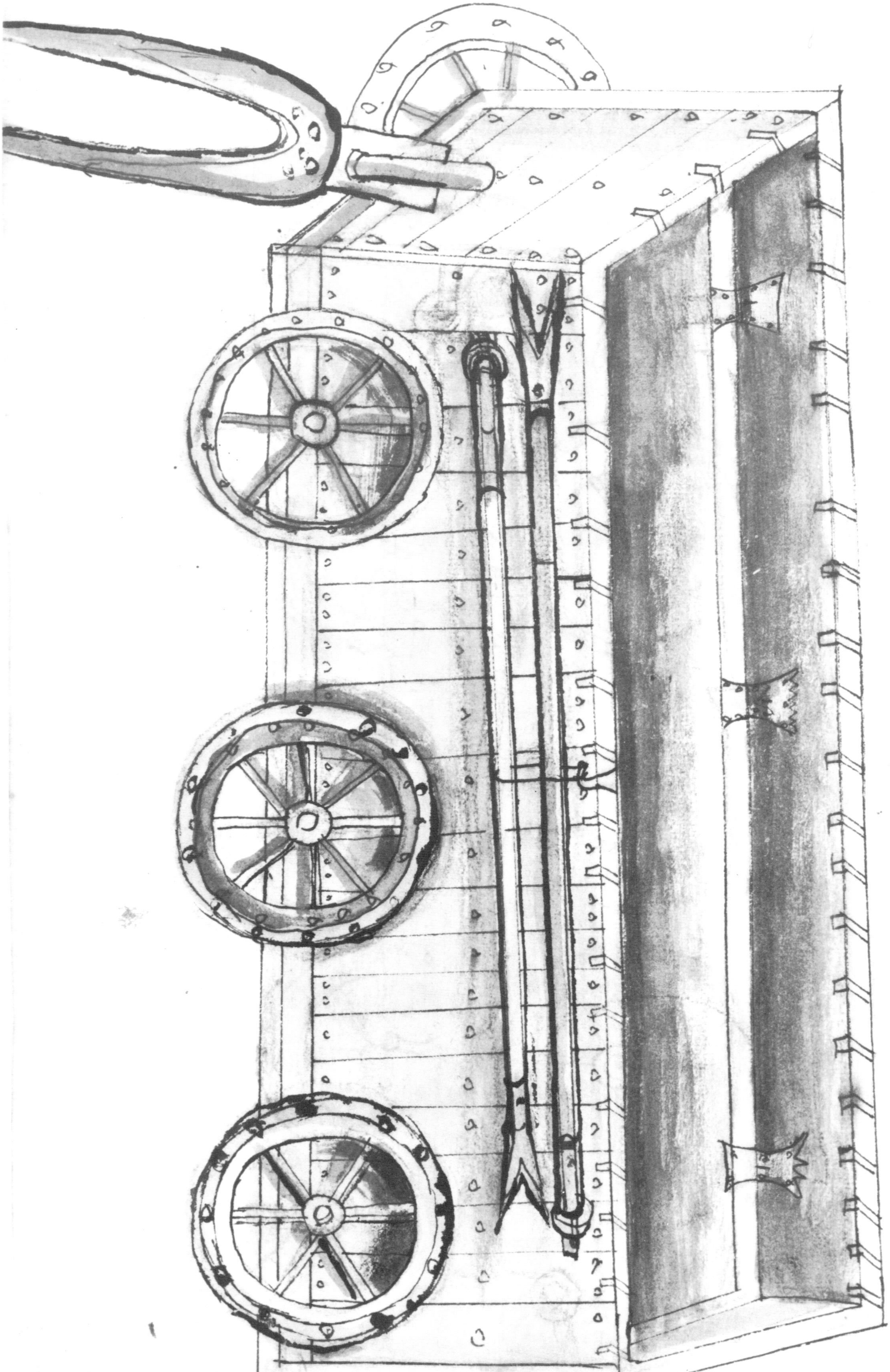

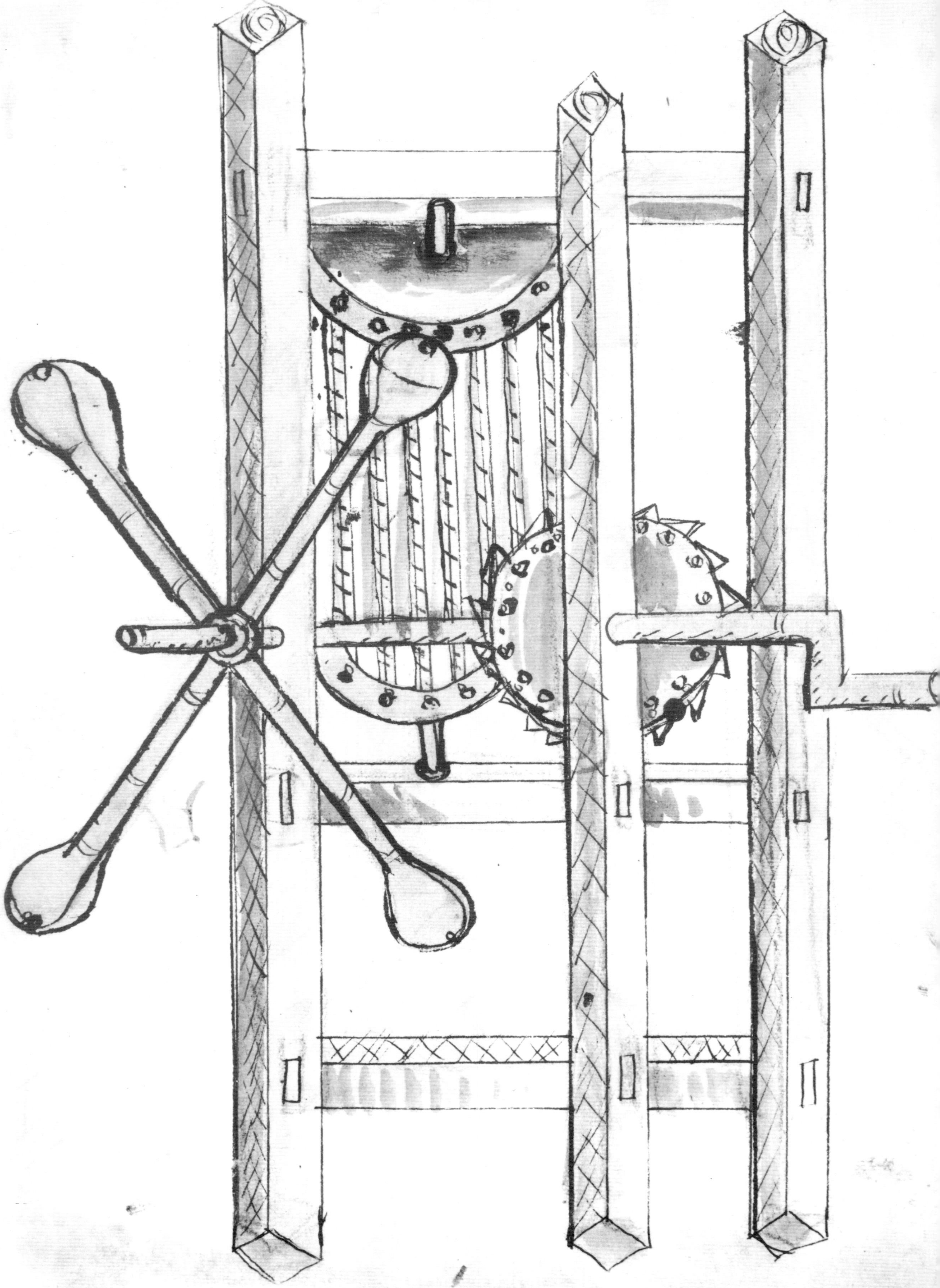

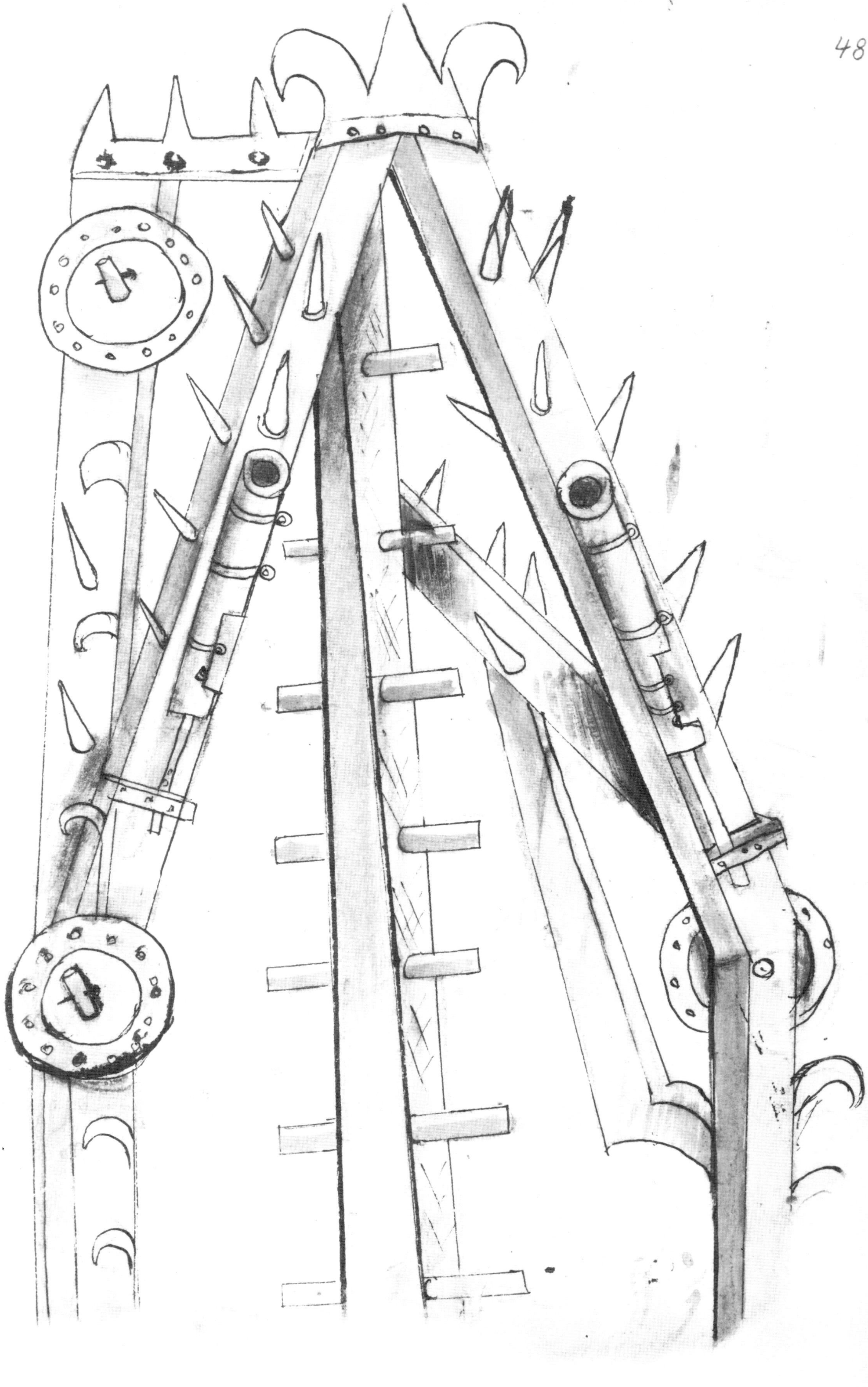

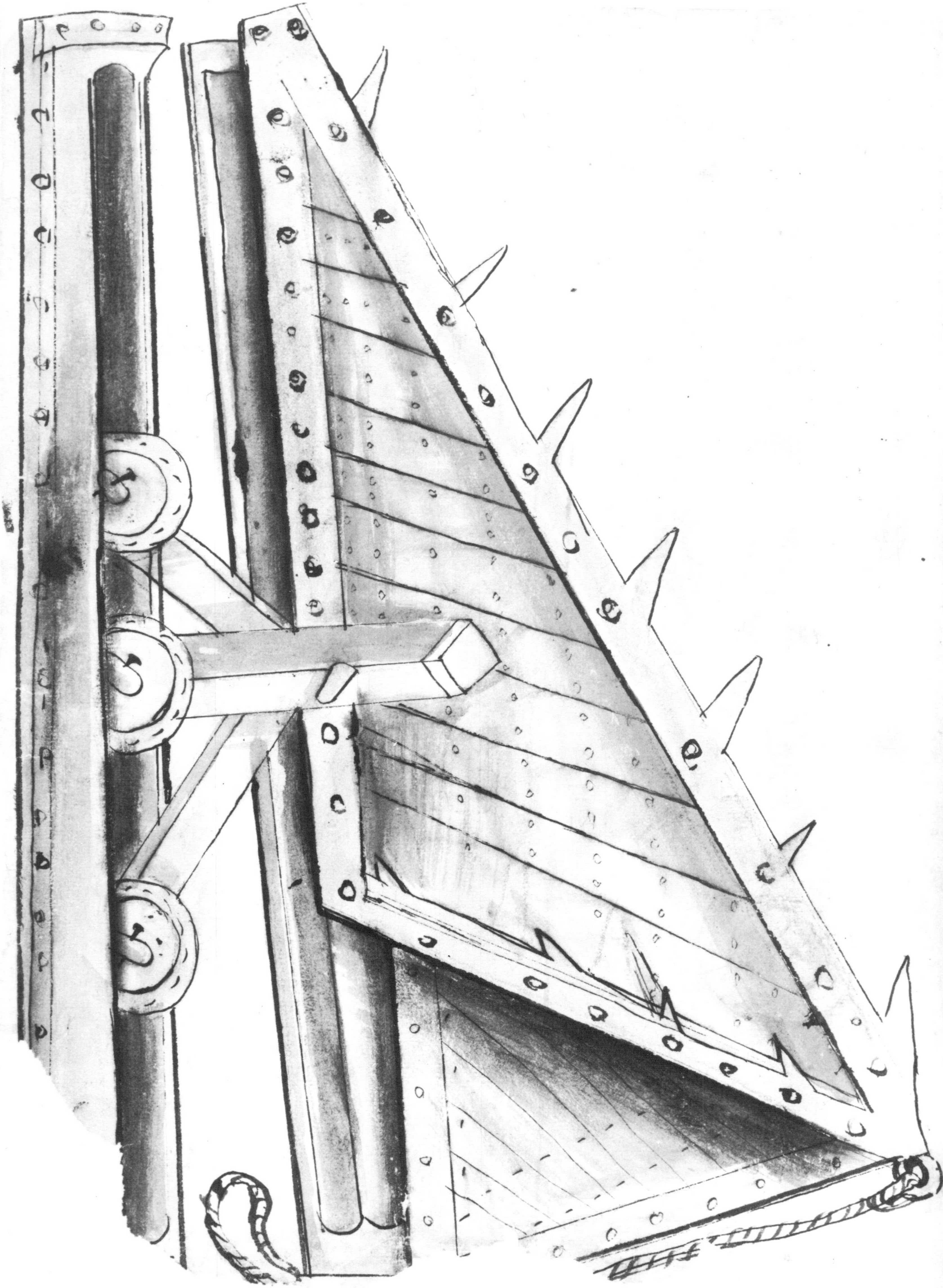